INDIGENOUS FOOD SOVEREIGNTY IN THE UNITED STATES

New Directions in Native American Studies
Colin G. Calloway and K. Tsianina Lomawaima, General Editors

INDIGENOUS FOOD SOVEREIGNTY IN THE UNITED STATES

Restoring Cultural Knowledge,
Protecting Environments,
and Regaining Health

Edited by
DEVON A. MIHESUAH and ELIZABETH HOOVER

Foreword by WINONA LaDUKE

UNIVERSITY OF OKLAHOMA PRESS : NORMAN

Portions of the introduction and chapter 2 appeared in Elizabeth Hoover, "'You can't say you're sovereign if you can't feed yourself': Defining and Enacting Food Sovereignty in American Indian Community Gardening," *American Indian Culture and Research Journal* 41, no. 3 (2017): 31–70. Reprinted by permission of the American Indian Studies Center, UCLA © 2016 Regents of the University of California.

Chapter 3 was previously published as Devon A. Mihesuah, "Searching for *Haknip Achukma* (Good Health): Challenges to Food Sovereignty Initiatives in Oklahoma," *American Indian Culture and Research Journal* 41, no. 3 (2017): 9–30. Reprinted by permission of the American Indian Studies Center, UCLA © 2016 Regents of the University of California.

Chapter 9 was previously published as Dennis Wall and Virgil Masayesva, "People of the Corn: Teachings in Hopi Traditional Agriculture, Spirituality, and Sustainability," *American Indian Quarterly* 28, nos. 3–4 (Summer–Fall 2004): 435–53. Reprinted by permission of the University of Nebraska Press.

Library of Congress Cataloging-in-Publication Data

Names: Mihesuah, Devon A., 1957– editor. | Hoover, Elizabeth, 1978– editor.
Title: Indigenous food sovereignty in the United States : restoring cultural knowledge, protecting environments, and regaining health / edited by Devon A. Mihesuah and Elizabeth Hoover ; foreword by Winona LaDuke.
Other titles: New directions in Native American studies ; v. 18.
Description: Norman : University of Oklahoma Press, [2019] | Series: New directions in Native American studies ; volume 18 | Includes bibliographical references and index.
Identifiers: LCCN 2018060695 | ISBN 978-0-8061-6321-5 (pbk. : alk. paper)
Subjects: LCSH: Indians of North America—Food. | Food sovereignty—United States. | Indians of North America—Social life and customs.
Classification: LCC E98.F7 I53 2019 | DDC 970.004/97—dc23
LC record available at https://lccn.loc.gov/2018060695

Indigenous Food Sovereignty in the United States: Restoring Cultural Knowledge, Protecting Environments, and Regaining Health is Volume 18 in the New Directions in Native American Studies series.

The paper in this book meets the guidelines for permanence and durability of the Committee on Production Guidelines for Book Longevity of the Council on Library Resources, Inc. ∞

Copyright © 2019 by the University of Oklahoma Press, Norman, Publishing Division of the University. Manufactured in the U.S.A.

All rights reserved. No part of this publication may be reproduced, stored in a retrieval system, or transmitted, in any form or by any means, electronic, mechanical, photocopying, recording, or otherwise—except as permitted under Section 107 or 108 of the United States Copyright Act—without the prior written permission of the University of Oklahoma Press. To request permission to reproduce selections from this book, write to Permissions, University of Oklahoma Press, 2800 Venture Drive, Norman OK 73069, or email rights.oupress@ou.edu.

To all those in the Indigenous Food Movement
And in honor of our ancestors who guide us

CONTENTS

List of Illustrations ▪ **ix**

List of Tables ▪ **xi**

Foreword: In Praise of Seeds and Hope ▪ **xiii**
 Winona LaDuke

Introduction ▪ **3**
 Elizabeth Hoover and Devon A. Mihesuah

1. Voices from the Indigenous Food Movement ▪ **26**
 Steven Bond-Hikatubbi, Carrie Calisay Cannon, Elizabeth Hoover, Laticia McNaughton, Devon A. Mihesuah, Brit Reed, Martin Reinhardt, Valerie Segrest, Sean Sherman, Chip Taylor, Elizabeth Kronk Warner, and Brian Yazzie

2. "You can't say you're sovereign if you can't feed yourself": Defining and Enacting Food Sovereignty in American Indian Community Gardening ▪ **57**
 Elizabeth Hoover

3. Searching for *Haknip Achukma* (Good Health): Challenges to Food Sovereignty Initiatives in Oklahoma ▪ **94**
 Devon A. Mihesuah

4. Kuaʻāina Ulu ʻAuamo: Grassroots Growing through Shared Responsibility ▪ **122**
 Kevin K. J. Chang, Charles K. H. Young, Brenda F. Asuncion, Wallace K. Ito, Kawika B. Winter, and Wayne C. Tanaka

5. Alaska Native Perceptions of Food, Health, and Community Well-Being: Challenging Nutritional Colonialism • 155
   Melanie M. Lindholm

6. Healthy Diné Nation Initiatives: Empowering Our Communities • 173
   Denisa Livingston

7. Planting Sacred Seeds in a Modern World: Restoring Indigenous Seed Sovereignty • 186
   Rowen White

8. What If the Seeds Do Not Sprout? The Cherokee Nation SeedBank & Native Plant Site • 198
   Pat Gwin

9. People of the Corn: Teachings in Hopi Traditional Agriculture, Spirituality, and Sustainability • 209
   Dennis Wall and Virgil Masayesva

10. Comanche Traditional Foodways and the Decline of Health • 223
    Devon A. Mihesuah

11. Bringing the Past to the Present: Traditional Indigenous Farming in Southern California • 253
    Gerald Clarke

12. On Intimacy with Soils: Indigenous Agroecology and Biodynamics • 276
    Devon G. Peña

13. Nephi Craig: Life in Second Sight • 300
    Devon A. Mihesuah

14. Indigenous Climate Justice and Food Sovereignty: Food, Climate, Continuance • 320
    Kyle Powys Whyte

    Conclusion: Food for Thought • 335
    Elizabeth Hoover and Devon A. Mihesuah

Study Questions • 339

List of Contributors • 347

Index • 355

ILLUSTRATIONS

Winona LaDuke in front of the Honor the Earth Bus, 2018 ▪ xv

Food sovereignty and good health ▪ 14

Sean Sherman at a Sioux Chef dinner at the James Beard House, 2017 ▪ 47

Brian Yazzie serving buffalo burgers at Standing Rock, 2016 ▪ 54

Braids of white corn hanging in the barn at Tsyunhehkwa ▪ 59

Hands-on corn-washing workshop at the Intertribal Food Sovereignty Summit ▪ 83

Prayer circle outside the public hearing for the Hāʻena CBSFA, 2014 ▪ 141

Limu lei weaving at the Global E Alu Pu in 2016 ▪ 145

Wild berries gathered in Alaska, 2015 ▪ 169

Rowen White holding a braid of white corn, 2016 ▪ 192

The mealing trough—Hopi, circa 1906 ▪ 212

Cornfield, Indian farm near Tuba City, Arizona, in rain, 1941 ▪ 215

Butchering beef and loading the meat, 1891 ▪ 239

Craig Nephi's dessert, 2015 ▪ 305

The Café Gozhóó logo ▪ 315

TABLES

2.1 Food Sovereignty Conferences and Events Attended ▪ 60

2.2 Sites Visited ▪ 61

5.1 Respondents' Demographic Information ▪ 157

12.1 Ethnomedical and Biodynamic/Agroecological Properties of Select Companion Plants in Traditional Milpas ▪ 286

WINONA LaDUKE

FOREWORD
In Praise of Seeds and Hope

At this moment, when food systems are being shaken by drought and industrial poisons, and our Indigenous communities suffer the ravages of colonial food, the voices in this book, strong and rooted, come together to tell their stories, tell our stories. For the past twenty years, a renaissance of Indigenous food systems has flourished. This is indeed the time of prophecies. We are told that in this time of the Sixth Fire, we will go and find ourselves, our foods, our songs, and our way. This remembrance of our foods and ourselves will make us stronger. This book records those voices—the voices of remembering, of returning, and of the songs for new seeds. For indeed, seeds are promise.

When I was a young woman at Harvard University, my father came to me one day and said, "I don't want to hear your philosophy if you cannot grow corn." In the simplicity of those words is not only the story of my life but a ground truth. If we are unable to feed ourselves, we will not survive; and if we lose our whole being to our minds, policy work, and scholarly discussions, we will have lost our direction. We need to strike a balance. Think of it this way: our ancestors navigated by stars, lakes, and trees; today, we navigate with a global positioning system. Due to pollution we can no longer even see many of the stars; that is, unless we return to the lands and the fields. Indeed, we must be conscious and work our way back to the soil. The soil and the seeds help us navigate the future. The chapters in this book seek to strike that balance between the documentation of history and the creation of policy versus the on-the-ground work and needs of indigenous communities.

Following the lead of my father, and so many others, I became a corn grower. First Ricardo Salvador gave me Bear Island Flint from the GRIN (Germplasm Resources Information Network) collection. I have seen the return of that variety—our seed from the island in the midst of Leech Lake Reservation—to gardens in our communities. I have seen even more varieties of *mandaamin*—corn, wondrous seed—return and flourish. Indeed, in a time of climate change, as described by Kyle White in this book, we have found that these seeds are resilient—never a crop failure, despite winds, droughts, and freezes. The corn has taught me hope, commitment, and a return to the craft of cooking—restoring hominy-making knowledge and recipes to our community. As my father surmised, corn teaches all of us.

At the first gatherings of Indigenous food producers at Slow Food International's Terra Madre event, we came to recognize our place in the international struggle of Indigenous peoples protecting and reclaiming the food the Creator gave us: our foods are who we are. Since then, and through subsequent Indigenous gatherings during which I have met many of the contributors to this book, our movement has grown and flourished as Indigenous food sovereignty takes root in our communities across Turtle Island. Our food and our future are related. And, with the emergence of the Turtle Island Association of Slow Food in 2016, we have taken our place at the international table of a movement for just, clean, and good foods worldwide. "You cannot say you are sovereign if you cannot feed yourself," Sugar Bear Smith of Oneida once said to me. *Clear and affirming.* We are the people growing and restoring the sovereign practice of food.

Despite the $13 billion corporate food industry, 70 percent of the world's food is grown by families, peasants, and Indigenous farmers. We are those people, and today when we return to our farms and our seeds, we take our place in history. In a time when agrobiodiversity has crashed and world food systems are filled with poisons, our seeds remain, and they return. These are our stories: stories of love and stories of hope.

Winona LaDuke in front of the Honor the Earth Bus, March 2018. *Photo by Elizabeth Hoover.*

It is said that in the time of the Sixth Fire we will go looking for much of what was stolen or lost, and we will recover those songs, medicine bundles, and seeds. Then we will come into the time of the Seventh Fire. In that time it is said we will make the choice between a scorched path and a path that is green. The essential part of following that green path is how we return to living here on this land. At the center is producing our food again, feeding our people. That is how we reaffirm our covenant and agreement with the Creator. We are becoming the people who no longer import our food from across the country, hemorrhaging our tribal budgets, but instead exercise and affirm the power of our relationship with our food.

To those who have collected these stories, I am grateful. And to those of us who have found our corn, I say find your courage in those seeds that you plant. Because those seeds are about hope, promise, commitment, the future. And in returning to our seeds and fields we are able to fulfill our responsibilities to all of our relatives, whether they have wings, fins, roots, or paws. Food sovereignty is an affirmation of who we are as Indigenous peoples, and a way, one of the most sure-footed ways, to restore our relationship with the world around us. That is the story of this book. These are stories of heroes of the time of the Seventh Fire. I am grateful to be present at this time.

INDIGENOUS FOOD SOVEREIGNTY IN THE UNITED STATES

ELIZABETH HOOVER
DEVON A. MIHESUAH

■ INTRODUCTION

The day's work began at dawn, the gourd rattle and songs ringing out in the cool Sonoran Desert air. Members of the Tohono O'odham Tribe, gathered for the annual TOCA (Tohono O'odham Community Action) *bahidaj* camp in Sells, Arizona, stirred from their tents and climbed into vans and pickup trucks, heading deep into the desert as the soon-to-be-blazing July sun crept up over the horizon. With *ku'ipad* in hand, long sticks fashioned from cactus ribs, groups of family and friends respectfully approached each towering saguaro cactus and knocked the ripe fruit, *bahidaj*, into a bucket below. The bright red fruits filled with tiny black seeds (about 2,000 per fruit) were plucked from their thick green skins, and the empty skins then placed with their red insides facing the sky, to call forth the clouds that bring the monsoon rains.

The gathered fruits were then brought back to the makeshift camp at the Alexander Pancho Memorial Learning Farm, where they were boiled in large pots over open fires, the seeds then painstakingly strained out, and the remaining juice boiled down to make syrup. Throughout the entire four-day camp, participants learned about O'odham language and culture, and the importance of continuing traditional foodways to maintain health. Later in the summer, O'odham people will utilize *akchin* (dryland) farming on the same land, relying on monsoon rains to grow hardy varieties of traditional tepary beans, sixty-day corn, and squash that they have adapted over generations to the arid environment, and which are also important for a healthy diet.[1]

Tohono O'odham people struggle to protect their fragile resources. In April 2018, they combined forces with the Hopi and Pascua Yaqui Tribes to file suit to halt the Rosemont Mine, an operation that would destroy a critical resource area south of Tucson. More controversial among tribal members is the simultaneously looming border wall, a physical boundary between the United States and Mexico and between tribal members who live north and south of the border divide. While some tribal members welcome the security from drug smugglers such a wall might bring, others have organized the O'odham Solidarity Project to protest this endeavor that will not only be a strike against their tribal sovereignty, but will also encroach on their sacred sites and damage the environment, preventing crucial animal migration and disrupting water flows during monsoon season.[2]

Similar traditions, concerns, and activist initiatives are being carried out all over Indian Country. In response to high rates of diabetes and obesity, environmental destruction, pollution, resource depletion, poverty, and general lack of access to healthy food, tribes and grassroots organizations of determined tribal members have initiated numerous projects, including seed distributions, food summits, farmers' markets, cattle and bison ranches, landscape restoration projects, community and school gardens, economic development initiatives, political activism, and legal actions.[3] These enterprises are steps toward achieving what many food activists refer to as "food sovereignty."

DISRUPTION OF FOOD SYSTEMS IN INDIAN COUNTRY

Over the past several centuries, colonialism has unleashed a series of factors that have disrupted Indigenous communities' ability to retain control of their food systems. In many cases, this interruption was intentional, with US settlers deliberately endorsing actions that would undermine Indigenous self-determination. As Potawatomi philosopher Kyle Powys Whyte argues, the goal was "erasing the capacities that the societies that were already there—Indigenous societies—rely on for the sake of exercising their own collective self-determination over their cultures, economies, health, and political order."[4] In American history this erasure includes actions ranging from deliberately destroying food in acts of war to interfering with the transfer of food-related knowledge from one generation to the next.

Scorched-earth battle tactics utilized against Native people in the eighteenth and nineteenth centuries destroyed food supplies and the land from which they came in order to force Native people to become reliant on the American government.[5] Indigenous communities have been pushed to marginal territories, and in many cases the treaty-making system alienated tribes from their traditional land.[6] Land bases were further diminished through the allotment system that allocated communal land to individuals and families. During the late nineteenth and the twentieth centuries, federal policies encouraged Native people on many reservations to farm on marginal lands, despite their histories of successful fishing and gathering practices. While some tribal communities traditionally practiced farming, others did not (such as Plains tribes and other communities across North America). The US and Canadian governments introduced farming projects to the latter in order to assimilate them, disrupt their hunting cultures, and expand the agricultural frontier—even as the best farmland was often usurped by non-Indians.[7]

During this era many Native youth were also forcibly sent to boarding schools, where they were often undernourished.[8] In these schools, youth were encouraged to forget their tribal connections and instead were forced to accept dietary staples that embodied Anglo ideals in which nutrition centered around starches and dairy—a shift for students accustomed to diets based on fresh and dried meats, fruits, and vegetables.[9] Following the boarding school era, urban relocation programs in the 1950s brought Native people from rural reservations to urban centers for employment opportunities, a move that often left families facing food insecurity.[10] Food insecurity refers to a state of having limited or uncertain access to food that is nutritionally adequate, culturally acceptable, and safe, or having an uncertain ability to acquire acceptable foods in socially acceptable ways.[11]

Environmental change has also hindered access to traditional foods, through both intentional reshaping of the landscape and climate change. For example, damming of the Missouri River in the 1940s and 1950s resulted in Native peoples losing most of their arable land on the Standing Rock, Cheyenne River, Crow Creek, and Fort Berthold Reservations in the Dakotas.[12] Similar dams built across the Northeast and Northwest have disrupted fisheries and flooded Indigenous homelands.[13] In addition, industrial contamination has damaged fishing in places like the Akwesasne Mohawk community on the New York–Canadian border and the Coast Salish Swinomish community in Washington State.[14] In the polar regions, persistent organic pollutants have made consuming the usual amounts of traditional foods hazardous to community health.[15] Climate change has led to declining sea ice in the Arctic, forced community relocations, shifts in plant and animal populations around North America, changes in river flows affecting water availability for crops, and broadening of the ranges of disease organisms.[16] All of these changes over the past century have impacted Indigenous food systems.

To stave off starvation and malnutrition that would have resulted from disrupted food systems, during the nineteenth century the US government distributed food rations on many Indian reservations, as stipulated in many treaties to make up for the loss of hunting, fishing, and agricultural lands. These rations consisted of foods that were foreign to Indian people: beef, bacon, flour, coffee, salt, and sugar.[17] The practice of the federal government providing food to American Indian communities has continued to the present day, now administered by the Food Distribution Program on Indian Reservations (FDPIR). This federal program provides foods sourced by the United States Department of Agriculture (USDA) to low-income households on or near Indian reservations and in designated areas of Oklahoma.[18] Although the USDA has been working to improve the quality of foods available to communities through this program, including making more fresh foods available, these programs have historically done little to reinforce the relational aspects that traditional food systems relied upon. This changed recently with efforts to include buffalo meat, blue cornmeal, wild rice, and frozen wild sockeye salmon in the FDPIR offerings. However, the inclusion of regionally relevant foods in the federal food distribution program has been slowed by the fact that a food must be available in quantities to supply all eligible participants in the United States; the FDPIR Food Package Working Group is currently seeking to resolve this issue.[19]

The disruption of traditional food systems has led to a number of health and social problems in Indigenous communities. American Indians have higher levels of food insecurity when compared to the US average.[20] In 2008, nearly 25 percent of American Indian and Alaska Native (AI/AN) households were food insecure, versus 15 percent of all US households. AI/AN children have approximately twice the levels of food insecurity, obesity, and type 2 diabetes relative to the average for all US children of similar ages.[21] Historically, Indigenous societies sometimes contended with seasonal and weather-related fluctuations of food sources and availability. But while hunger is still a problem in some households, it is the increased consumption of processed foods that has contributed to an elevation in diet-related health issues among Native peoples. This trend of increased consumption of sugary and starchy foods, paired with decreased consumption of garden produce and game meats, began for some Indigenous communities in the mid-nineteenth century and has only increased over time.[22] Currently, AI/AN adults (16.1%) are more likely than black adults (12.6%), Hispanic adults (11.8%), Asian adults (8.4%), or white adults (7.1%) to have ever been told they had diabetes. These rates vary by region, from 5.5 percent among Alaska Native adults to 33.5 percent among American Indian adults in southern Arizona.[23]

More than physical health problems resulted from the disruption of traditional food systems: as the availability of foods declined, so too have the stories, languages, cultural practices, interpersonal relationships, and outdoor activities implicated in those food systems. A tribal community's capacity for "collective continuance" and "comprehensive aims at robust living" are hindered when the relationships that are part of traditional food cultures and economies are disrupted.[24] Tristan Reader and Terrol Johnson, who worked together to form TOCA, describe how "the endangerment of Tohono O'odham symbolic culture followed directly the decline in material culture. People did not stop planting the fields because the ceremonies were dying out; the ceremonies began to die out when people stopped planting their fields. After all, if you never plant crops, the importance of rain is diminished."[25]

DEFINING FOOD SOVEREIGNTY

While the issues laid out so far are specific to the Indigenous nations in what is now the United States, other marginalized people around the globe have similarly experienced and resisted attacks on their food systems, coalescing in a multinational movement for food sovereignty. The term "food sovereignty" was first defined in 1996 by La Via Campesina, an international group of peasant and small-scale farmers who sought to articulate a common response to neoliberalism and the dominant market economy and to defend their rights to land and seeds.[26] The term was refined and brought to the world stage at the 2007 Forum for Food Sovereignty in Sélingué, Mali, during which five hundred delegates from more than eighty countries adopted the Declaration of Nyéléni. According to the declaration, "Food sovereignty is the right of peoples to healthy and culturally appropriate food produced through ecologically sound and sustainable methods, and their right to define their own food and agriculture systems."[27] The declaration highlights the importance of putting food producers and consumers, rather than corporations, at the heart of food systems policies; the need to include the next generation in food production, as well as to empower food producers and artisans; the importance of environmental, social, and economic sustainability; and the need for transparent trade as well as equality between genders, racial groups, and social classes.[28] Everyone in the food chain is positioned as a potentially powerful actor.[29]

The food sovereignty movement has grown out of, and pushed back against, definitions of food security that activists and scholars have criticized as simply addressing an adequacy of food supply without specifying the means of food acquisition, definitions which have led to correspondingly limited efforts to address the problem. For example, according to the United Nations Food and Agriculture Organization, food security describes "a situation that exists when all people, at all times, have physical, social and economic access to sufficient, safe and nutritious food that meets their dietary needs and food preferences for an active and healthy life."[30] This definition does not specify how, where, and by whom the food that all people should have access to should be produced, contributing to a focus on food-related policies that emphasize maximizing food production and give inadequate attention to who exactly will benefit from where and how that food is produced. Accordingly, efforts toward developing global food security have promoted the liberalization of agricultural trade and the concentration of food production, both of which have benefited multinational agribusiness corporations. This lack of specification as to the source of food promotes the dumping of agricultural commodities at below-market prices and the use of genetically modified seeds and other expensive agricultural inputs. These developments have devastated domestic agricultural systems, undermining the economic position of small farmers and reinforcing power differentials by promoting multinational corporations, rather than putting resources back into the hands of those who would produce food for themselves.[31] In the specific context of North American food security studies, Indigenous scholars have argued that a single focus on the supply end of food procurement does not adequately address the food conditions, histories, and relationships of Indigenous peoples, even if the intention is to document and address hunger in individual households.[32]

The food sovereignty movement, in contrast, seeks to address intersecting issues of hunger, environmentally unsustainable production, economic inequality, and social justice on a political level. The goal is to democratize food production, distribution, and consumption, shifting "the focus from the right to access food to the right to produce it."[33] This movement is seen as an alternative to neoliberal economic development and industrial agriculture, which have devastated the livelihoods of peasant and small-scale farmers in favor of large agribusiness and have contributed to economic and environmental crises.[34]

By supporting more environmentally sustainable production and greater reliance on smaller producers, food sovereignty also seeks to address what sociologist Philip McMichael has labeled the "triple crisis": displaced local food production for almost 50 percent of humanity, deepening fossil fuel dependency in an age of "peak oil," and industrial agriculture that generates roughly a quarter of the greenhouse gas emissions which are contributing to global climate change.[35] Yet the imperative of food sovereignty is not simply to add social justice components to an environmentally sustainable food system. Rather, social justice is the foundation from which a food system that works to correct historical and structural injustices must be built.[36] The production, consumption, and distribution of culturally appropriate food must be accomplished while strengthening community, livelihoods, and environmental sustainability.[37] The food sovereignty movement highlights the social connections inherent in the production and consumption of food, demanding that we not treat food as a mere commodity.[38]

Whereas "sovereignty" conventionally refers to the sovereignty of the state over its territory and its right to enact policies without external interference, the food sovereignty movement defines food sovereignty as a "right of the peoples," adopting a pluralistic concept that attributes sovereignty not only to states but to non-state actors such as cultural and ethnic communities.[39] For communities that experience nested layers of sovereignty, the situation can be complicated—the nation-state might seek to be sovereign over food production and distribution free of the interference of multinational entities, making a particular reading of food sovereignty "attractive to national governments advocating for strong state regulation of food chains."[40] Therefore, the focus on "peoples" is "not just a semantic move to make food sovereignty feel inclusive; it indicates a focus on collective action to assert and maintain political autonomy at multiple scales" since, particularly in colonized societies, peoples' and countries' rights are not necessarily the same thing.[41] In many Indigenous communities, food sovereignty is a continuation of anticolonial struggles; the politics employed by Indigenous people engaged in the food sovereignty movement are "not only a politics moored in both space and place, but a politics developed as part of longer struggles against exploitation and colonization of that place."[42]

For Indigenous communities who experience nested layers of sovereignty, food sovereignty as a term and concept can take on different layers of meaning compared to either the broader peasant struggle or urban communities.[43] In the Native American context, whether as sovereign nations or "domestic dependents" (as Chief Justice John Marshall labeled them in the precedent-setting *Cherokee Nation v. State of Georgia* [1831]), tribes have been integrating the struggle for food sovereignty into broader efforts of self-determination.

TRIBAL SOVEREIGNTY

Defining what exactly "sovereignty" means for Native communities has proven challenging. When Elizabeth Hoover asked Anishinaabe (Ojibway) scholar, activist, community organizer, and economist Winona LaDuke to define food sovereignty, she highlighted one part of the debate for many Indigenous people: "What is food sovereignty? You know I'm going to be honest with you, I actually have problems with the word 'sovereignty,' because sovereignty is a definition that comes from a European governance system based on monarchy and empire. And I'm really not interested in monarchy and empire. They have no resilience, they have really nothing to do with who we are."[44] Mohawk scholar Taiaiake Alfred notes that the concept of sovereignty originated in Europe and denoted a single divine ruler. Describing sovereignty as "an exclusionary concept rooted in an adversarial and coercive Western notion of power," he wonders why more people have not questioned how a European idea and term became so central to the political agenda of Native peoples.[45] On the other hand Joanne Barker has argued that "sovereignty" has no fixed meaning because the term is embedded within the specific social relations in which it is invoked and given meaning, determined by political agendas and cultural perspectives. She argues that in its links to concepts of self-determination and self-government, sovereignty insists on the recognition of rights to political institutions that are historically and culturally located.[46]

Contemporary understandings of sovereignty have included understandings that nations are autonomous and independent, self-governing, and generally free of external interference.[47] Tribal sovereignty has come to include the authority of tribal governments to engage in a range of activities, including determining citizenship, regulating on-reservation commercial activities, deploying varying levels of criminal jurisdiction, overseeing natural resource management, providing child welfare and social services, and more. Moreover, sovereignty serves as the legal framework for most American Indian rights claims.[48] Advocates have operationalized the term food sovereignty as a means of leveraging this cultural, political, and economic autonomy for the purposes of revitalizing food systems.

INDIGENOUS FOOD SOVEREIGNTY

In the context of the very specific meanings of the term "sovereignty" for many Indigenous individuals, and specifically for Native American and First Nations communities, as well as because of both the very particular cultural connections to land and the political relationships to settler-colonial governments, scholars and activists (and scholar-activists) have worked to define a specifically Indigenous food sovereignty.[49] These definitions are constructed within a framework that recognizes the social, cultural, and economic relationships that underlie community food sharing and seek to stress the importance of communal culture, decolonization, and self-determination, as well as the inclusion of fishing, hunting, and gathering—not just agriculture—as key elements of a food sovereignty approach.[50] Put simply, Indigenous food sovereignty "refers to a re-connection to land-based food and political systems," and seeks to uphold "sacred responsibilities to nurture relationships with our land, culture, spirituality, and future generations."[51]

Kyle Whyte explains that the "indigenous food systems" at the center of these definitions "refer to specific collective capacities of particular indigenous peoples to cultivate and tend, produce, distribute, and consume their own foods, recirculate refuse, and acquire trusted foods and ingredients from other populations." He specifies that the concept of "collective capacities" describes "an ecological system, of interacting humans, nonhuman beings (animals, plants, etc.) and entities (spiritual, inanimate, etc.) and landscapes (climate regions, boreal zones, etc.) that are conceptualized and operate purposefully to facilitate a collective's (such as an indigenous people's) adaptation to metascale forces."[52] As described earlier, Indigenous communities' abilities to adapt to these forces were intentionally and unintentionally disrupted through the establishment of settler colonial nations. As Karlah Rae Rudolph and Stéphane McLachlan describe, "An indigenous food sovereignty framework explicitly connects the health of food with the health of the land and identifies a history of social injustice as having radically reduced indigenous food sovereignty in colonized nations."[53]

The concept of Indigenous food sovereignty is not focused only on *rights to* land, food, and the ability to control a production system, but also *responsibilities to* and culturally, ecologically, and spiritually appropriate *relationships with* elements of those systems. This concept entails emphasizing reciprocal relationships with aspects of the landscape and the entities on it, "rather than asserting rights over particular resources as a means of controlling production and access."[54] Secwepemc scholar Dawn Morrison describes Indigenous food sovereignty as a framework for exploring the right conditions for "reclaiming the social, political, and personal health we once experienced prior to colonization. But the framework itself does not resolve where the responsibility for it lies." The responsibility lies with Indigenous people to participate in traditional food-related activities on a daily basis, to build coalitions with friends and allies, and to assert and insist on the utilization of Indigenous values, ethics, and principles in making decisions that impact "forest and rangeland, fisheries, environment, agriculture, community development, and health."[55] Because of this focus on cultural relevancy and specific relationships to food systems, cultural restoration is imperative for Indigenous food sovereignty, "generally more so than to non-indigenous food sovereignty."[56]

While some of the previous arguments about the use of terms like sovereignty could apply to the struggle of defining and enacting Indigenous food sovereignty, Indigenous people are defining food sovereignty to their advantage. In their work with the O-Pipon-Na-Piwin Cree Nation, Asfia Gulrukh Kamal and colleagues point out that this community uses the term "food sovereignty" in a way where neither "food" nor "sovereignty" retains its classical meaning. Food, which is typically framed as "consumable commodities," is instead framed under its cultural meaning as the bond between people, health, and land. Sovereignty, rather than being understood as control over land, water, or wildlife is instead framed by this community as a relationship with these entities that allows for the mutual benefit of all parties.[57]

To summarize, Dawn Morrison and the Working Group on Indigenous Food Sovereignty developed four principles of Indigenous food sovereignty: (1) the recognition that the right to food is *sacred*, and food sovereignty is achieved by upholding sacred responsibilities to nurture relationships with the land, plants, and animals that provide food; (2) day-to-day *participation* in Indigenous food-related action at all of the levels of individual, family, community, and region is fundamental to maintaining Indigenous food sovereignty; (3) *self-determination*, or the ability of communities and families to respond to their needs for culturally relevant foods and to have the freedom to make decisions over the amount and quality of food they hunt, fish, gather, grow, and eat; and (4) *legislation and policy support* to reconcile Indigenous food and cultural values with colonialist laws, policies, and mainstream economic activities.[58]

This book opens with "Voices from the Indigenous Food Movement," a collection of statements from scholars, activist cooks, scientists, and community activists who have dedicated their lives to this movement. Indeed, a common thread throughout the following chapters is that food choices, environmental protection, and cultural relationships are deeply personal. Ethnobotanists Steven Bond-Hikatubbi and Carrie Cannon; anthropologist Elizabeth Hoover; doctoral candidate in Native American and American Studies Laticia McNaughton, historian and gardener Devon Mihesuah, cooks Brit Reed, Sean Sherman, and Brian Yazzie; nutritionist Valerie Segrest; insect ecologist and biologist Chip Taylor; and federal Indian law professor Elizabeth Kronk Warner share their personal ruminations about their relationships with foods, the environment, and why they do what they do. They are only a few of the participants in this growing movement.

These personal narratives are followed by substantive chapters, starting with "'You can't say you're sovereign if you can't feed yourself,'" where Elizabeth Hoover expands on this discussion of a distinct Indigenous food sovereignty by exploring how "food sovereignty" is defined and utilized by participants in Indigenous farming initiatives across the United States. Through interviews conducted with participants in thirty-eight tribal community farming projects, Hoover delves into how issues of cultural identity; access to food, land, and information; relationships to the environment, food sources, and other people; and the need for independence to make choices about defining food systems, dovetail with a focus on education, health, and heritage seeds.

Food production and distribution, as well as maintaining healthy environments for farming, hunting, fishing, and gathering involve a complex meshing of social, political, religious, economic, and environmental concerns. The goals of this anthology are both to identify the challenges facing Indigenous communities in revitalizing and maintaining traditional food systems and also to highlight inspiring and successful food and health initiatives in Indian Country. How these challenges affect Oklahoma are discussed in Devon Mihesuah's chapter, "Searching for *Haknip Achukma* (Good Health)." The state is home to thirty-eight tribal nations, and its multifaceted history presents opportunities for discussion about the plethora of political, economic, and social obstacles tribal members face in creating and maintaining food sovereignty and environmental initiatives.

Almost four thousand miles to the west are the Hawaiian Islands, an archipelago that conjures images of white sandy beaches, fresh fish, and readily

PERSONAL AND GRASSROOTS

- Desire for healthy lifestyle
- Activism around food justice
- Security in identity
- Cultural pride
- Reverence for natural world
- Adequate health care
- Backyard gardens
- Relationships
- Agreed-on definition of "traditional foods"

HEALTHY ENVIRONMENT

- Pollinators
- Healthy soil
- Free-flowing clean water
- Clean air
- Protective environmental standards upheld
- Invasive species managed
- Concern for seventh generation
- Animal welfarew

ACCESS TO LOCAL, CULTURALLY RELEVANT FOODS

- Support of tribal leadership, policies, and funds
- Community support
- Grants to tribal organic farmers and ranchers
- Right to control food production
- Job opportunities
- Access to medicinal plants
- Access to land and waterways
- Fair trade
- Tools for food production
- Long-range planning

KNOWLEDGE

- Traditional Indigenous knowledge
- Tribal language
- Elder knowledge
- Formal education
- Tribal schools
- Seed saving
- Community-based and historical research
- Culinary skills
- Tool making
- Nutritional education
- Gardening skills
- Hunting and fishing skills
- Ethnobotanical (or foraging) knowledge

Achieving food sovereignty and good health entails a complex mesh of social, political, religious, economic, and environmental concerns. *Text by authors.*

available tropical fruit. Indeed, prior to colonization, Hawaiians had a multitude of resources and were food self-sufficient. Today, the development of beachfront tourism infrastructure is disrupting Hawaiian Natives' ability to procure traditional foods. Invasive flora and fauna, overbuilding of waterfront hotels and resorts, and environmental contamination have also caused resource depletion. Now Hawaiian Natives import 90 percent of their foods, contributing to high rates of obesity and diabetes. In "Kua'āina Ulu 'Auamo," Kevin Chang, Charles Young, Brenda Asuncion, Wallace Ito, Kawika Winter, and Wayne Tanaka detail efforts to restore Hawai'i's social-ecological systems by concentrating on humans' relationships with each other and with the land and water. E Alu Pū, which means "move forward together," is a network focused on creating and empowering community-based biocultural resource management under the umbrella of the backbone organization, Kua'āina Ulu 'Auamo (KUA).

Alaskan Natives have undergone a similar transformation, from procuring their own foods through hunting and fishing to depending on imported products. This transition from consuming unprocessed foods to daily use of market, processed foods has resulted in high rates of cancer, obesity, high blood pressure, and diabetes. For Alaskans, the loss of their subsistence lifestyles has caused a decrease in cultural knowledge, including traditional hunting skills, that should be passed to the younger generations. Melanie Lindholm documents in "Alaska Native Perceptions of Food, Health, and Community Well-Being" that the Alaskans she interviewed also expressed concern at the mental illness that results from becoming dependent on another culture's food. Foreign food items are readily available, yet are nutritionally inadequate and culturally inappropriate, a situation that Lindholm refers to as "nutritional colonialism." Lindholm outlines how Alaska Native individuals and communities have adapted to changes in subsistence availability, changes that many of them call cultural genocide.

The Diné (Navajo people) of the Four Corners region of the Southwest were hunters, gatherers, farmers, and traders with access to a variety of plant and animal foods. They moved with the seasons and by necessity stayed physically active. Around 1900, the Diné steadily moved away from garden produce and began consuming packaged goods they procured from trading posts, such as white flour, sugar, candy, cookies, and sugared canned fruit. Serious food-related health problems appeared a mere decade later and exploded in the late 1980s. The Navajo Nation attempted to deal with the burgeoning health crisis by creating the Diné Community Advocacy Alliance (DCAA) in 2012, but that has not been enough. Denisa Livingston and a few others decided to attempt something more radical to combat diabetes and obesity (what she calls "diabesity"): a controversial sales tax on junk food and a reduction of taxes on healthful foods. In "Healthy Diné Nation Initiatives," Livingston chronicles a familiar tale of overcoming numerous challenges, from intratribal disagreements to defining what exactly is "junk food." The goal that most Diné seem to agree on is to focus on achieving a state of good health for tribal members rather than to assign negative connotations to foods many tribal members still choose to consume.

For agricultural tribes, seed saving was crucial for survival. Since contact, flora and fauna have disappeared; ecosystems have been altered or destroyed; and because of boarding schools, missionaries, intermarriage, and economic disparities, many tribespeople have lost cultural knowledge about ancestral foodways, how to save seeds, and how to cultivate, hunt, and gather. Representatives of tribes in North, Central, and South America are currently searching to reclaim their heirloom seeds, and many are fighting cross-pollination with genetically modified seed plants. Protecting tribal heirloom seeds against companies intent on taking the seeds and patenting them is a real concern for many tribes. For example in New Mexico, the Traditional Native American Farmers Association has been lobbying for legislation to protect Indigenous heritage seeds, and the Tesuque Pueblo maintain a robust seed bank to both house traditional seeds and keep them available to local communities.[59]

Seed sovereignty activist Rowen White chronicles her journey of hearing tribal food stories, learning how to plant, and locating Haudenosaunee heirloom seeds among elder farmers, universities, and seed companies in "Planting Sacred Seeds in a Modern World," a chapter threaded through with her expressions of love for seeds, which she calls "my trellis of hope" and the "Earth Apothecary." She describes seeds as "life capsules of memory," a sentiment similarly expressed by White Mountain Apache and Diné chef Nephi Craig, who says in chapter 14 that a corn kernel "is an Indian microchip." One of the founders of the Indigenous SeedKeepers Network (ISKN) as well as the Haudenosaunee Seedkeepers Society, White has been instrumental in the resurgence of interest in traditional Indigenous seed-keeping and agricultural practices.

Traditionally, Cherokees also were agriculturalists. Today, Oklahoma Cherokees wrestle with many of the same food, environmental, health, and socioeconomic issues that other tribes deal with. The Cherokees were removed from the Southeast to Indian Territory during the 1830s, and in 2006, no one could locate a single crop plant that Cherokees cultivated prior to the Trail of Tears, much less prior to 1492. It was not until Pat Gwin, director of the Cherokee Nation SeedBank, visited the home of "Corn Man" Carl Barnes that a few varieties of corn were located to provide the first specimens for the seed bank. Another meeting with individuals working with the Eastern Band of Cherokees provided him with more varieties of corn, squash, and beans, and the Science Museum of Minnesota donated Cherokee ceremonial tobacco.[60] These collective efforts, and the overcoming of myriad challenges similar to those faced by other food projects across Indian Country, culminated in the Cherokee Nation seed distribution program, chronicled by Gwin in "What If the Seeds Do Not Sprout?"

Other tribes also are revitalizing or attempting to maintain their agricultural and foraging traditions. Hopis in the high desert of Arizona continue to dry-farm as they have for millennia. "People of the Corn," by Dennis Wall and the late Virgil Masayesva, first appeared in the *American Indian Quarterly* in 2004, but the content is timeless. Hopis have raised corn by relying on precipitation, runoff water, and prayer for nearly a thousand years. This article describes aspects of a unique relationship between Hopis and the ancient agricultural practice of dry farming that sustains them. Aside from the sustenance it provides, corn enters into nearly every aspect of traditional Hopi life, contributing to development of values, sharing and passing on of tradition, and celebration of and connection with the Great Mystery.

Conversely, some tribes had no agricultural tradition, no farms to remember, no seeds to rediscover. In contrast to what Rowen White and Pat Gwin discuss about seed saving, in "Comanche Traditional Foodways and the Decline of Health," Devon Mihesuah discusses the health ramifications of the Comanche Nation having no agricultural tradition to revitalize. This poses serious problems for those tribes who relied mainly on game meat and are desirous of reconnecting to their traditional ways of eating. Comanches, for example, historically roamed over a vast area of various ecosystems with myriad resources. They consumed mainly game meats (hunting most notably bison and antelope), but also relied on a variety of trade items (as well as food they stole, such as corn, squash, and sheep), gathered wild plants, and by necessity stayed physically active, at least until confinement to a reservation at Fort Sill. Today, few members of the Comanche Nation of Oklahoma have access to wild game and most suffer from obesity, diabetes, and high blood pressure brought on by a sedentary lifestyle and a diet rich in starch, sugar, and fat. It could be that the absence of food initiatives among the Comanche is a cultural issue.

Gerald Clarke writes in "Bringing the Past to the Present" that anthropologists and historians traditionally defined "farming" as the act of owning and controlling land, and the establishment of agriculture as marking the beginning of "civilization." Conversely, California tribes' place-based knowledge, respect for the natural world, and understanding of the concept of sustainability enabled them to cultivate crops and raise animals more efficiently than most European farmers, thus making California "one big farm." Native Californians are restoring the environment through organizing workshops on building bridges among governmental agencies, tribal leaders, and environmental advocates. Backyard planting of traditional flora is one strategy used to avoid run-ins with Forest Service personnel who often disallow the gathering of cultural materials.

Devon Peña focuses on historic Indigenous permaculture practices and how these traditional ways of knowing, combined with necessary adaptions, can restore and transform the land in "On Intimacy with Soil." Peña recalls his grandmother's relationship to the soil she so devotedly tended in order to convey the basic principle of Indigenous agroecological practices: soil is alive. For flora to thrive, the soil must contain healthy microbial life, adequate nutrients, and sufficient moisture, and in order for these conditions to happen, Indigenous farmers "must begin with our standing ground in solidarity with healthy soil." That necessitates the interweaving of environmental wellness, spiritual integrity, cultural resilience, and community health.

There may be thousands of Indigenous peoples who garden and cook according to their tribal traditions, but chefs are among the most visible agents of the Indigenous Food Movement (IFM). To succeed in the competitive restaurant business, and especially to ensure that one's eatery is successful, a chef must create signature dishes and capitalize on a trend. Preparing lovely dishes that taste good is one thing, but fancy plating and Michelin-starred restaurants overlook the component of cultural meanings behind the food items. Also absent from the multitude of headline-grabbing stories about Indigenous foods these days is what chef Nephi Craig calls "the colonial reality": poverty, food-related maladies, environmental destruction, and loss of culture. In "Nephi Craig: Life in Second Sight," Mihesuah recounts conversations with Craig, a White Mountain Apache and Diné chef and founder of the nonprofit Native American Culinary Association (NACA). Although Craig is profiled in prominent publications and has worked in a five-star restaurant, he has no interest in what he calls "The Chefy Phenomenon." At his brainchild, Café Gozhóó, which will open in 2019 on the outskirts of Whiteriver, Arizona, he intends to offer dishes made from local produce geared to his fellow Apaches, arts and crafts, and a full-service gas station. The café is also open to interested outsiders. Most importantly to Nephi, Café Gozhóó will serve as part of the Nutritional Recovery Department for outpatients at the Rainbow Treatment Center, the White Mountain Apache Tribe's addition treatment facility.

Another goal of this anthology is to illustrate that for Native people, discussions about the many facets of Indigenous foodways are subjective—rooted in home, family, health, and culture. A happy, healthy life depends on having available culturally connected resources and the ability to hunt and fish in traditional ways. In "Indigenous Climate Justice and Food Sovereignty," Kyle Whyte revisits his experiences and work across North America to make the case that Indigenous food sovereignty aspires to increase the degree of what he calls "collective continuance" of our peoples. In the collective Native way of understanding our place in the world, our peoples' capacities to adapt to changes, whether natural or human caused, are based on our relationships with human and nonhuman relatives. One of the distinctive aspects of Indigenous climate justice movements is their focus on food as a vector for understanding the relationship between environmental change and climate justice. Today's climate change ordeal represents a particular kind of challenge for Indigenous collective continuance. Environmental changes such as sea-level rise and increases in severe weather, are more difficult to endure because of discrimination against our sovereignty, cultural integrity, and economic vitality by nation-states and corporations.

We are not on the outside looking in at what many non-Natives refer to as a fascinating food trend, like kale chips and pea milk. On the contrary, cultivating, hunting, and gathering our traditional foods, and respecting the land that sustains the flora and fauna we depend on, in large measure makes us who we are.[61] Growing and gathering the same plants as our ancestors, saving seeds, protecting the Earth, and fighting for economic and social equality and justice does more than just honor our forebears. Raising our own food and using fresh, unprocessed ingredients is not only greatly empowering, it also represents an attempt to mend our disconnect to our cultures and to the land that sustains us.[62] Eating healthfully also means we become aware of where our food comes from, how it is prepared, and by whom.

There are hundreds of Indigenous food and health initiatives in Canada, but in this book we focus on the United States.[63] We have chosen specific themes, geographical areas, and activist voices to illustrate the complexity of food sovereignty and health initiatives. Our initial objective was to include a variety of US geographical regions, ranging from Alaska to Hawaiʻi, the Northeast, Southwest, Great Plains, Southeast, and California. However, each of these is a region of myriad ecosystems, weather patterns, and populations. Obviously, we could not focus on them all. There are many more voices and initiatives we could have included and had hoped to represent. Not every cultural group is amenable to spreading their cultural knowledge, however, including how they manage and utilize their resources, such as methods for preparing plants for medicine. Biopiracy, the appropriation of cultural knowledge for profit, is a real threat, as is the theft of intellectual property such as ceremonies involving planting and harvesting. We respect those who prefer to keep such knowledge private.

Addressing treaty abrogations, environmental destruction, cultural food appropriation, racism, food deserts, educational deficits, and poverty takes energy, determination, political expertise, and thick skin. Tribes are indeed different, yet many of their concerns are similar. Native writers, scholars, educators, lawyers, economists, farmers, seed savers, activists, cooks, chefs, and environmentalists with vested interests in Indigenous foodways all contribute to the goal of ensuring that tribes have access to healthy food and resources. Networking to share knowledge is crucial because the challenges of revitalizing and sustaining cultural knowledge and foodways and protecting the natural world are formidable. Our various skill sets, tribal cultures and worldviews, and access to resources and funding, are all valued and needed.

These papers cover a spectrum of topics, but they represent only the tip of the proverbial iceberg of what the more than five hundred recognized tribes in the United States face. The Indigenous food movement is developing so rapidly, and the political climate today is so volatile, that delineating definitive strategies for achieving food sovereignty is premature. It is our hope that these papers will fuel discussions and inspire initiatives.

NOTES

1. For more information see Elizabeth Hoover, *Tohono O'odham Community Action (TOCA)*, September 17, 2014, https://gardenwarriorsgoodseeds.com/2014/09/17/toho-oodham-community-action-toca-sells-az/; Tristan Reader and Terrol Dew Johnson, "Tohono O'odham *Himdag* and Agri/culture," in *Religion and Sustainable Agriculture: World Spiritual Traditions and Food Ethics*, edited by Todd LeVasseur, Pramod Parajuli, and Norman Wirzba (Lexington: University of Kentucky Press, 2016), 315–36.

2. O'odham Solidarity Project, http://www.solidarity-project.org/.

3. Indigenous Food and Agriculture Initiative, *A National Intertribal Survey and Report: Intertribal Food Systems* (Fayetteville, AR: Indigenous Food and Agriculture Initiative, 2015).

4. Kyle Powys White, "Food Sovereignty, Justice and Indigenous Peoples: An Essay on Settler Colonialism and Collective Continuance," in Anne Barnhill, Tyler Doggett, and Mark Budolfson, eds., *Oxford Handbook of Food Ethics* (Oxford: Oxford University Press 2018), 349.

5. For example, see the description of Sullivan's campaign against the Haudenosaunee in Jane Mt. Pleasant, "The Paradox of Plows and Productivity: An Agronomic Comparison of Cereal Grain Production under Iroquois Hoe Culture and European Plow Culture," *Agricultural History* 85, no. 4 (2011): 460–92; and in the nineteenth century, see the stories collected and recorded by the Diné of the Eastern Region of the Navajo Reservation, Patty Chee and Title VII bilingual staff, *Oral Histories of the Long Walk = Hwéeldi Baa Hané* (Crown Point, NM: Lake Valley Navajo School, 1990).

6. Nicholas James Reo and Angela K. Parker, "Re-thinking Colonialism to Prepare for the Impacts of Rapid Environmental Change," *Climatic Change* 120 (2013): 671–82.

7. Karlah Rae Rudolph and Stéphane M. McLachlan, "Seeking Indigenous Food Sovereignty: Origins of and Responses to the Food Crisis in Northern Manitoba, Canada," *Local Environment* 18, no. 9 (2013): 1082.

8. Charlotte Coté, "'Indigenizing' Food Sovereignty: Revitalizing Indigenous Food Practices and Ecological Knowledges in Canada and the United States," *Humanities* 5, no. 3 (2016): 57, https://doi.org/10.3390/h5030057.

9. Jennifer Bess, "More Than a Food Fight: Intellectual Traditions and Cultural Continuity in Cholocco's Indian School Journal 1902–1918," *American Indian Quarterly* 37, nos. 1–2 (2013): 77–110.

10. Valarie Blue Bird Jernigan, "Addressing Food Security and Food Sovereignty in Native American Communities," in *Health and Social Issues of Native American Women*, edited by Jennie Joe and Francine Guachupin (Santa Barbara, CA: Praeger, 2012), 113–32.

11. Katherine W. Bauer, Rachel Widome, John H. Himes, Mary Smyth, Bonnie Holy Rock, Peter J. Hannan, and Mary Story, "High Food Insecurity and Its Correlates among Families Living on a Rural American Indian Reservation," *American Journal of Public Health* 102, no. 7 (2012): 1346, https://www.ncbi.nlm.nih.gov/pmc/articles/PMC3477997/pdf/AJPH.2011.300522.pdf.

12. Richard White and William Cronon, "Ecological Change and Indian-White Relations," in *Handbook of the North American Indian*, vol. 4, *History of Indian-White Relations*, edited by Wilcomb E. Washburn (Washington, DC: Smithsonian Institution, 1988), 417–29.

13. For the Northeast see Elizabeth Hoover, "Cultural and Health Implications of Fish Advisories in a Native American Community," *Ecological Processes* 2, no. 4 (2013): np, https://doi.org/10.1186/2192-1709-2-4; Laurence M. Hauptman, *In the Shadow of Kinzua: The Seneca Nation of Indians Since WWII* (Syracuse, NY: Syracuse University

Press, 2013). On the Northwest, see Kari Marie Norgaard, Ron Reed, and Carolina Van Horn, "A Continuing Legacy: Institutional Racism, Hunger, and Nutritional Justice on the Klamath," in *Cultivating Food Justice: Race, Class, and Sustainability*, edited by Alison Hope Alkon and Julian Agyeman (Boston: Massachusetts Institute of Technology Press, 2011), 23–46.

14. Elizabeth Hoover, *The River Is in Us: Fighting Toxics in a Mohawk Community* (Minneapolis: University of Minnesota Press, 2017); Jamie L. Donatuto, Terre A. Satterfield, and Robin Gregory, "Poisoning the Body to Nourish the Soul: Prioritizing Health Risks and Impacts in a Native American Community," *Health, Risk and Society* 13, no. 2 (2011): 103–27, https://doi.org/10.1080/13698575.2011.556186.

15. Pamela K. Miller, Viola Waghiyi, Gretchen Welfinger-Smith, Samuel Carter Byrne, Jane Kava, Jesse Gologergen, Lorraine Eckstein, Ronald Scrudato, Jeff Chiarenzelli, David O. Carpenter, and Samarys Seguinot-Medina, "Community-Based Participatory Research Projects and Policy Engagement to Protect Environmental Health on St. Lawrence Island, Alaska," *International Journal of Circumpolar Health* 72, no. 1 (2013), https://doi.org/10.3402/ijch.v72i0.21656.

16. Bob Weinhold, "Climate Change and Health: A Native American Perspective," *Environmental Health Perspectives* 118, no. 2 (2010): A64–A65; Kathy Lynn, John Daigle, Jennie Hoffman, Frank Lake, Natalie Michelle, Darren Ranco, Carson Viles, Garrit Voggesser, and Paul Williams, "The Impacts of Climate Change on Tribal Traditional Foods," *Climate Change* 120, no. 3 (2013): 545–56, https://doi.org/10.1007/s10584-013-0736-1.

17. Dennis Wiedman, "Native American Embodiment of the Chronicities of Modernity: Reservation Food, Diabetes, and the Metabolic Syndrome among the Kiowa, Comanche, and Apache," *Medical Anthropology Quarterly* 26, no. 4 (2012): 595–612, https://doi.org/10.1111/maq.12009.

18. US Department of Agriculture Food and Nutrition Service, "Food Distribution Program on Indian Reservations," 2018, www.fns.usda.gov/fdpir/food-distribution-program-indian-reservations-fdpir.

19. Email correspondence between Elizabeth Hoover and Joe Van Alstine, president of NAFDPIR (National Association of Food Distribution Programs on Indian Reservations), January 2, 2017.

20. Rachel M. Gurney, Beth Schaefer Caniglia, Tamara L. Mix, and Kristen A. Baum, "Native American Food Security and Traditional Foods: A Review of the Literature," *Sociology Compass* 9, no. 8 (2015): 681–93, https://doi.org/10.1111/soc4.12284.

21. USDA Food and Nutrition Service, "Addressing Child Hunger and Obesity in Indian Country: Report to Congress," January 2, 2012, https://www.fns.usda.gov/addressing-child-hunger-and-obesity-indian-country-report-congress.

22. Devon A. Mihesuah, "Historical Research and Diabetes in Indian Territory: Revisiting Kelly M. West's Theory of 1940," *American Indian Culture and Research Journal* 40, no. 4 (2016), 1–21.

23. Centers for Disease Control and Prevention, "National Diabetes Fact Sheet, 2011," https://www.cdc.gov/diabetes/pubs/pdf/ndfs_2011.pdf.

24. Kyle Powys Whyte, "Justice Forward: Tribes, Climate Adaptation and Responsibility in Indian Country," *Climatic Change* 120, no. 3 (2013): 518, https://doi.org/10.1007/s10584-013-0743-2.

25. Reader and Johnson, "Tohono O'odham Himdag and Agri/Culture," 329.

26. See, for example, La Via Campesina's website at https://viacampesina.org/en/food-sovereignty/.

27. "Declaration of Nyéléni," February 2, 2007, Sélingué, Mali, http://nyeleni.org/spip.php?article290.

28. "Declaration of Nyéléni."

29. Bina Agarwal, "Food Sovereignty, Food Security, and Democratic Choice: Critical Contradictions, Difficult Conciliations," *Journal of Peasant Studies* 41, no. 6 (2014): 1247–68, https://doi.org/10.1080/03066150.2013.876996.

30. United Nations Food and Agriculture Organization 2001, cited in Hannah Wittman, Annette Aurélie Desmarais, and Nettie Wiebe, "The Origins and Potential of Food Sovereignty," in *Food Sovereignty: Reconnecting Food, Nature, and Community*, edited by Hannah Wittman, Annette Aurélie Desmarais, and Nettie Wiebe (Halifax, NS: Fernwood, 2010), 1–14.

31. Michael Menser, "The Territory of Self-Determination: Social Reproduction, Agroecology, and the Role of the State," in *Globalization and Food Sovereignty: Global and Local Change in New Politics of Food*, edited by Jeffrey McKelvey Ayres, Peter Andree, Michael J. Bosia, and Marie-Josee Massicotte (Toronto: University of Toronto Press, 2014), 53–83; Wittman, Desmarais, and Wiehe, "Origins and Potential of Food Sovereignty."

32. Tabitha Martens, Jaime Cidro, Michael Anthony Hart, and Stéphane McLachlan, "Understanding Indigenous Food Sovereignty through an Indigenous Research Paradigm," *Journal of Indigenous Social Development* 5, no. 1 (2016): 18–37.

33. Menser, "Territory of Self-Determination," 59.

34. Priscilla Claeys, "Food Sovereignty and the Recognition of New Rights for Peasants at the UN: A Critical Overview of La Via Campesina's Rights Claims over the Last 20 Years," *Globalizations* 12, no. 4 (2015): 452–65, https://doi.org/10.1080/14747731.2014.957929. Wittman, Desmarais, and Wiehe, "Origins and Potential of Food Sovereignty."

35. Philip McMichael, "Food Sovereignty in Movement: Addressing the Triple Crisis," in Wittman, Desmarais, and Wiehe, *Food Sovereignty*, 168–85.

36. Raj Patel, "What Does Food Sovereignty Look Like?" in Wittman, Desmarais, and Wiehe, *Food Sovereignty*, 186–95; Meleiza Figueroa, "Food Sovereignty in Everyday Life: Toward a People-Centered Approach to Food Systems," *Globalizations* 12, no. 4 (2015): 498–512.

37. Annette Aurélie Desmarais and Hannah Wittman, "Farmers, Foodies, and First Nations: Getting to Food Sovereignty in Canada," *Journal of Peasant Studies* 41, no. 6 (2014): 1153–73, https://doi.org/10.1080/03066150.2013.876623.

38. See Wittman, Desmarais, and Wiehe, "Origins and Potential of Food Sovereignty"; Figueroa, "Food Sovereignty in Everyday Life"; Madeleine Fairbairn, "Framing

Resistance: International Food Regimes and the Roots of Food Sovereignty," in *Food Sovereignty*, edited by Wittman, Desmarais, and Wiehe (Halifax and Winnipeg: Fernwood, 2010), 15–32.

39. Judith Ehlert and Christiane Vossemer, "Food Sovereignty and Conceptualization of Agency: A Methodological Discussion," *Austrian Journal of South-East Asian Studies* 8, no. 1 (2015): 9.

40. Ehlert and Vossemer, "Food Sovereignty and Conceptualization of Agency," 9.

41. Amy Trauger, "Putting Food Sovereignty in Place," in *Food Sovereignty in International Context: Discourse, Politics, and Practice of Place*, edited by Amy Trauger (London and New York: Routledge, 2015), 1–12.

42. Sam Grey and Raj Patel, "Food Sovereignty as Decolonization: Some Contributions from Indigenous Movements to Food System and Development Politics," *Agriculture and Human Values* 32, no. 3 (2015): 434, https://doi.org/10.1007/s10460-014-9548-9.

43. See Figueroa, "Food Sovereignty in Everyday Life."

44. Winona LaDuke, interview with Elizabeth Hoover, August 29, 2014, White Earth Reservation.

45. Taiaiake Alfred, "Sovereignty," in *A Companion to American Indian History*, edited by Philip Deloria and Neil Salisbury (Malden, MA: Blackwell, 2002), 460–74.

46. Joanne Barker, "For Whom Sovereignty Matters," in *Sovereignty Matters*, edited by Joanne Barker (Lincoln: University of Nebraska Press, 2005), 1–50.

47. Heidi Kiiwetinepinesiik Stark, "Nenabozho's Smart Berries: Rethinking Tribal Sovereignty and Accountability," *Michigan State Law Review* 339: 339–54.

48. Jessica Cattelino, "The Double Bind of American Indian Need-Based Sovereignty," *Cultural Anthropology* 25, no. 2 (2010): 235–62, https://doi.org/10.1111/j.1548-1360.2010.01058.x.

49. The terms "scholar" and "activist" (as well as "community member") are not meant to be mutually exclusive; an actor could ascribe to multiple or all of these designations.

50. Desmarais and Wittman, "Farmers, Foodies and First Nations," 1154–55.

51. Quotations, respectively, from Martens et al., "Understanding Indigenous Food Sovereignty," 21; and Dawn Morrison, "Indigenous Food Sovereignty: A Model for Social Learning," in *Food Sovereignty in Canada: Creating Just and Sustainable Food Systems*, edited by Hannah Wittman, Annette Aurélie Desmarais, and Nettie Wiebe (Halifax, NS: Fernwood, 2011), 97–113.

52. Kyle Powys Whyte, "Food Justice and Collective Food Relations," in *The Ethics of Food: An Introductory Textbook*, edited by Anne Barnhill, Mark Budolfson, and Tyler Doggett (New York: Oxford University Press 2015), 7.

53. Rudolph and McLachlan, "Seeking Indigenous Food Sovereignty," 1081.

54. Amanda Raster and Christina Gish Hill, "The Dispute over Wild Rice: An Investigation of Treaty Agreements and Ojibwa Food Sovereignty," *Agriculture and Human Values* 34 (2017): 268.

55. Morrison, "Indigenous Food Sovereignty: A Model."

56. Asfia Gulrukh Kamal, Rene Linklater, Shirley Thompson, Joseph Dipple, and Ithinto Mechisowin Committee, "A Recipe for Change: Reclamation of Indigenous Food Sovereignty in O-Pipon-Na-Piwin Cree Nation for Decolonization, Resource Sharing, and Cultural Restoration," *Globalizations* 12, no. 4 (2015): 565.

57. Kamal et al., "Recipe for Change."

58. Morrison, "Indigenous Food Sovereignty: A Model."

59. See Elizabeth M. Hoover, "Tesuque Pueblo Farm, NM," July 27, 2014, https://gardenwarriorsgoodseeds.com/2014/07/30/traditional-native-american-farmers-association-tnafa/ and https://gardenwarriorsgoodseeds.com/2014/07/27/tesuque-pueblo-farm-nm/.

60. For information about the Eastern Band of Cherokees' seed-saving efforts, see Jordan Wright, "An Eastern Band of Cherokee Farmer Fosters 'Memory Banking' and Growing of Heirloom Seeds," *Indian Country Media Network*, February 11, 2013, https://newsmaven.io/indiancountrytoday/archive/an-eastern-band-of-cherokee-farmer-fosters-memory-banking-and-growing-of-seeds-qfTMh_tLtk-4OB4MM4cwQg/.

61. Robin Wall Kimmerer, *Braiding Sweetgrass: Indigenous Wisdom, Scientific Knowledge, and the Teachings of Plants* (Minneapolis, MN: Milkweed, 2015); Enrique Salmon, *Eating the Landscape: American Indian Stories of Food, Identity, and Resilience* (Tucson: University of Arizona Press, 2012); Melissa K. Nelson, ed., *Original Instructions: Indigenous Teachings for a Sustainable Future* (Rochester, VT: Inner Traditions International, 2008).

62. Devon A. Mihesuah, *Recovering Our Ancestors' Gardens* (Lincoln: University of Nebraska Press, 2005); Winona LaDuke, "Food as Medicine: The Recovery of Traditional Foods to Heal the People," in *Recovering the Sacred: The Power of Naming and Claiming* (Cambridge, MA: South End Press, 2005), 191–212.

63. See, for example, Food Secure Canada, https://foodsecurecanada.org/; Tides Canada, "Sustainable Food Systems," http://tidescanada.org/focus/sustainable-food-systems/.

1
VOICES FROM THE INDIGENOUS FOOD MOVEMENT

Food choices, environmental protection, and cultural relationships are deeply personal. In this chapter, ethnobotanists, an anthropologist, a doctoral candidate in Native American and American Studies, a historian and gardener, activist cooks, a nutritionist, an insect ecologist and biologist, and a federal Indian law professor share their personal ruminations about their relationships with foods, the environment, and why it is they do what they do.

STEVEN BOND-HIKATUBBI

It is clear to me we need a balance of the old lifeway and the benefits modern technology affords to truly be sovereign. We are coexisting in both worlds, as modern traditional people it is important to overlay not to split our efforts. This fuels my passion as an ethnobotanist and environmental scientist. Sharing these skills with tribal communities to navigate the changes in growing conditions, commerce, and interpersonal relationships can be very technical and necessary. Our families before us had their own set of issues to address during a rapid push for nationalism and integration, struggles where much was lost. Ethnobotany was originally developed to document our cultures as they were disappearing, whereas I embrace the discipline to strengthen our knowledge and embolden traditional practices.

My maternal great-grandparents Gracie (Choctaw) and Orn (Scandinavian) were migratory farmworkers for most their lives, retiring to a subsistence farm and home they built from scratch near Wister, Oklahoma, where my passion for agriculture was nurtured as a child. My paternal great-grandparents Mary (German) and Raymond (Chickasaw and original allottee) purchased a home in Ada, Oklahoma, the first home my family did not build from materials gathered in the woods. Raymond's son became an engineer and pilot in the Air Force, and my dad is a lineman. It was the stories I received on Raymond's couch and the family outings into the woods that nurtured my passion for Chickasaw culture, flora, and fauna, leading to my pursuit of knowledge and development as a scientist. We remain Chickasaw and I remain determined to share, cultivate, and disseminate the cultural elements our ancestors held dear; our ancestral foods and material culture are the embodiment of these values.

CARRIE CALISAY CANNON

To the untrained observer traveling the landscape of northwestern Arizona, the land may appear dry, desolate, and devoid of life. The Hualapai people, however, know the true bounty locked inside the dry desert façade. The Hualapai Indian Reservation runs along 108 miles of the South Rim of the Grand Canyon. Here is a landscape rich in edible, nutritious, life-giving foods. Over the last decade I have had the honor and pleasure to work with the Hualapai Tribe to establish the ongoing Hualapai Ethnobotany Youth Project. The work has been as rewarding as it has been challenging. The Hualapai, like many tribes, lost a significant amount of their traditional homelands and lifeways through the colonization efforts of the last two centuries. Our program is funded through charitable grant donations, and we offer harvesting activities throughout the year. Hualapai tribal elders teach twenty of the tribal youth the traditional plant knowledge throughout the year.

In the spring we travel to the canyon rim to harvest the mescal agave plants and pit-roast the agave hearts for twenty-four hours in an earthen oven layered with rocks heated by a juniper wood fire. Slices of barrel cactus are placed atop the coals to prevent the mescal from scorching while it bakes overnight. This traditional food source yields a high content of calcium, which sustained the people in a precontact diet where dairy was unavailable. In the summer, Ethnobotany Youth Project participants gather the prickly pear cactus fruit, mesquite bean pods, and banana yucca fruits. These are traditional foods Western science is only recently beginning to study for their nutritional value to diabetics. In the fall and winter we follow the pinyon jays into the forest to collect pinyon nuts, acorns, and manzanita berries. In a good hunting season, a hunter will donate elk or deer meat to our program, and we will demonstrate traditional food preservation methods, drying the meat into jerky cured with the local salt harvested deep within the Grand Canyon from ancient salt mines.

Tribal ethnobotany project elder Jorigine Paya recalls eating the traditional foods: "I remember when growing up with my paternal grandparents, we harvested a lot of the traditional food plants, such as the prickly pear, mescal, banana yucca, sumac berries, and the Indian tea. We prepared it in different ways depending on the plant, and we would eat these foods, and never did we have a problem with diabetes." These days Hualapai youth live in a completely different, modern era severed from the traditional lifestyles of their grandparents' memories. During our ethnobotany project harvesting field trips, the distractions of text messages, iPods, and YouTube videos are dialed down. The favorite snacks of hot Cheetos and Mountain Dew are put away, and the attention is brought back to an ancient plant knowledge that sustained generations for millennia, and brought power and life to the people.

The fate of the Hualapai Ethnobotany Youth Project, like so many tribal programs, rests in the hands of funding sources to keep the project going. What does the future hold for the project? Strength in numbers and tribal solidarity among affiliated tribes is one answer. The Hualapai language is a member of the Yuman language family, which includes fourteen distinct tribes that are ancestrally tied to the Colorado River spanning from the Grand Canyon to the Baja Peninsula of Mexico. Besides the Hualapai, among these fourteen Yuman-speaking tribes are several other closely related Pai people, including the Havasupai, Yavapai, and the Paipai of Baja California del Norte, Mexico. The Pai tribes speak closely related languages, share a common religious heritage, and all trace their origin story to a sacred mountain known as Spirit Mountain towering above the Colorado River near Laughlin, Nevada.

Collectively, the Pai tribes' land base lies within a region of the world that is botanically distinctive and rare. The ancestral land base encompasses both Mojave and Sonoran Desert types. The Hualapai Ethnobotany Youth Project endeavors to collaborate with these closely related Pai peoples. Efforts are currently underway for the Hualapai Cultural Center to spearhead the creation of a "Pai-wide" ethnobotany archive. In an age of continuing globalization, human cultures are subject to ongoing homogenization. This collaboration has the potential to restore knowledge from the edge of extinction. Documenting, preserving, and passing on Pai ethnobotanical and land-based knowledge to the next generation are vital steps toward ensuring the survival of a unique and irreplaceable local biological and cultural knowledge.

Transferring ethnobotanical knowledge takes time. It is an all-day effort to harvest enough mescal agave to do a traditional roast. When you talk to a tribal elder they will tell you, "You don't just harvest it and you're done." After you pry the agave rosettes from the ground, you take them back to the community, and it is another daylong effort to remove all the green, thorny leaves to prepare the agave hearts for roasting. Then you prepare the pit, and gather the rocks, the juniper wood, and the barrel cactus, which you de-thorn, slice, and place in the pit. When the fire is lit and all the coals have burned down, the mescal is placed in the earthen oven and buried overnight; it is unearthed the very next day with great anticipation. Students and elders of the ethnobotany project are involved in each stage of this process. The effort is rewarded with a taste of sweet, juicy, calcium-rich roasted mescal that has a flavor like nothing else in the world. In each stage of the process, the elders speak in the tribal language, they say prayers, and they instruct tribal youth on the traditional tribal philosophies.

When asked the significance of the ethnobotany project, tribal elder Malinda Powskey had this to share:

> When we teach the plant knowledge, we do it in the language; *a'ha*, that is "cottonwood," and the name in Hualapai tells you that this plant grows by the water, *hamsi'iv*, that is "cattail," part of that word means "star" for the knowledge that when the plant goes to seed and the fluff blows away in a breeze, the shape of that seed fluff, it resembles a star. If our children lose their heritage language, they lose part of their history, who they are, where they come from. In the teaching of the ethnobotany to the children, we believe it is important for them to think in Hualapai, and connect to this land that is their heritage.

ELIZABETH HOOVER

I grew up with my hands in the soil, turning fresh earth every spring; poking in seeds with my father, mother, grandmother, and sisters; spending all summer weeding and picking off bugs; and then harvesting in fall. I grew up in upstate New York, in the Helderberg Mountains, surrounded by love and chickens and pigs and goats and turkeys and gardens and wild berries. These filled our table all summer, and our freezers and pickle jars for the winter. The production and processing of food symbolized to me what it meant to be self-sufficient, and how people expressed love—through the constant need and desire to feed each other.

Our mountains were adjacent to the Mohawk Valley, the original home of some of my mother's Mohawk ancestors before settlers pushed them north. Now, aside from the gardens planted by the Kanatsiohareke Mohawk community in Fonda, the fields that once supported entwined Three Sisters crops of Haudenosaunee corn, beans, and squash now sprout every spring with uniform rows of field corn and soybeans and gardens of conventional market vegetable varieties. In our home, aside from the wild berries we ate, our seeds and livestock came to us through the local ag store, nondescript breeds and varieties that promised uniform performance and productivity based on our growing zone. It was not until I was living in the Mohawk community of Akwesasne, working with the farming and gardening group Kanenhi:io Ionkwaienthon:hakie, and having discussions with members of the Akwesasne Task Force on the Environment that I discovered the wonders of heritage seeds—treasures passed down among relatives and shared among friends and community members. Some lost, some who found their way home through seed keepers and collectors, and some who had been continuously planted, harvested, and cooked for centuries or more. These seeds embodied connections to the land, the seasons, and the generations of hands who had planted, harvested, selected, and saved those that were best suited for their region and tastes.

But access to traditional foods in Akwesasne was made challenging by a number of factors, not least of which were one federal and two state Superfund sites directly upstream from the community, which destroyed dairy farms and gardens in the 1970s, and then made some fish hazardous to eat.[1] This, combined with a number of other factors—including limited land, a loss of gardening knowledge, a lack of time on the part of community members who had to work within the wage labor market, and a general increased reliance on prepackaged foods—had diminished the number of family and community gardens in

Akwesasne.[2] Groups like the Akwesasne Task Force on the Environment and Kanenhi:io and, more recently, the horticultural arm of the Á:se Tsi Tewá:ton Akwesasne Cultural Restoration Program (which came out of the Natural Resources Damage Assessment), as well as the Oherokon Rites of Passage, have been working hard to restore horticultural knowledge to their fellow community members, to encourage more people to garden and eat healthy food. Conversations at night over cups of coffee after long days of pulling weeds or braiding corn centered around how to fund these garden projects, how to get more community members interested, how to involve more youth.

These experiences led me to become interested in other Native community gardening projects, to find out more about how they were addressing these challenges and working to get healthy food onto dinner tables in their communities. I took to the road, traveling twenty thousand miles around the country to visit with friends I had made over the years through powwows (I have been a fancy shawl dancer for the past three decades), and through food sovereignty summits like the one hosted by the Oneida Nation of Wisconsin and the Indigenous Farming Conference at White Earth in Minnesota. I was able to see firsthand how forty different community farming projects were working to address issues of health disparities, food insecurity, and cultural attrition in their nations. It was inspiring; Indigenous people are resilient, even in the face of impossible odds coming from outside and from within. I now work as part of the executive boards of the Native American Food Sovereignty Alliance and the Slow Food Turtle Island regional association to support these kinds of projects and to create and strengthen broader networks between and among these community food projects. I still keep a little garden, here on the campus of Brown University where I now work as a professor. But now I cultivate beautiful heritage seed varieties—plump scarlet runner beans, climbing dark red cranberry beans, blackish purple Haudenosaunee sweet corn, towering Seneca stripe sunflowers. Seeds are now passed among friends and relatives rather than purchased from a conglomerate. And this is how we will persist.

LATICIA MCNAUGHTON

I am Kanien'kehà:ka/Six Nations Mohawk, and six years ago, I found myself sick, overweight, diabetic, and facing numerous metabolic and autoimmune health problems. I decided I no longer wanted to feel bad, and I would do everything in my power to change that. With the help of a naturopathic doctor, I began a process of elimination in my diet that was not easy. I immediately removed the harmful "colonizers' foods" from my plate, including refined sugars, dairy, white flour and wheat/gluten, white potatoes, and processed chemicals like artificial sweeteners. Furthermore, I eliminated foods that would affect my diabetes negatively, like high-glycemic fruits, dried sweetened fruits, juices, refined starches, and processed foods in general. Today, I am seventy-five pounds lighter than I was and have seen improved health. My Hemoglobin A1c (a measure of blood sugar control over the past three months) and other health markers have improved remarkably.

This was just the beginning of my quest for health. My interests in food grew deeper along this journey. I tried to become more conscious of where my food came from, whether it was organic or genetically modified or grown sustainably. I thought about who produced the food, the labor and hands that prepared that food, whether products were "fair trade," whether the animals were raised without cruelty. I searched labels for terms previously foreign to me like "grass-fed," "free-range," or "cage-free." I eventually questioned the act of relying on a grocery store for sustenance and began to explore gardening more seriously, connected with my community for traditional and wild-caught foods, looked into local growers' products, and dehydrated and preserved my own food at times.

This experience ignited a wellness journey that has changed my ideas about and relationship with food. With some help along the way, I am making my best attempt to restore my diet based on ancestral and Indigenous forms of sustenance adapted to modern-world realities. A friend and I had a conversation years ago, and we concluded that in order to take good care of ourselves these days, we need to be "food revolutionaries" and fight for the appropriate sustenance for our bodies because the industrial food system is not designed to be in our best interests or best for Yonkhinisténha Owéntsyia (Mother Earth). Supporting food sovereignty is about being advocates for better health for our bodies and nations by eating the foods we were meant to and revitalizing traditional practices. It is not merely decolonizing, but re-indigenizing and breathing life back into the traditions that sustain us.

Currently, I am a PhD candidate writing a dissertation titled, "Tetewatskàn:hons ne Sewatokwà:tshera (We Eat from the 'Dish with One Spoon'): Reclaiming Haudenosaunee Food Traditions." I explore the food traditions, diet, and practices of the Native American Haudenosaunee nation, more commonly known as the Iroquois. Located in the Great Lakes region and beyond, the Haudenosaunee Confederacy communities have a rich history of Indigenous foods and food sovereignty practices. At the same time, US Indian policy history has had a tremendous colonizing impact on Native American diets and bodies. My research examines the ways Haudenosaunee people today negotiate these traditional food practices with contemporary realities and the potential for reconnecting to Indigenous knowledge, foods, and language through community-engaged activism. My personal health story enriches my research in many ways.

Before my diet evolved, I was inspired by Seneca scholar John Mohawk's work calling for the decolonization of the Native diet to address alarming diabetes rates and health issues in Indigenous communities. He emphasized that to promote health, "exercise is important, food is important; self-esteem and a long list of other healthy things are important; and sharing a path toward healing is the most important of all."[3] It is rooted in this last idea that the Indigenous Food Revolutionary blog was started . . . we are stronger by sharing a path toward healing together. In our own actions individually and as a community, we can choose to heal and prevent the diseases associated with colonizing foods. This revolution happens in the little backyard garden producing tomatoes and peppers. It is in the maple syrup boiled down from the sap that rose earlier in the spring. It happens with the community hunts and feasts sharing venison and rabbit. Or it is in the simple *nia:wenh* (thank you), expressing gratitude, said after finishing a meal. There is a power of healing through food that connects us in so many deep and complex ways to each other and all of creation. Food sovereignty work is a mode of healing, a rebuilding of nationhood through individual bodies and nations. The contributors to this volume are doing this work in tremendous, multifaceted ways. I share my story alongside them with great respect and gratitude. *Niawenhkó:wa*.

DEVON A. MIHESUAH

My *apokni* (grandmother) Eula was tall and quiet, with heavily hooded eyelids she passed on to her descendants. Her great-grandfather was of the Okla Hannali clan and one of the Choctaw headmen who felt he had little choice but to sign the infamous removal treaty, also known as the 1830 Treaty with the Choctaws, or the Treaty of Dancing Rabbit Creek. Eula's grandfather was a Nanulhtoka (lighthorseman), the sheriff of Sugar Loaf County, Moshulatubbee District; her grandmother was Chickasaw, while her great-uncle was the son of the Chickasaw Chief Ochantubby.[4] Nana belonged to the Indian Grandmothers' Club in Muskogee, Oklahoma, and the Pocahontas Club, whatever that was.

According to my childhood interpretation of life, she seemed the opposite of her boisterous Irish and staunchly Catholic husband, Tom.

I spent a lot of time in my grandparents' kitchen, either eating or sitting at the table playing with something until meals were ready. I never waited long. Nana cooked her dishes, especially the ones made in summer, quickly from fresh ingredients straight from her large garden. If she needed an onion or an herb, it was easy enough to walk outside and gather more or to reach into her pantry of dried goods. I never saw her eat meat, but she often cooked my favorite fish—*nakishtalali* (catfish).

I dutifully followed Big Tom through his straight rows of *tanchi* (corn), *isito* (squash), *ahe* (potatoes), and *tobi* (beans). Thick *bissa vpi* (blackberry bushes) grew around the fence line. Peach trees stood outside the garden perimeter, as did a small tool shed. Okra stalks loved the heat and grew more than six feet tall. One morning in early summer he gently pushed his pitchfork into the soil next to some wilted greenery and he pried up a pile of potatoes covered in dirt. I was only around four years old, but I recall my astonishment at learning potatoes grew underground and not on trees. Later in the summer, Nana cooked green pole and bush beans, crookneck and zucchini squashes, sweet corn, okra, purple and red tomatoes, and watermelon, and transformed them into a plate of food that I prefer over any other. I cannot think of another meal that evokes in me such a physical and emotional reaction as the distinctive and earthy flavors of that garden plate.

These foods—along with *nita* (bear), *issi* (deer), and *fukit* (turkey)—are the same foods my ancestors cultivated and hunted. Almost every Choctaw family made the staple *tamfula*, created from finely ground and sifted corn, water, and wood ash lye. One might add walnuts, pecans, squirrel, or turkey to the mixture and allow it to cook all day. Another dish I often prepare instead of bread is *banaha*, made of cornmeal, hickory oil, and meats boiled in corn shucks. It looks and tastes similar to Mexican tamales, but because there is no lard in the mixture, it has a coarser texture.

My first novel, *Roads of My Relations*, is based on seven generations of my family stories, with the family garden serving as a symbol of cultural continuity and a source of emotional strength through each generation, especially after the arduous and horrifying ordeal of removal from the Southeast to Indian Territory in the 1830s. At my homes in Flagstaff, where the growing season is short and the soil is ashy, and now in Kansas, where summers might be intensely hot or too wet for corn and squash, I have attempted to emulate my ancestors' gardens. I grow plants in raised beds, a large in-ground garden, and a modest greenhouse where I can control the moisture and sun intensity with makeshift shades of old bedsheets and curtains. Smaller cold frames are constructed of rebar, PVC pipes, binder clips, and painter's plastic. If you want to grow something, you will find a way.

I am not trained in culinary arts, but I can cook simple and nutritious meals with produce from our garden and fruit trees; fish from our pond; and deer, turkeys, and quail my husband, Josh, hunts on the family allotment outside Duncan, Oklahoma. Josh is Comanche and has no agricultural legacy. His ancestral memories, and the memories of his father and grandfather, are of meat, not of corn, squash, or beans. His great-grandfather Mihesuah hunted bison, and his grandfather, Joshaway, was forced to attend boarding school, where he learned farming as a way to survive. Ultimately, he was one the few Comanches who found success as a farmer, but he still hunted—mainly turkeys, squirrels, and deer.

Our tribes are facing unprecedented health problems directly related to diet, poverty, and lack of knowledge about nutrition. Why this happened is complicated. After my Choctaw ancestors settled in the Kully Chaha (High Spring) township in the shadow of Nvnih Chufvk (Sugar Loaf Mountain), the environment deteriorated from overhunting and overgrazing. The tribe fractured along party lines, and an economic class system developed that continues until today. My ancestors are buried amidst the pastures where cows graze. It is painful to know that this ecosystem was once lush with flora and fauna. Now it is stark and stripped of its beauty. That is a typical scenario throughout the Choctaw Nation and, indeed, throughout Oklahoma. Resources have been depleted; Indigenous knowledge has been lost and disparaged because of the influence of missionaries, boarding schools, and intermarriage with non-Natives.

Through my cultural and familial knowledge, scholarship, food photography, speaking, and teaching, I try to educate others about the consequences of inactivity and poor diets and the physical, emotional, and spiritual benefits of revitalizing our traditional ways of eating. My book, *Recovering Our Ancestors' Gardens: A Guide to Diet and Fitness* (2005) won the Special Award of the Jury of the Gourmand World Cookbook Awards and was finalist for Best in the World Cookbook. Despite those accolades, that book has been forgotten amidst the voices that have recently emerged in the food movement. I acknowledge that the book was lacking, and merely a baby step in chronicling the realities of the immense health and environmental challenges we are facing. It was only the precursor of what I wanted to do. Nevertheless, I continue my message by managing the website *American Indian Health and Diet Project* and the Facebook page *Indigenous Eating,* and try to show by example that even home cooks with modest kitchen skills can create nutritious and appetizing meals.[5] Each fall, as an alternative to the American Thanksgiving holiday, I facilitate the "Week of Indigenous Eating" event that challenges Natives to consume only foods of this hemisphere. I advocate for traditional and manageable backyard patches, the small family gardens that were cultivated by many tribes. In addition, through my other writings about racism, stereotyping, violence in Indian Country, misogyny, and inter- and intratribal factionalism, I try to impart that any food sovereignty initiative is a multifaceted process and that reaching that goal will not be easy.

BRIT REED

A major part of my development while growing up centered on time spent in the kitchen with my family—cooking with my parents and eating meals together at the kitchen table. Through these acts, I grew to learn how food has the ability to bring people together and strengthen relationships. I carried that lesson forward through my education and into a career as a community-based cook.

Evergreen State College's Reservation-Based, Community-Determined Program teaches upcoming leaders how to guide their communities in a good way, through learning about tribal economics, leadership, health, and more. As a student there, I began learning about tribal food (in)security and sovereignty as it relates to the health of communities, and the importance of ensuring community access to traditional foods. While in the Master in Public Administration (MPA) Tribal Governance concentration at Evergreen State College, I continued my focus on tribal health, food (in)security, and sovereignty.

The Facebook group Food Sovereignty is Tribal Sovereignty (FSiTS) began as an assignment in a food policy class. It started with a few invited people and has grown to more than seven thousand members from all parts of the Indigenous food system and movement—people who are gathering foods, hunting, or growing traditional foods in their gardens or in agricultural settings, as well as academics, policy makers, people running programs in their communities, and Native chefs and cooks. It is both a database and a talking platform—there is a really beautiful dialogue that happens with the group, and it is a great resource for any Indigenous person interested in Indigenous goods, whether you are looking for a recipe or you have questions about particular ingredients or food policy. Although FSiTS is a great place to share information, there is also however growing dialogue regarding the challenges of colonized ways of thinking regarding how information is shared, conflicting with traditional ways of thinking that rather than traditional food knowledge being shared on an internet platform, people need to be going back to communities and asking permission from people for that information. Going forward we will continue to learn about different protocols as they develop for different communities regarding sharing traditional food knowledge and intellectual property, and what communities themselves will see as the best ways to preserve, utilize, and pass down information.

Having spent much of my time writing and researching about food as a part of my academic life, while also working with Indigenous women in a culturally traditional kitchen, I felt that I should move beyond simply theorizing and writing about food and health and transition to gain real-world skills and knowledge to more fully serve the community. With this in mind, I enrolled in the Seattle Culinary Academy. During my MPA capstone I spoke with professional Native chefs about how food policies are affecting their ability to serve traditional foods to the public—this experience gave me a heads up to what it is like to be a Native in a professional kitchen. While culinary school was a great place of learning and mentorship, it also became, for me, a battleground to ensure that Native contributions to national and international cuisines were recognized alongside European contributions. Additionally, informing future and current culinarians about sourcing foods such as salmon and shellfish from Nisqually, Muckleshoot, Lummi, and other tribes in the Coast Salish region became important.

I have been blessed to have cooked for participants in the Yappalli Choctaw Road to Health project, run by Dr. Karina Walters (Choctaw, University of Washington), Dr. Michelle Jennings (Choctaw, RICH Program), and others. This project gathers cohorts of women from the Choctaw Nation of Oklahoma and works with them to address different health issues, such as diabetes, heart disease, and addiction in part through a ten-day walk retracing the Trail of Tears walked by their ancestors.[6] I am also currently working with the Tulalip Health Clinic diabetes program, where I teach healthy recipes and cooking classes. This gives me the opportunity to talk to elders to learn about their food systems to ensure that the program is culturally relevant.

Bringing recognition to traditional southeastern Indigenous foods has been an important aspect of my work. Well before I stepped through the doors of a professional kitchen, I was inspired by Cherokee chefs Taelor Barton and

Bradley James Dry, who learned the art of Cherokee traditional cooking from their families. Chef Taelor Barton is the granddaughter of Edith Knight, who was declared a Cherokee Nation National Treasure. Taelor and Bradley have both been working to put Cherokee dishes on restaurant, pop-up, and catering menus in Tulsa and across the Cherokee Nation for a number of years now. Additionally, I have been blessed to share the kitchen and conversations with fellow I-Collective member David Rico, a member of the Choctaw Nation of Oklahoma. David spent his time at Yale studying the link between chronic diseases and land destruction following the removal of Native people from traditional foods. He is an incredible cook who has been dedicated to perfecting his craft on the restaurant line and is in constant pursuit of learning more about Choctaw traditional foods.

My newest culinary endeavor is as a member of the I-collective, a group of Indigenous chefs, activists, herbalists, seed keepers, and knowledge keepers who are working to create a new narrative that not only highlights historical Indigenous contributions, but also promotes their community's resilience and innovations in gastronomy, agriculture, the arts, and society at large.[7] I took part in a four-day series of pop-ups in New York City in an effort to flip the script of the myth that Americans have about Thanksgiving, as well as to educate people about issues of culinary appropriation and its impact on Indigenous people. In addition, while we were there we worked in solidarity with people from Palestine to talk about the struggles they are facing as a result of US and international policy, as well in support of workers from Tomcat Bakery who were handed over to ICE by their own employers. We are working to employ gastro-diplomacy to educate the public about Native people through food.

Going forward, one of the most important issues we need to consider is how to cook with and share Indigenous foods in a way that is sustainable. This includes not overharvesting or overusing resources, especially by non-Native people who are just learning about these foods that Natives have had in their meals forever and that Native communities need to sustain themselves. This also includes the sustainability of knowledge transmission and use: in a society where information is considered up-for-grabs, it is important to transmit to the general public the problems of culinary appropriation and the importance of Indigenous people maintaining their teachings and taking this opportunity to tell their own stories. It is also important to be able to take care of the relatives that Native people have agreements with—the animals, the plants, the water, whatever or whoever it is. I hope to continue to support Native businesses and projects that sustain people working with traditional foods. I want to make sure community food producers are as well known as the chefs in this movement. I also hope to continue to address issues of health through healthy traditional cooking, and to help bolster coming generations of people who want to be cooks.

MARTIN REINHARDT

At the 2017 Indigenous Farming Conference, I proposed that Indigenous foods are often identified in at least two important ways: in relation to themselves and their surroundings, and in relation to Indigenous peoples. For Indigenous peoples, I proposed that we may have deep historical, shared, spiritual connections to the foods through ancestral food experience, and that they are cultural relatives and part of Indigenous peoples' traditional identities.

Some of my favorite and earliest memories of Indigenous foods are of fishing and gathering berries on Sugar Island, a place we call Ziisbaakwat Minis in our Native language—Anishinaabemowin. The island is located in the St. Marys River in an area known by our people as Baweting. Although the United States claims the island as part of the Upper Peninsula of Michigan, many Anishinaabe people believe that we never ceded it and that it remains part of our sovereign territory.

My Anishinaabe Ojibway grandmother would spend countless hours with us grandkids, picking berries, helping us bait hooks, and teaching us how to cast when we were little. As we got a little older, our uncles and older cousins would take us out spear fishing in Baraga Bay and on Bay de Wasie.

It never occurred to me that we were exercising our aboriginal and treaty rights to fish until I got a little older and I experienced it firsthand when the US Coast Guard approached our boat as we were out spearing. My uncles threw our equipment into the water. As they moved the boat away from that location, they told us younger guys to remember where it was because we were going to dive in and get it after the "Coasties" left. I am not going to lie, it was cold and scary to dive into those deep, dark waters and feel around for the equipment.

When I returned from my military experience in South Korea in 1990, my cousins and I decided to go out spearing one night in Bay de Wasie. As we trolled around near a small island out near the ship channel looking for walleye, I heard a shot ring out. I was startled at first, then my military training kicked in and I unplugged the headlight from the car battery and told my cousins to get low in the boat as another shot whizzed by, closer to our boat. Thank goodness, whoever shot at us either was just trying to scare us, or was a really bad shot, because we did not get hit. We floated around for about a half hour in the dark and then made our way back to shore without the light.

No one in our family ever bothered getting fishing licenses or asking permission to hunt or gather berries. This was just something we did because we liked to do it. It provided our family with foods that we liked to eat. It was a way for us to access food that we could not afford if we had to purchase it in a store. Meat and fresh fruits were expensive, and we were poor. Fresh fish, venison, and berries were gifts of food from Mother Earth that did not require us to have cash.

As I look back now, I realize how important those days were in shaping my relationships with Indigenous foods. I think about how lucky I was to have had those experiences, and how they have helped shape my identity as a mixed-ancestry Anishinaabe Ojibway person. I also think about the non-Indigenous foods that played a significant role in my early years, like commodity foods that we got from the USDA.

Oh, how I loved commodity cheese! The very unhealthy meals we made using ingredients like white flour, sugar, salt, mystery meat, and cheese were all too common and filled in the gaps between the more nutritious meals we made with wild game, fish, and fresh berries. Although "commods" were comfort food that reminded us of family and home, they were also taking their toll on our health. Some of the worst health conditions for Indigenous people in the United States are rooted in poor nutrition. The connection between commods and poor health is so apparent that people jokingly call obesity commod-bod. I have witnessed the negative impact of commods and other junk food on my family and community over the years in the forms of heart disease, high cholesterol, obesity, and diabetes.

Fast-forward to 2010 and the preparation for the First Nations Food Taster at Northern Michigan University. As we worked to prepare a meal for the community that reflected Indigenous culture, I really began to ponder the relationship between the foods we commonly eat today and those that were eaten by our ancestors prior to colonization. Commods were never part of a precolonial diet. The ever-popular fry bread was also not part of our traditional Indigenous food patterns. This line of thinking eventually led me to ask a question that would ultimately grow into a full-blown research study, "If I wanted to eat the foods my Native ancestors ate, what would I have to know and do?"

We called the research study the Decolonizing Diet Project (DDP), and began researching other similar studies that had been conducted previously. I asked Devon Mihesuah and Jim St. Arnold to act as advisors for the project, and we began the planning phase. The research team developed a master food list and investigated ways to access those foods on a local level. We recruited twenty-five research participants, including both Native and non-Native people, and launched our implementation phase in March 2012.

I, along with one other person, ate Indigenous foods from the Great Lakes Region at a 100 percent commitment level, and the remainder of the participants ate them at 25–99 percent levels. We all kept an online journal of our experiences, and many took photos and made videos regarding certain aspects. We were all expected to increase our physical activity to better match a precolonial level of daily exercise. We also all got regularly scheduled annual physicals and quarterly checkups.

I examined the results of the DDP in a multidimensional fashion, focusing on biological, cultural, and legal-political outcomes. My daughter Nim also did a three-year follow-up study. The data showed significant results on both group and individual levels. On a biological level, we were very happy to report that group data showed statistically significant reductions in weight, girth, and BMI. Individuals also experienced noteworthy or significant reductions in blood pressure, cholesterol, and blood glucose levels. As participants drifted away from the DDP commitment, however, there was a decrease in positive biological outcomes. On a cultural level, we concluded that family and community support were very significant to the sustainability of these efforts, and that convenience and price played major roles in which Indigenous foods participants selected, both during and after the implementation phase. Most of the participants reported that they learned a lot about Indigenous foods during the DDP, and that this continued to help them in making better food choices afterward.

On a legal and political level, treaty rights and boundaries seemed to make a big difference in how participants accessed certain Indigenous foods. Some participants were citizens of federally recognized tribes, and some were not. Northern Michigan University sits near the boundary between the Treaties of 1836 and 1842, and there were participants who had treaty rights on one side or the other. In a subsequent study, I found that 295 of the treaties signed between the American Indian tribes and the United States had food-related provisions in them.

I am now preparing for another study called Indigenous Foods TEKnology in the Great Lakes Region. The purpose of this study is to document ten to twelve Indigenous food activities with individuals who are willing to share their traditional knowledge with the general public. The TEK part of the title refers to traditional ecological knowledge. TEK has been defined in various ways, but essentially it refers to the type of knowledge that Indigenous people inherit from our ancestors about our surroundings. In this case, I am focusing on how that knowledge can inform us about food.

It is my belief that we are witnessing what our ancestors prophesized as a time of great struggle. We will have to make big changes in the way we live (or do not live) as humans on Mother Earth. We can and must help those who have oppressed us, and the nonhuman peoples, learn a better way of living that is sustainable and loving to Mother Earth. This is the only way we will be able to return to a state of true food security and sovereignty.

VALERIE SEGREST

The Puget Sound is blessed with majestic mountain ranges that rapidly descend to the whitecaps of the Salish Sea. Our mountain ranges produce swift-running rivers, feeding ancient food forests, calling on wild salmon nations. The ceremonious return of the coho, chinook, chum, pink, and sockeye salmon feeds the soil, the animals, and the very waters of the Salish Sea. On a clear day, when Mount Rainier is in its full magnificence, I find myself observing its bighearted presence and I think, "I'm hungry." I crave the taste of that mountain and daydream about putting it in my mouth and eating it whole. Does that sound crazy?

Up until a few years ago I thought it was pretty crazy and would keep thoughts like that to myself. Then during a routine mountain huckleberry harvest, it occurred to me that I was, in fact, eating the mountain. You see, we have a teaching in my culture that the harvester is required to perform a taste test on each bush we harvest from. I like to sing a little song to myself I learned from a story about a chipmunk girl, and it goes, "One for me, one for the basket, two for me, two for the basket." While harvesting and singing, I was also eating the mountain.

Not only are huckleberries a sacred food that is crucial to the Coast Salish Native diet, they are among our most highly esteemed teachers and produce a tremendous amount of cultural pride. In our oral traditions, the huckleberries are the blood of the Earth and are meant to feed our spirits health and joy. This makes sense to me, because I have never felt anything less than joy while harvesting, eating, and sharing these precious gems, or any berry for that matter.

For the Coast Salish people, the berries, the wild game, the salmon, and the roots and plant people are our living heritage and our true local foods. These foods are our absolute link to the land, gifting us with a sense of place and weaving together our social fabric. So, for me, eating the mountain is about a deep food culture, one that provides an understanding and a cultural realization sent to us by the ancestors and delivered directly to our sense of taste through simply eating ancient huckleberries.

For me, participatory food practices like these make eating a profoundly essential and transformational experience. Think of the most precious thing in your life, that one thing you wholeheartedly adore, that you cannot resist wrapping your love and fierce protection around. That is what food sovereignty means to us. We are actively praying, loving, tending, cultivating, eating, and protecting what we love—our ancestral foods. We are inviting that food and all of its ancient nourishment to come home to our hearts and feed our spirit. Why not? This is what our ancestors have done since time began so that we can uphold our health. Is it that much of a surprise we would not just abandon our foods for the mystery of supermarket offerings?

In the process we animate a culture of health that empowers our traditional ecological knowledge, raises up our food producers and all the wisdom they carry, and ultimately strengthens our civic responsibilities. One thing that restoring a traditional food culture unapologetically requires is that we abandon our consumer mindset and become citizens of our community. It means we are not in a grocery store buying blueberries and making an impersonal transaction. We are in high mountain meadows eating those huckleberries, storing them up for ourselves and our families, and making sure we harvest enough to share with precious people in our lives who are not able to make it out there. Ultimately, we encourage the health of others by participating in this way and really becoming a health practitioner. That is the prayer answered. That is what it will take to strengthen our sovereignty, and that is why we do this work.

For the past ten years, I have been honored to witness the collaborative efforts throughout our region, and really the world, to strengthen food sovereignty. In Muckleshoot, my community, we have partnered with many stakeholders across tribal programs. At our tribal college we facilitate community-based research, which has led to developing a nutrition curriculum that uplifts our traditional food culture and empowers tribal members to eat a traditional-foods diet in a modern world. We have installed a native berry teaching garden, which expands itself every year. We have worked with our tribal cooks to develop the Traditional Foods Protocols of the Muckleshoot Kitchens, ensuring that traditional foods are offered to more than one thousand community members once per week. The rest of the menu is based on seasonal and local inspiration. We have launched Native

Infusion: Rethink Your Drink, a campaign that encourages our people to put down the sugary beverages that are directly linked to many health disparities, like diabetes and tooth decay, and to pick up their ancestors' beverages: clean spring water, nettle tea, rosehip tea, Douglas fir tree tips, and the list goes on.

We have developed traveling curriculum kits called Cedar Box Teaching Boxes that focus on fourteen quintessential Northwest Native foods and are meant for use in schools and other learning groups. Youth are learning to be leaders in our natural resource fields as we bring more resources into our classrooms and bring youth into field courses out on our traditional territories and cultural ecosystems. They learn about healthy ecosystems, the hundreds of different foods to harvest, and they make medicine for arthritis and blend nutritive teas, then take those ancient remedies to our tribal elders.

For us, food sovereignty looks a lot like people being active on the land, moving through each food source as the season changes, but along the way tending the wild so that it in turn will continue to bless us with its rich medicine. Two hundred years ago, my ancestors were witnessing an apocalypse. I imagine that time looked a lot like the end of the world to them. Pandemics had just spread through the land, wiping out thousands and striking down entire nations of families—something that would be comparable to burning down all libraries and losing the world wide web. Knowledge keepers became scarce, then their knowledge became outlawed and went underground in order to retain our identity. Then this remarkable thing happened, which is the ultimate act of resilience in our known history. Leaders from throughout the Salish Sea gathered in 1855, deliberated, and shared their vision for the future, for all of us. In our territories we call it the Treaty of Medicine Creek and the Treaty of Point Elliot, to this day the most powerful piece of environmental legislation of this land. It ensured that we obtained the right to fish, hunt, and harvest our foods and medicines. In all their wisdom, they wrote the kind of policy that upholds our existence and provides for us the remedy we need to face the problems we are carrying throughout Indian Country. Our teachings remind us that if these foods cease to exist, so do we as a people—we may move and breathe on this land, but we are nobody without our foods.

SEAN SHERMAN

America and Canada have, for the most part, ignored the Indigenous history of the land that they sit on today. In most cities, you can walk a few blocks and find foods from countries around the world. But there is literally nothing representing the people that have lived here for thousands of years. It is really important to showcase that our cultures are strong and resilient, and food is a sign of that resiliency. My goal is to help Indigenous restaurants pop up across North America to celebrate Indigenous knowledge and food systems.

I am a member of the Oglala Lakota Nation and was born and raised on the Pine Ridge Reservation in South Dakota, until my mom moved my sister and me to the town of Spearfish in the Black Hills. She was going back to school and working three jobs to support us, so I started working in tourist restaurants at the age of thirteen to help support the family. I continued working in restaurants all through high school and college, starting at the bottom—washing dishes, bussing tables, prepping in the back—and worked my way up through hard work and self-education. After college I moved to Minneapolis where, four years later, I got my first chef position and jumped into the farm-to-table movement that was up-and-coming at the time. A few years into my chef career, after surviving hundred-hour workweeks and a lot of pressure, I bought a one-way ticket to Mexico and moved to a quiet beach town in the state of Nayarit on the Pacific Coast. There, I was inspired by the local Indigenous people, the Huichol, who have amazing beadwork and stories, and who have shared many colonial struggles and atrocities like our northern tribes, but have really been able to maintain much of their traditional knowledge—making their own clothes and gathering a lot of their own food and medicine. As I became interested in their foods, it hit me that I should be focused on trying to figure out the foods of my own heritage. I had been studying cuisine from other cultures around the world—Japanese, French, Spanish—and I realized that I did not have that much knowledge about my own heritage food. Growing up in the 1970s and '80s on Pine Ridge we had some traditional foods—we hunted a lot and we gathered things like timpsala, chokecherry, and juniper. But we also ate a lot of commodity foods. Since the beginning of colonization, the US government has worked actively to wipe out Native American culinary traditions. I have grandparents and great-grandparents who went to boarding schools where they were forced to assimilate and were introduced to commodity foods. When I first started searching for Native American recipe books, the ones I found included traditional foods like wild rice and buffalo, mixed with things like canned cream of mushroom soup. I began the search to figure out what my Lakota ancestors were eating before the introduction of European foods.

I moved from Mexico to the Red Lodge area in Montana, where I spent a lot of time outdoors, getting to know plants and how to utilize them, while also talking to chefs and elders, and reading a lot of books on history, ethnobotany, anthropology, and archaeology—digging through everything with a culinary lens to try to understand it all. I couldn't just order *The Joy of Native American Cooking* cookbook, so I had to dig deep and gather that knowledge. I was picking up the pieces, teaching myself about wild foods and learning about Native American farming, food preservation, and salt-making techniques—all sorts of processes and technologies that were being utilized here, and then piecing them all together so that I had a foundation where I was able to remove all European influence and focus on what was indigenous to here.

I moved back to Minnesota in 2011 and began to cook dinners using this knowledge. In the fall of 2014 I started the Sioux Chef with barely $4,000 in my pocket, and I have been able to grow the business slowly over a few years until now I have a crew of twelve people who travel across the Midwest and around the country cooking Indigenous foods for events. The Sioux Chef name started as an email handle—the word "Sioux" was created by the French and other communities to designate us, and does not mean anything in our Oglala Lakota language. But as a play on words, "the Sioux Chef" took off. It was never intended to be a moniker for myself, even though people have called me the Sioux Chef; it is really about the whole business. And the Sioux Chef is more than just a culinary business; it is about community building and capacity building. We have studied wild foods; permaculture; Native American agriculture; seed saving; seasonal lifestyles; ethno-oceanography; hunting, fishing, and butchery; salt, sugar, and fat production; crafting; cooking techniques; regional histories; Indigenous traditional medicines, food preservation; fermentation; nutrition; health; and spirituality. All those pieces are things that we think about when we are learning about the cuisines of different Indigenous areas. In addition, our goal is to embrace Native values and ignore the very militaristic kind of hierarchy you see in kitchens designed by European people. We are not like that, we are a team and a family, and we are having fun making dishes taste like where we are.

Aside from the Sioux Chef catering, our first business was called Tatanka Truck, which was a food truck we started with the Little Earth of United Tribes—the Native urban community housing project in Minneapolis. For the menu, we cut out all colonial ingredients and challenged ourselves to cook with only Indigenous foods. Instead of beef, pork, or chicken, we served turkey, bison, duck, venison, rabbit, game birds, and local fish.

Our first cookbook, *The Sioux Chef's Indigenous Kitchen,* was published by the University of Minnesota Press in October 2017. The cookbook, which has

Sean Sherman caramelizing maple sugar on a dessert at a Sioux Chef dinner at the James Beard House, October 2017. *Photo by Elizabeth Hoover.*

more than one hundred recipes for healthy foods, is designed to be an educational tool to share with communities, to give them ideas of how they can apply this approach to their own region. We want to encourage them to create and use similar recipes, using ingredients that they have around them—whether those are heirloom agricultural crops or wild foods—to help improve and maintain their community's health.

We also just founded our nonprofit called NATIFS, or North American Traditional Indigenous Food Systems, which has two main goals: Indigenous culinary education and Indigenous food access. We will be opening a culinary hub in Minneapolis, the Indigenous Food Lab, which will be a restaurant and educational center. The restaurant will act as a live training center where we can bring in people to learn hands-on how to work with traditional foods in a modern way. The second phase of the project will be to reach out to tribal communities to help them develop satellite food businesses in their own communities, offering them a full business model and plan and giving them the tools they will need to create small food businesses while utilizing our main training hub. The third phase of the program will be to replicate this model, creating food hubs in large metropolitan areas across the country, with satellite businesses in surrounding tribal communities.

Our goal has always been to support micro-regional economic systems by prioritizing Indigenous vendors for all of our food purchases. In Minneapolis, we have been buying wild rice from the White Earth Reservation, fish from the Red Lake Reservation fisheries, and heirloom varieties of produce from two local Indigenous farms, Dream of Wild Health and Wozupi. Wherever we serve a dinner, we try to search out Indigenous communities selling food, and we prioritize those purchases and put those pieces on our menus first. Our business plan, developed through NATIFS, is to continue keeping food dollars within Native communities and to make food systems sustainable for everybody.

It is time that people living in North America wake up and know that they no longer get the colonial privilege of ignoring the Indigenous histories and the spirit of this land that was taken. We are all on Indigenous land, and we all have much to learn when it comes to understanding the Indigenous ways of connecting to and caring for this land for the sake of our future generations. As Indigenous peoples, we should be the answer to our ancestors' prayers, to be healthy and to carry on a lot of this knowledge and to bring a lot of the education and knowledge that they had through to future generations, to really make a difference and to really stand strong.

CHIP TAYLOR

I am working on my fourth career. As a trained insect ecologist, I started my career studying hybridization of two species of sulphur butterflies, both common in alfalfa fields in the Southwest. It was fascinating work, but it came to an end when I became progressively allergic to these butterflies. From there, I reinvented myself as a honeybee expert and for twenty-two years led research on the so-called killer bees as they advanced through the Americas. That, too, came to an end as funding priorities shifted in the early 1990s. Again, I reinvented myself, this time as a monarch butterfly biologist through the creation of a program known as Monarch Watch at the University of Kansas.[8] I am continuing in that role, but I have been forced to change once more. Of necessity, at the young age of eighty, I am endeavoring to become a restoration ecologist.

Pollinators—and that includes butterflies, bees, moths, flies, beetles, wasps, and hummingbirds—collect nectar, and often pollen, from flowering plants. These visitations spread pollen, resulting in the production of the fruits, nuts, berries, foliage, and roots and shoots that sustain literally millions of species. Pollinators' key role is sustaining plant diversity, which in turn sustains insect and seed diversity, which sustains insect- and seed-eating birds and small mammals, as well as birds of prey and larger mammals. There is a web of life out there, and we need to sustain and restore it. This is why I am trying to become a restoration ecologist.

When I started Monarch Watch, the monarch population was robust, with perhaps a billion monarchs overwintering in Mexico in the winter of 1996–97. Things have changed, and monarch numbers are down, with the overwintering population averaging less than 20 percent of the numbers in the 1990s. The reasons for the decline are many. Land use across the monarchs' breeding range, especially in the Upper Midwest and the grasslands that were originally home to this species, has changed dramatically over the last two decades. The adoption of herbicide-tolerant soybean and corn varieties, which allowed the crops to be sprayed with glyphosate to eliminate weeds, all but eliminated milkweed from row crops by 2005. This loss of milkweed was significant since a substantial proportion of the adult monarch population each summer originated from larvae that had fed on milkweeds in these fields. The renewable fuel standard, which was signed into law in 2007 with the goal of producing ethanol from corn to reduce our dependency on foreign oil, had the effect of placing a premium on lands that could be used to grow corn. This policy had the effect of driving grain production out of the Upper Midwest into grassland areas. In addition to these losses, land development increases the annual rate of habitat loss for monarchs by 1–1.5 million acres a year. Thus, for every year we do not offset the loss of at least a million acres of habitat, we can expect the monarch population to decline further.[9] The monarch decline is a signal of larger issues and greater losses. The vast prairies and grasslands that were once home to monarchs have been replaced by farms, rangeland, and cities. There are connections here that need to be understood and appreciated.

While I focus on restoring the milkweeds monarchs depend on as larvae through our milkweed programs at Monarch Watch, the issue is much larger than milkweeds. It is about restoring native flowering plants and grasses in general, and that is not easy. For the purpose of restoring milkweeds, we have reached out to numerous individuals and groups, but our most ambitious and challenging project has been working with the Chickasaw, Citizen Potawatomie, Miami, Muscogee-Creek, Osage and Seminole Nation, and Eastern Shawnee Tribes in eastern Oklahoma to help them restore milkweeds and other native plants to some of their degraded landscapes. This effort, known as TEAM

(Tribal Environmental Action for Monarchs), is basic technology transfer. We provide the training on everything from identifying species of native plants to seed harvesting, storage, stratification, germination, growing out, and planting. We are learning as we go and trying to profit from both our successes and our failures. We are working hard to continue this program and also to build on it by reaching out to other tribes through the creation of another program, the Tribal Alliance for Pollinators (TAP). It's our hope that other tribes will see the value of learning how to restore milkweeds and other native plants to their lands, a process that can help maintain or restore connections with plants of medicinal and cultural value.

Along with this work comes a lot of inquiries from the press. Most of the reporters who call me do not have a background in science or even basic natural history. Some ask questions like, "Why should we care about losing the monarch migration?" One reporter framed the question even more bluntly: "Why do you environmentalists protect animals rather than people?" Implicit in this question, and many others, is the notion that by protecting wildlife we are expending resources that could be better spent on people.

The bottom line is that it is in our self-interest to save the monarch migration, pollinators' habitats, and all the life-forms that share the same ecosystems. This issue is not just about insects. It is also about us and the world we want to leave for future generations. And, if you think about all that is implied by such actions, it means that we have to address the even larger issues—the increasing emissions of greenhouse gases, acidic oceans, rising sea levels, and other threats to life as we know it. Everything is connected; there are no separate parts. As monarchs go, so go many, many other things. There are seven and a half billion of us, and we are having a massive impact on a relatively small planet. The pace of change is exceeding expectations. The monarch decline—among many other signals—is telling us we need to slow down and put the brakes on the processes that are leading to degradation of the systems that support life. Saving wildlife is all about saving ourselves from ourselves. It's about the future. KEEP IT CONNECTED, PLANT MILKWEED!

ELIZABETH KRONK WARNER

A sense of place and a rootedness in the environment of my tribal ancestors have always brought me happiness. Growing up in the Upper Peninsula of Michigan near my tribal reservation, the Sault Ste. Marie Tribe of Chippewa Indians, I was surrounded by the bounty and beauty of nature—Gitche Gumee (Lake Superior), white birch, every type of berry, moose, deer, turkey, and vegetation galore. I loved picking vegetables out of my Nokomis' (grandma's) garden. Even to this day, a profound sense of joy spreads over me when I cross over the Mackinac Bridge and enter the Upper Peninsula, seeing my beloved evergreens sprouting from limestone cliffs—friends welcoming me home.

The ensuing four decades, however, have brought noticeable environmental changes, most of which have been to the detriment of my beloved environment. These developments set me on the path of becoming a lawyer and law professor specializing in environmental law within Indian Country. It is not uncommon for my students to become disheartened, overwhelmed by the scope of environmental degradation. But I always remind them that the law does provide options and, in some cases, solutions. As inherent sovereigns preexisting the formation of the United States, tribes possess the power to develop laws regulating and protecting their environments, so long as such laws do not directly conflict with federal laws of general application, such as the Clean Water Act and Clean Air Act.

Utilizing their inherent sovereignty, many tribes are innovating in new and interesting ways within the field of tribal environmental law. For example, several tribes have adopted tribal regulations and policy statements aimed at assisting tribal communities in their adaptation to climate change. The Climate Change Strategic Plan of the Confederated Salish and Kootenai Tribes is an excellent example of tribes leveraging their inherent authority to protect the environment. The Salish and Kootenai developed a plan that incorporates traditional ecological knowledge and also recognizes that climate change may have an impact on culture as well as negative social environmental and economic consequences. The focus on culture in the strategic plan is consistent with the tribes' overall incorporation of cultural considerations for natural resources in land-use planning. In fact, the strategic plan places a special emphasis on the importance of protecting tribal culture and traditional environmental knowledge, as Section 2.3 focuses on both extensively, in addition to providing excerpts of tribal elders' observations related to climate change. In this way, the tribes utilized their inherent sovereign authority to protect their tribal environment in a culturally appropriate way.

Treaties also provide a potent tool for protecting tribal environments. Tribes have often turned to their treaties with the United States as a way of protecting valuable rights. For example, in the 1970s, the Swinomish Indian Tribal Community successfully asserted its treaty rights to fish, a cultural keystone for the tribe. Historically, federal courts have interpreted treaties in expansive and progressive ways. For example, in 1908, the United States Supreme Court in *Winters v. United States*, 207 U.S. 564, determined that tribal treaties, even those that made no explicit mention of water rights, still reserved water rights sufficient for the primary purposes of a reservation. Similarly, in 1974, the Western District of Washington district court in *United States v. Washington*, 384 F. Supp. 312, determined that tribal treaties provided for a reserved right of tribes to be co-managers of fisheries along with the states, despite the fact that the treaties involved did not explicitly reference such a right. These decisions and others demonstrate the capacity for federal courts to interpret treaties in broad ways in order to protect tribal resources. Moreover, such decisions also demonstrate federal courts' willingness to demand specific action from the federal government on the basis of implicit treaty provisions.

Beyond inherent sovereign authority and treaties, tribes may also turn to federal environmental law to help protect their tribal environments. For example, several federal environmental statutes, such as the Clean Water Act and Clean Air Act, have Treatment as State (TAS) provisions allowing tribes that meet certain requirements to adopt environmental regulations just as states can. TAS status provides tribes with legal authority to adopt environmental standards more stringent than federal or state standards, which in turn can allow tribes to be protective of their tribal environments. For example, Isleta Pueblo, which is downstream from Albuquerque, was one of the first applicants for TAS status under the Clean Water Act. Under its TAS status, Isleta Pueblo promulgated more stringent water quality standards than those adopted by the City of Albuquerque, and the Environmental Protection Agency (EPA) approved the more stringent standards. The City of Albuquerque challenged the EPA approval in federal court. In *City of Albuquerque v. Browner* (1996) the court held that tribes with TAS status could adopt more stringent water quality standards because such actions were consistent with "powers inherent in Indian tribal sovereignty."[10] Further, at least one tribe, the Navajo Nation, has determined that taking advantage of the TAS provisions helps to promote tribal sovereignty, as TAS allows tribes to enact laws that are not identical to the federal standards. Further, Professor Alex Tallchief Skibine also concluded that TAS status may benefit tribes, as it "allows Indian tribes to extend the reach of their sovereignty beyond the reservation borders."[11]

In short, tribes have many legal options, including those mentioned above and others under domestic and international law, which they can utilize to protect the tribal environment. For this reason, I encourage my students to be hopeful and passionate advocates.

BRIAN YAZZIE

My name is Brian Yazzie, and I'm a Diné (Navajo) chef/cook from Dennehotso, Arizona, now living in St. Paul, Minnesota. I started cooking at the age of seven, helping my mother in the kitchen with basic stuff like chopping up vegetables or stirring soup or making salad. My mother was a single parent who worked double shifts to support us, so as the youngest in the family, I worked to help out through cooking. Preparing a meal for the family and seeing their enjoyment was basically what got me into cooking—being in a kitchen, lost in my own world, creating and plating a work of art is my medicine to comfort and heal. In 2013, after moving to the Twin Cities with my partner, Danielle, I decided to go back to school, and attended Saint Paul College where I earned an associate's degree in culinary arts. As part of my education we learned about different cuisines from across the world, but not about Native American food. When I questioned my chefs and instructors, the answer was that there was no one out there who was putting out cookbooks or talking about these types of foods. Even when we learned about Minnesota staples like walleye and wild rice, the culinary school was sourcing farmed walleye from Canada and paddy-grown wild rice from California. I challenged them on this, and started giving them information about walleye from Red Lake fisheries and wild rice from White Earth. Some of the instructors were offended at having a student bring them this information, but one chef, the leader of the culinary program, understood and has encouraged me ever since.

Starting in 2014 after I met Sean Sherman, I began focusing more on Indigenous foods. Danielle and I had started a Native American club on campus at St. Paul College, and we were looking for a Native caterer—something besides the usual Indian tacos and fry bread. We connected with Sean Sherman; he catered our event, then asked if I wanted to work for him. I now work as the chef de cuisine for the Sioux Chef enterprise, catering meals based on the ingredients available to Native peoples in the upper Midwest region prior to contact, in addition to working on my own projects under the brand Yazzie the Chef.

While there are some Indigenous chefs who are working regionally within their own communities, I am currently expanding on my work as a traveling chef, working with different Indigenous foods in different regions in collaboration with those communities. This can be tricky work, not overstepping my

Brian Yazzie serving buffalo burgers to water protectors at the Oceti Sakowin camp at Standing Rock, November 2016. *Photo by Elizabeth Hoover.*

boundaries as an Indigenous chef from another tribe working with other communities' food and foraging on their land. It is my goal to serve as a resource to these different communities and encourage people to explore their own local, seasonal foods.

Speaking out about the cultural appropriation of food has also been a focus of mine, based on concerns about non-Native chefs profiting off of Native foods. Everyone has to eat throughout the day, and as people are starting to learn more about their food, they are starting to have more interest in where their food is coming from. People are talking about farm-to-table and sustainability, and as part of that they should focus on the revitalization and stories of their own ancestral foods, as opposed to "saving" the foods of others, even if they are well meaning. There are Native community members and chefs and farmers and seed savers who are working hard on revitalizing Indigenous foods. When other people are stepping in to talk about saving or featuring Native foods, they are taking that opportunity away from Native chefs who are trying to do that for their own communities. People who are interested in Native foods should support Native producers and Native chefs, as opposed to profiting from those foods themselves.

I have also used my experience with healthy, traditional foods to support people who are fighting for their Indigenous rights. In 2016 I traveled to the Oceti Sakowin camp at Standing Rock to feed water protectors out of the main kitchen. I first traveled to the camp in August and found that the main kitchen was being run by a vegan chef. With the permission of local elders, I stepped in and cooked bison and wild rice and other traditional foods. Some of the diners were upset that I did not want to cook fry bread for the meals, but the path that I am on is serving healthy Indigenous food. The wheat and lard that make up fry bread are not indigenous to North America. When I entered the pantry tent by the main kitchen, part of one wall had nothing but flour, lard, and baking powder—all the stuff to make fry bread. The moment I stepped into that pantry, it reminded me of what our ancestors went through during the Long Walk, when they were introduced to those ingredients they did not know, basically how fry bread came about. Stepping into that pantry brought back for me negative memories of what our ancestors had to go through. Fry bread is survival food, and there are still people in certain areas who are making fry bread not just as a traditional food, but out of poverty—that is all they have. But for me, I am going to focus on healthy Indigenous foods, and fry bread is one of the breads I do not serve. Danielle and I returned to the camp in November, with a twenty-seven-foot-long U-Haul filled with donations of food and kitchen supplies from co-ops and individuals in the Twin Cities. These supplies went to the main kitchen and the various smaller tribal camp kitchens. They also went toward a Thanksgiving feast we cooked in collaboration with Winona's kitchen and the California kitchen and several other kitchens at camp. Everything came together on eighteen eight-foot tables set up in the giant dome, where we fed a two-hour-long buffet line of people. At that time my way of contributing to the cause and helping with the movement was by cooking. With this pipeline going through, and with the possibility of it leaking, not only will it affect the tribes in the area and the millions of people who reside downriver, but it will also affect our animals and plants and those things that we depend on for food. That was my focus in being here, to bring awareness about the food we have that we have to protect and fight for.

For Native people who have a passion for cooking, I would encourage you to focus on your tribe's traditional foods. Start from where you are at and learn about what grows in your area—what kind of wild game and wild plants you have in your area, and focus on the four seasons, or however many seasons you have in your area. Focus on what grows and the time frame that it grows in. Being a Native chef is not just about being in the kitchen and getting food delivered to you. It's all about learning the foods in your area, what's around you and growing with the seasons.

NOTES

1. Elizabeth Hoover, "Cultural and Health Implications of Fish Advisories in a Native American Community" *Ecological Processes* 2, no. 4 (2013), https://doi.org/10.1186/2192-1709-2-4.

2. Elizabeth Hoover, *The River Is in Us: Fighting Toxics in a Mohawk Community* (Minneapolis: University of Minnesota Press, 2017).

3. John Mohawk and José Barreiro, *Thinking in Indian: A John Mohawk Reader* (Golden, CO: Fulcrum, 2010), 22–23.

4. Devon A. Mihesuah, *Choctaw Crime and Punishment, 1884–1907* (Norman: University of Oklahoma Press, 2009).

5. See Devon A. Mihesuah, *Roads of My Relations* (Tucson: University of Arizona Press, 2000), and *Recovering Our Ancestors' Gardens* (Lincoln: University of Nebraska Press, 2005); American Indian Health and Diet Project: http://www.aihd.ku.edu/; and Indigenous Eating: https://www.facebook.com/IndigenousEating/.

6. Yappalli—The Road to Choctaw Health, March 27, 2015, https://clinicaltrials.gov/ct2/show/NCT02400554.

7. I-Collective, https://www.icollectiveinc.org/home.

8. For all the projects, blogs, and educational activities initiated by Monarch Watch, see http://www.monarchwatch.org/.

9. For texts and references dealing with habitat loss, see the Monarch Butterfly Recovery Plan: Part Two (monarchwatch.org/blog/2015/12/01/creating-a-monarch-highway/) and Monarch Population Status (monarchwatch.org/blog/2014/01/29/monarch-population-status-20/).

10. *City of Albuquerque v. Browner*, 97 F.3d 415 (10th Cir. 1996).

11. "Tribal Sovereign Interests beyond the Reservation Borders," *Lewis and Clark Law Review* 12 (2008): 1022.

ELIZABETH HOOVER

2

"YOU CAN'T SAY YOU'RE SOVEREIGN IF YOU CAN'T FEED YOURSELF"

Defining and Enacting Food Sovereignty
in American Indian Community Gardening

A sign at the entrance to the long, narrow driveway proclaimed that this space was home to Tsyunhehkwa (Life Sustenance), a certified organic farm and program of the Oneida Nation.[1] The shuttle turned down the driveway, past the small yellow farmhouse that serves as the farm's office, and pulled up in front of the large red barn. Climbing down the shuttle steps, representatives from tribal projects in California, Saskatchewan, the Cherokee Nation, the Navajo Nation, and various Ojibway communities from across the Great Lakes region were promptly greeted by Don, one of the farm's employees. Each of these representatives had traveled from his or her own corner of Indian Country to the Oneida Nation of Wisconsin for the Food Sovereignty Summit, hosted by the Oneidas along with the First Nations Development Institute and the Intertribal Agriculture Council. Along with employees Ted and Jeff, Don took the visitors through the barn filled with braids of white corn hanging from the rafters, past pastures filled with chickens and grass-fed cows, through acres of Iroquois white corn; and through the greenhouse and chicken-processing facility. Then everyone reboarded the shuttle for a tour of the tribal cannery. In addition to these tribal representatives, hundreds of other Indigenous gardeners, farmers, ranchers, seed savers, fishers, foragers, hunters, community organizers, educators, and chefs attended the summit, all seeking to better connect with others in the movement and to envision what food sovereignty could look like in their communities.

Taken up by activists and academics alike, "food sovereignty" has now become a rallying cry for both established tribal programs and grassroots projects across Indian Country. However, what is meant by the term often varies considerably. In this chapter I expand on the various definitions of food sovereignty discussed in the introduction, exploring in detail how Native American community farmers and gardeners describe and define food sovereignty as both concept and method, and examining how these definitions are being operationalized in pursuit of community goals of promoting health and reclaiming and maintaining tribal culture.

METHOD

I first became engaged in conversations around food sovereignty through volunteering with the Akwesasne Mohawk community-based organization Kanenhi:io Ionkwaienthon:hakie (We Are Planting Good Seeds), with which I have been involved since 2007. Located in a community that has been contending with environmental contamination and an overall diminishment of farming and gardening, the goal of Kanenhi:io is to boost local food production by helping Mohawk people access land, equipment, funds, and a community of fellow gardeners.[2] Conversations with fellow project participants on how to increase community involvement and access to funds then led me to take part in twenty-five different food sovereignty summits and Indigenous farming conferences, hosted by tribal nations such as the Oneida Nation of Wisconsin, community groups such as the White Earth Land Recovery Project, and organizations including the First Nations Development Institute, the Intertribal Agriculture Council, and the Native American Food Sovereignty Alliance (see table 2.1). In an effort to learn more about the Indigenous community-based farming and gardening projects I was hearing about during these conferences, I drove twenty thousand miles around the United States to visit thirty-nine of these projects in person in the summer of 2014 (see table 2.2).

In the process, I conducted fifty-two formal interviews, and recorded thirty-four conversations and farm tours.[3] Of the fifty-two interviews, forty-six were

Braids of white corn hanging in the barn at Tsyunhehkwa.
Photo by Elizabeth Hoover.

with individuals who identify as Native American, and six were with project staff who do not identify as Native but have been working closely for a number of years with the Native communities that hired them to run these projects. Interviews were transcribed, then coded in NVivo 8, based on themes presented in the interview questions and those that arose organically through the interviews.[4] I also wrote a blog post about each of the communities I visited, which featured the story of each project and accompanying photographs.[5] Project participants were asked to approve each post to ensure that they were portrayed in a manner they felt was appropriate. The blog served two purposes: it gave these projects a

TABLE 2.1

FOOD SOVEREIGNTY CONFERENCES (25) AND EVENTS (3) ATTENDED

CONFERENCES

- First Nations Food Sovereignty Summit
  Oneida, WI (October 2013, 2014, 2015)
- Indigenous Farming Conference
  White Earth, MN (March 2011, 2012, 2013, 2014, 2015, 2017)
- Native American Culinary Association Conference
  Tucson, AZ (December 2013, November 2015)
- Intertribal Agriculture Council Conference
  Las Vegas, NV (December 2013)
- Terra Madre panels with Indigenous seed keepers
  Turin, Italy (October 2014, September 2016)
- Slow Food Turtle Island Planning Meeting
  Taos, NM (February 2016)
- Slow Food Turtle Island Delegate Meeting at Slow Food Nations
  Denver CO (July 2017)
- Mohawk Seed Keepers Meeting
  Six Nations, Ontario (April 2016)
- Great Lakes Food Sovereignty Summit
  Oneida, WI (April 2013, 2014, 2015)
  Jijak Camp, MI (April 2016, 2017)
- Food Sovereignty Summit
  Red Lake Reservation, MN (September 2016)
- Native American Food Sovereignty Alliance Board Meeting
  Albuquerque, NM (October 2016)
- Native American Food Sovereignty Alliance Storytelling Project Meeting
  Tucson, AZ (April 2017)

EVENTS

- Ponca Corn Planting in Nebraska
  (June 2014)
- Tohono O'odham Community Action Bahidaj Cam
  Arizona (July 2014)
- Honor the Earth pipeline horse ride
  Minnesota (August 2014)

TABLE 2.2
SITES VISITED

- Kanenhi:io Ionkwaienthon:hakie, Akwesasne Mohawk (NY)
- Coushatta Tribe Hydroponic Program (LA)
- Mvskoke Food Sovereignty Initiative (OK)
- Ponca Agricultural Program (OK)
- Cherokee Heritage Seed Project (OK)
- Tesuque Pueblo Farm (NM)
- Pueblo of Nambé Community Farm (NM)
- Traditional Native American Farmers Association (NM)
- Native Food Sovereignty Alliance, Taos County Economic Development Corporation (NM)
- Red Willow Farm, Taos Pueblo (NM)
- Cochiti Youth Experience (NM)
- Zuni Youth Enrichment Project (NM)
- Cultural Conservancy (CA)
- Tohono O'odham Community Action (AZ)
- San Xavier Co-op Farm (AZ)
- People's Farm, Ndée Bikíyaa (AZ)
- Hopi Tutskwa Permaculture Institute (AZ)
- Black Mesa Water Coalition, Pinon (AZ)
- Big Pine Paiute Tribe Sustainable Food System Development Project (CA)
- Bishop Paiute Aquaculture Project (CA)
- Sierra Seeds Cooperative (CA)
- Native American Youth and Family Center (OR)
- Muckleshoot Food Sovereignty Project (WA)
- Nisqually Community Garden (WA)
- Cheyenne River Youth Project (SD)
- Slim Buttes Agricultural Development Program (SD)
- White Earth Land Recovery Project (MN)
- Wozupi Farm, Shakopee (MN)
- Spirit Lake Native Farm, Fond du Lac (MN)
- Little Earth Urban Farm (MN)
- Mashkiikii Gitigan Medicine Garden (MN)
- Dream of Wild Health (MN)
- Bad River Gitiganing Community Garden (WI)
- Tsyunhehkwa, Oneida Reservation (WI)
- Ho-Chunk Nation Gardening Cooperative, Whirling Thunder Farm (WI)
- Minnesota Museum of Science, Scott Shoemaker, indigenous seed curator (MN)
- Food Is Our Medicine, Seneca Nation (NY)
- Crandall Minacommuck Farm, Narragansett Food Sovereignty Initiative (RI)

web presence, which was later helpful for some in acquiring additional funding and recognition, and it also helped these community organizations to learn more about one another.

During the interviews, I asked participants to describe the history of their project, some of their successes and challenges, and advice they would give to new Indigenous food projects. I also asked them all to define food sovereignty and to describe how this term or concept fit into their own work or was utilized in their own communities. In the results section I break down elements of the definitions they provided, put them in conversation with each other, and highlight where aspects of these definitions converge and diverge with other food sovereignty definitions.

RESULTS: COMMUNITY DEFINITIONS OF FOOD SOVEREIGNTY

When participants were asked to define food sovereignty, their answers coalesced into a number of themes—namely, the importance of food to cultural identity; relationships to the environment, food sources, and other people; and the need for independence to make choices around how to define food systems and what exactly to eat, at the tribal and community levels as well as an individual level. Additionally, the importance of access to food, land, and information was also raised, as was the role of the tribe in providing this access. Participants also focused on the issue of control—over what they put in their mouths, what seeds are planted, and how their tribes should take back control of their land and food systems from outside influences. Participants further raised the importance of providing education, improving health, and focusing on the youth and future generations. Heritage seeds—most passed down through generations of Indigenous gardeners, with some reacquired from seed banks or ally seed savers—were often discussed as the foundation of the movement, as living relatives to be protected from patent or modification, but also as tools for education and reclaiming health.

To conclude this chapter, I discuss the assertion made by many of the participants that in order for tribes to properly assert that they are fully sovereign, they need to work toward achieving food sovereignty first. With that in mind, many saw food sovereignty not as a final state that could be achieved, but rather as a process, a method, and a movement.

Health

Driving much of their work on issues related to food sovereignty, some of the main motivators that participants described for taking part in community-based projects included concern about poor health in their community, anxiety about the grave health statistics described in the introduction, and a desire to try to rectify this situation. Food sovereignty was described as a necessary tool to solve existing health problems, as well as to promote better health in the future. In many communities traditional foods have become less available, and processed, packaged foods have become more available, contributing to poor health. As Julie Garreau (Lakota), director of the Cheyenne River Youth Project, explained,

> For so long we ate those foods that weren't good for us, and we didn't know. ... We struggle with diabetes, it's just rampant in our communities, so we just need to change our diet.... Because in the end, if you don't have a healthy population, you don't have anything. They're not going to get to school, they're not going to have long, healthy lives, they're not going to be able to raise their children. You need healthy people. And who doesn't want a healthy nation? We want that for us. As parents and grandparents, we want our kids to have long, productive lives. So food is a part of it.

These concerns motivated Garreau to incorporate a garden and kitchen into her youth program. Similarly, Dan Powless (Ojibway) described the purpose of the Bad River Gitiganing Community Garden project as "to regain the health that we need. We've got a lot of health problems, nutrition problems on reservations and things like that, so I think that's the first thing we're looking at is health of the people ... the main focus that we kind of think of is the health first." George Toya (Jemez Pueblo), who runs the garden for Nambé Pueblo, also connected diet change to addressing health problems and saw it as part of his mission to work toward gradually reversing the situation.

> Our diet has changed so much, and the evidence is in the health of the people. They're not as healthy as they used to be even a few generations ago. It's really changed. If it took that long to change us, it might take that long for us to get back to that point where we're healthy people again, and this is kind of our attempt to do that. Being sovereign is not just about being a totally isolated nation, it's about being able to really feed—make your people well, and feed them again.

This, he recognized, was not going to happen quickly, but is an important goal to work toward.

For Kenny Perkins (Mohawk) from Kanenhi:io, working toward food sovereignty will result in good health for himself as well as future generations:

> I believe that food sovereignty means that we're able to feed ourselves, and by feeding ourselves we know what's going into our body. And when we know what's going into our body and we're healthy, we are able to make

better decisions, especially for those future generations that're coming up. And if we can show them the right way the first time, they won't know any other way. And so in turn, they'll become healthy.

Speaking to an audience full of Native gardeners, foragers, chefs, and others interested in food in 2013, Valerie Segrest (Muckleshoot) advocated that

> when we follow our traditional diets we're healthier people. Our immune systems can stand up to the seasons. One hundred years ago diabetes and heart disease were nonexistent in our communities. We know what we need to do to be able to solve our health crisis. Telling people what to eat is not the root cause of our problem; it's access to our traditional foods. Preventable diseases rise when we don't have access to traditional foods.[6]

Segrest carried this thought into our interview in July 2014: "The reason why we have a lot of diabetes and heart disease in our community is because we've been taken away from our traditional food system and have experienced the effects of a superimposed diet on people. When I talk to my leaders they know and they preach about how if we ate our traditional foods we wouldn't be sick." For these reasons, Segrest and others working on similar projects are promoting a shift in diet specifically to culturally important health foods.

Food as a Component of Culture

Grim public health statistics reflect the effects on physical health caused by disrupted food sources. But it is also important to note the impacts of these disruptions to cultural and spiritual health, which are reliant on important cultural connections to food. It is notable that many of the projects I visited are not trying to grow just any nutritious food—in many cases, they are seeking to restore culturally relevant food. Guaranteed access to "culturally appropriate foods" is a central tenet of the most basic definition of food sovereignty. This phrase was reflected in many of the definitions provided by participants. For example, Diane Wilson (Dakota), director of the Dream of Wild Health program, defines food sovereignty as "having access to healthy, affordable, culturally appropriate food." She describes that "part of this cultural recovery process," which many tribes as well as urban Indian communities are undertaking, "is the idea that you have control over your own food." Dream of Wild Health seeks to provide that access and control to urban Native youth and their families through internships at the farm as well as cooking classes. Similarly, Scott Shoemaker

(Miami) defines food sovereignty as "the ability to seed your own community with cultural appropriate foods." As a curator of an Indigenous seed collection at the Minnesota Museum of Science, he worked to do that through collaborations with nearly a dozen Indigenous community projects that partnered to form the Indigenous Seed Keepers Network and are now growing out seeds from that collection and sharing them with other community members.[7] Even if these foods are already available, Tom Cook (Mohawk), who directed the Slim Buttes Agricultural Development Program on Pine Ridge for more than two decades, described food sovereignty as "the expansion of local, culturally produced foodstuffs."

These culturally appropriate foods are seen as providing more than just nourishment for the physical body. Roberto Nutlouis (Navajo), who runs food and farming projects through the Black Mesa Water Coalition, explains that "corn isn't just corn for our people, it has so much spiritual significance. It's a biological and spiritual nourishment to our people." Nutlouis works with youth to maintain fields of Navajo heritage corn using traditional dry-farming methods, then uses that corn to feed youth and elders.

Several cultural programs in Native communities focus on food as an important vehicle for delivering cultural information. Kenny Perkins recalled that a major focus of the Á:se Tsi Tewá:ton Akwesasne Cultural Restoration Program was to restore traditional foodways disrupted by environmental contamination. On the opposite coast, in the state of Washington, Romajean Thomas (Muckleshoot) described a "cultural sovereignty" class that focuses on food culture, held at Muckleshoot Tribal College. "Food sovereignty is really at the root of cultural sovereignty. It's what our treaties are for and what our ancestors fought for." Similarly, Bob Shimek (Ojibway), the current director of the White Earth Land Recovery Project in the Ojibway community of White Earth in Minnesota, is focused on

> using Ojibway food systems as the vehicle for cultural restoration and revitalization. Those little creation stories that come with each one of our relatives, whether they be the fish or the birds or the plants or the insects or the frogs or turtles or whatever, so many of those have a little story about how they got here. Inside those words that tell that story, that's where the true meaning and value of our culture is stored, in our languages that tell those stories. So that's the effort I'm making right now—it's to not only keep building on our physical health, improving our physical health by

teaching people not only about gardening and small-scale farming but also all the wild plants, the wild foods that are out there, and packaging those up in the historical, cultural, and spiritual context which is part of the original understanding in terms of our role here on this turtle island.... Food sovereignty means that we're taking care of that cultural and spiritual relationship with our food.

In this way, food sovereignty is not just a goal in and of itself but a tool to achieve other aspects of cultural restoration that are connected to health and language. The Mvskoke Food Sovereignty Initiative (MFSI) in Oklahoma partners with the Euchee language immersion program every summer, helping the students to plant a garden at the school. Stephanie Berryhill (Mvskoke Creek), who worked as a youth programs coordinator for MFSI, asserted, "Language is the most critical marker of the health, and the cultural health, of the community. It's an important mark of the sovereignty of each of our Indigenous nations." As we stood and watched half a dozen girls from the summer program tuck corn seeds into freshly tilled soil, Richard Grounds (Yuchi/Seminole), the director of the Euchee language project, described the garden as the perfect place to learn and practice language "because you're physically doing what's being said [and] that helps you to remember and learn and associate the meaning with the activity and that has all that repetition built in. So we can view the language in a natural way," and in addition have food to show for it.

Food was also described as a core and necessary component of culture. Cassius Spears Sr. (Narragansett), who is heading up the Narragansett Food Sovereignty Initiative, stated that food "is to me what identifies your culture, your traditions, basically who you are. And it brings people together, it's like the kitchen of the house," playing on the image of the kitchen as not just the place where food is prepared, but also the central gathering place in many homes, the place where people often receive wisdom from elder women culture-bearers. For Spears food is the central hearth, the foundation, of culture and overall tribal sovereignty, the same way a kitchen is to a home.

Without this core component of food, Indigenous cultures are compromised. Valerie Segrest explains, "When our foods cease to exist so do we as a people. They're there to remind us who we are and where we come from." She goes on to say that tribal "creation stories tell us that we are to commit ourselves to ceremonies around food. Food is our greatest teacher—without a spoken word." On the other hand, she believes that "culture repression" impedes her community's ability to access teachings from fish, cedar trees, and other elements from their environment in the Pacific Northwest. Access to traditional foods and the practice of ceremony around those foods are necessary for the continued survival and growth of Coast Salish tribal culture.[8]

Alan Bacock (Paiute), from the Big Pine Paiute permaculture project in California, described what happened when Indigenous people were denied access to traditional foods: "we saw through our history that when we lacked the ability to provide food for our people is when our culture started to decline... if we were able to maintain our local food control, we would still have a strong cultural identity, strong cultural heritage." For these reasons, the reconstruction of traditional food systems is seen as imperative to cultural restoration and health.

Relationships

These cultural practices are in many ways centered on relationships—with food and with other tribal members around food—as opposed to considering food simply as a commodity. Jeremy McClain (Ojibway), formerly with the Bad River Gitiganing project, described "that symbiotic relationship with your environment. To me that's food sovereignty: if you take care of your environment it will take care of you." Within that context, he also mentioned the importance of the different Anishinaabe nations maintaining relationships with each other, and of different programs and departments within the tribal government establishing relationships in order to foster the success of food sovereignty projects. Similarly, Lannesse Baker (Ojibway), with the Mashkiikii Gitigan Medicine Garden project in Minneapolis, described food sovereignty as being "about that relationship we have with food and our ability to feed ourselves and sustain ourselves." Because she works with urban Indigenous populations, she described the importance of projects like Mashkiikii Gitigan in "facilitating that relationship to the earth and the environment and food, the healthy foods, the original foods."

Some foods were described as actual relatives with whom positive relationships needed to be maintained. After we returned from a four-day *bahidaj* (saguaro cactus fruit) picking camp hosted by Tohono O'odham Community Action, Terrol Dew Johnson (Tohono O'odham) described how "the raw food that was harvested this weekend, in our traditional songs is referred to as being a little girl, a person, a woman." He went on to tell the story about a little girl who was neglected by her mother and, despite the help of different birds and animals, became so sad she sank into the ground and grew into a cactus bud.[9] As described in greater detail below, seed keepers I interviewed repeatedly highlighted the importance of *relationships* to seeds, as opposed to *ownership* of them. Clayton Brascoupe (Mohawk), director of the Traditional Native American Farmers Association for more than twenty years, repeatedly described seeds as "our living relatives," who need to be cared for and protected from people who would treat them as commodities.

These relationships between human communities and the other communities that make up a tribal nation's food system are reflected in what Mariaelena Huambachano describes as Indigenous "good living" philosophies within which food sovereignty and food security should be framed, because these philosophies do "not solely focus on economic growth but rather place an emphasis on Indigenous peoples' tenets of duality, equilibrium and reciprocity in order to enjoy and preserve the bounties of Pachamama to safeguard food security." She goes on to argue that these philosophies "offer models for promoting biodiversity, social equity and economic growth without agrochemicals, and preserving Mother Earth."[10] Maintaining these philosophies, specifically the Anishinaabe concept of *Mino Bimadiziiwin*, was described as the key to a healthy productive community. According to Winona LaDuke (Ojibway),

> The Creator gave us instructions, *Mino Bimadiziiwin*, about how to lead a good life. And the Creator gave us this land, *Oma akiing*, here upon which to live. Our instructions were to take care of each other—take care of all of our relatives, whether they had wings or fins or roots or paws—to be respectful, and to live that life. That's what I want to do. In that life, we feed ourselves—our food does not come from Walmart, our food does not come from fast food, we are not engaged in an industrial era. We are people that live from the gifts here.

Food sovereignty is the process of nurturing the proper relationships with food elements. As Bob Shimek put it:

> Food sovereignty also means that we're taking care of that cultural and spiritual relationship with our food. This is not by any means a one-way thing. I mean, it's not like we can just go out there and keep taking and taking from all that was put here for us without properly taking care of that land and those relatives of ours that were put here for our use, benefit, and enjoyment. So I think the true measure of food sovereignty is when you have that reciprocal relationship where Anishinaabeg is thriving, as

are all our plant, animal, bird, fish relatives, etc. That's food sovereignty, when it's all lock-stepping together in what we call *Mino Bimadiziiwin*, the good life.

This good life philosophy encapsulates the harmony that is established when symbiosis is maintained through respectful relationships between humans and the other communities that contribute to their food systems.

Independence

Many of the participants equated food sovereignty with a level of independence from outside forces when it came to sourcing food—on an individual level, as a community, and as a sovereign tribal nation. Being able to feed oneself was at the root of an individual's responsibility toward broader food sovereignty; as Milo Yellowhair (Lakota), from the Slim Buttes Agricultural Development Program, reflected, "Sovereignty is an issue that's rooted in the ability to feed one's self." Woodrow White (Ho-Chunk), formerly from the Whirling Thunder garden project, described food sovereignty as follows: "You can grow lock, stock, and barrel all of your own food . . . if you can feed yourself, that's a giant step. No dependency out there. That's the sovereignty you're talking about, and there's not that many of us to take care of our own." Looking around at the collection of individual raised garden beds in the community garden he had helped establish, Woodrow went on to describe how once individuals became independent, they could then contribute to feeding an entire community.

For other interviewees, it was the ability to rely on their fellow community members, rather than outside companies, for inputs in running the local food system that constituted the necessary independence for food sovereignty. Angelo McHorse (Pueblo), who ran the Red Willow farm at Taos Pueblo, defined food sovereignty as "you don't have to depend on any companies for your seed or your fertilizer, even big tractors or oil much less. We have all our own ditches, we have all our own seed. We have all our own energy—our own hands." Similarly, Jayson Romero (Pueblo), who apprentices young farmers through the Cochiti Youth Experience, defines food sovereignty as "not having to go outside of ourselves to get the things that we need and use." Looking out over his field of knee-high corn plants sprouting up out of the sand, he described all of the special occasions that require traditional Pueblo corn. Food sovereignty, he decided, would be accomplished when "the ladies here do not have to go anywhere else to find the stuff that they need," because farmers would be able to provide all of the corn necessary for these occasions. Sitting in his adobe home in his wife's community of Tesuque Pueblo, Clayton Brascoupe also showed me pictures of the cornfields maintained by his entire family. He stated that their community will have achieved food sovereignty when "we have the ability to provide for ourselves, our children, our neighbors, within our community."

Food sovereignty entails the ability not only for individual members of the community to rely on each other rather than multinational corporations, but also specifically for the tribal community to provide for its members. Grace Ann Byrd (Nisqually), who works for the Nisqually Community Garden, defines food sovereignty as "being able to provide for your own people, to work the land, to have that garden stand. . . . So being able to provide for our own . . . we like to provide for our own people because that's what I believe is sovereignty, is providing for the tribal members, the community members that reside here, and our elders in the diabetes program."

A third level of independence described by participants was on a tribal level—the tribe as both a community and a government becoming less dependent on outside entities to provide food for its constituents. As Jeremy McClain (Ojibway) described, food sovereignty is "reducing our tribe's dependence on the mainstream food production system and distribution." Chuck Hoskins, the Cherokee Nation secretary of state, described food sovereignty as "Cherokees producing for themselves, producing food for our families and for our people. Not being dependent on outside state influences." Amos Hinton (Ponca), who started an agriculture program for his tribe, explained

> Food sovereignty is the ability to take care of yourself without input from outside forces. If I as a department head can produce all of the food for my tribe that they need, then not only are we food sovereign, we are indeed sovereign. You look at a tribe who says, "We are a sovereign nation." Where do you get your food from? Do you buy it from an outside source? If you buy it from somebody else, then you are not a sovereign nation, because you're dependent on somebody else for your food. To me, if you're growing all of your own food, then you are a sovereign nation. At one time all Native American tribes were sovereign nations. They are not now.

Within these levels of independence, participants recognized that in many cases the support of tribal government went a long way toward supporting food sovereignty for individuals as well as the community as a whole.

Economics

The focus on more equitable economic systems, which constitutes much of the focus of global food sovereignty definitions, also surfaced in these participants' definitions, not only in the context of keeping food dollars within the community to support tribal food producers, but also in efforts to make these nonprofit organizations sustainable. Winona LaDuke described a survey of the White Earth Reservation, conducted by the White Earth Land Recovery Project, which found that their community was spending one-quarter of its economy on food, a majority of the money being spent off-reservation. She has since expressed her determination to direct more of those food dollars to support on-reservation food producers.[11] Stephanie Berryhill from MFSI reflected on the quantity of food being served at the tribe's casino, all of which is "purchased from outside vendors when we should be producing it ourselves. We should be providing jobs and keeping this money in our communities."

To remedy situations like this, some of the participants are developing payment systems or buying practices that seek to keep food dollars within Native communities. Director Garreau made arrangements for the Cheyenne River Youth Project to accept bank cards at their little store and farmers' market, as a way of directing federal government dollars provided to community members toward supporting local food producers. Lilian Hill (Hopi) described her efforts to help create markets for farmers and local producers, in order to support them as well as promote the sale of healthy food. Hopi Tutskwa Permaculture, which Hill runs, partnered with the Natwani Coalition, the Hopi Food Co-op, and the Hopi Special Diabetes Program to create the Hopi Farmers Market & Exchange on Second Mesa, which provides an opportunity for food producers and consumers to connect directly. The market "provides a venue for local farmers and gardeners to sell or exchange their fresh, seasonal produce directly with the Hopi community," and in a feature unusual to farmers' markets but in line with a traditional Hopi economy, it encourages "community members to bring fresh produce, vegetables, home-prepared foods, and crafts to trade/barter/exchange with farmers market vendors."[12]

Native chefs and restaurant owners have also become involved in promoting Native food producers. Sean Sherman (Lakota), the chef behind the Sioux Chef enterprise, described his efforts to "really try to use as many Native producers as possible—so keeping a lot of these food dollars within the Native communities will be a thing in making these food systems sustainable for everyone—the farmers, the wild rice harvesters, the people foraging and just gathering stuff that can be sold, or people raising animals." Similarly, in a conversation in his Denver-based restaurant, Tocabe, chef Ben Jacobs (Osage) described his buying practices as "Native first," purchasing first from Indigenous food producers even if they are outside of Colorado, and purchasing second from local non-Native food producers.

Because the majority of the participants I spoke with worked for nonprofit organizations centered on food, they described the struggle to make their projects more economically self-sufficient. Diane Wilson's goal is to make the Dream of Wild Health "farm ultimately become economically independent" through their farmers' market and other programs, rather than relying solely on grants and gifts. Romajean Thomas similarly reflected on the struggle to find the necessary funds to keep their community-based programs running. "It's a sustainability question: How do we keep funds coming in?" Many other project leaders expressed similar concerns about how to keep the necessary staff to run these projects without hindering the livelihoods of those staff. For example, at the time of my interview with them, two project co-directors had not been paid in several months, but continued coming to work—sacrificing their personal economic well-being in order to keep their organization afloat.

But in addition to conventional monetary exchanges—and beyond the notion that projects and communities should be economically independent and supply all of their own food to be considered "truly food sovereign"—participants also highlighted the important role of trade, historically and in the present. For example, in their research with Indigenous communities in what is now British Columbia, Nancy Turner and Dawn Loewen describe archaeologically documented extensive trade networks that specifically brought plant products to the Pacific Northwest, in order to obtain plants which were unavailable or difficult to access locally, as well as products that required specialized skills. They argue that few if any environments provide all of the resources a group needs at a given time, so trade has long been used to counter instabilities in resource supply and to provide variety. With this in mind, they suggest that rather than building a strictly local food system as an alternative to the global industrial system (the language of the local food movement), Indigenous communities are in many ways seeking to protect traditional food practices and networks.[13] Similarly, Scott Shoemaker states that he prefers the term "interdependence" to "independence," arguing that tribal communities have always relied on trade and reciprocity.

Josh Sargent (Mohawk), with the group Kanenhi:io, worked to unpack what his fellow group members mean by "food sovereignty" and "what we mean when we say 'independence,'" positing that "I know we're not going to go 100 percent because, honestly, no one ever has. Trade has always been a trait that humans do, they don't live in bubbles." He describes the need for people to "at least make your own basic needs, you have to be able to do that," to be considered sovereign, but beyond that he sees trade as having always been important. This sentiment was also embodied in a call to reconnect or reestablish trade routes between Native communities as a form of economic and cultural support and revitalization. Seed keeper Rowen White (Mohawk) directs the Sierra Seeds cooperative and currently serves as the chair of the board for Seed Savers Exchange. She believes that seed keeping and seed exchanges are part of planning for the future: "I think also in times of global climate change, we will be reestablishing trade routes, we will be connecting with other tribes and other people because I think that was always happening in the first place: corn went from this tribe to this other tribe, and we mixed it with ours and made something new." Similarly, Pati Martinson (Lakota/French), in describing what constituents are asking for from the still-developing Native American Food Sovereignty Alliance (NAFSA), mentioned that "people have said they're really interested in bringing back those trade routes. And part of that could be a big economic development, community development, piece as well." Dan Cornelius (Oneida) and the Intertribal Agriculture Council worked to enact this, beginning the Mobile Farmers Market Reconnecting the Tribal Trade Routes Roadtrip in 2014, on which participants collected, purchased, and exchanged foodstuffs from tribal communities across the United States.[14] The market continues, with a brick-and-mortar store recently opened in Madison, Wisconsin, stocked with food products and nonfood items that have been traded for food from across Indian Country.

Access

Many of the food sovereignty definitions centered on forms of access—to food, land, and knowledge. Darlene Fairbanks (Ojibway), from the Little Earth housing project in Minneapolis, defined this simply as "just having access to healthy food," a description echoed by several others. Melissa Nelson (Turtle Mountain Chippewa), director of the Cultural Conservancy, framed her definition of food sovereignty around terms of access at the individual, community, and political levels. On an individual level, she noted, "We don't always have control over what we have access to," being limited by factors like affordability and availability of foods, "but what we put in our mouths we really do." Jeff Metoxen (Oneida) from Tsyunhehkwa similarly pointed out that their project can grow and package white corn and other traditional foods, but it is up to individuals "whether you decide to access it." Nelson further described, "on a community level, it's a community's ability to determine the foods that they have access to, and that they can utilize for the health and well-being of the whole, so it's really about access and sharing." Then, on a political level, "Indian nations really have a legal obligation to their citizenship about what foods they grow and make accessible for their larger nations."

Barriers to accessing healthy, culturally appropriate food include the cost of these foods for consumers, and access to land for growers. As Lannesse Baker at the Mashkiikii Gitigan garden in Minneapolis reports, "People talk within the urban community about a lot of challenges related to access to healthy food, whether there are barriers to access because of affordability issues or challenges with transportation." Similarly, Keith Glidewell (Paiute) of the Bishop Paiute Tribe defines the challenges to food sovereignty, even with "access to food funds," as "having food availability. Is there food within the reservation that can be accessed? . . . Can you afford to buy it?" Especially in a resort town like Bishop, California, "they gouge us for everything here. So that's my main thing, is it available, is it affordable?" Similarly, Amos Hinton of the Ponca Tribe in Oklahoma stated, "Our average household income is $7,000 a year. You can tell them all day long to eat healthy, but they can't afford it."

Access to land—whether for farming or hunting and harvesting wild foods— was cited as another challenge in achieving food sovereignty. Valerie Segrest reported the Muckleshoot Tribe is looking to address this issue through the recent purchase of 96,000 acres of land from a timber company. Stephanie Berryhill reports that some Mvskoke Creek tribal members moved to urban areas "because of access to jobs," but their families lost land in the process. Berryhill cites lack of access to land as a major factor in limiting the nation's ability to be food sovereign. To address this "means that we promote policy, and ultimately tribal laws, that will enable citizens to have access to land to grow food." Otherwise, as Lori Watso from the Shakopee Mdewakanton Sioux describes, "if there comes a time where we don't have access to clean food, or any food, what difference does everything else that you've built [make]—it doesn't matter . . . if we're able to develop or reach a point of food sovereignty, we'll be OK." Access to land is seen as imperative in many cases to accessing sufficient culturally relevant food.

Decisions and Choices

Having the freedom to make decisions about food and the ability to make good choices were both cited as integral parts of what it means to have food sovereignty. Julie Garreau outlined food sovereignty as "being able to decide what we eat, grow what we want." Similarly, Zach Paige, who works with the White Earth Land Recovery Project, described food sovereignty as a "form of freedom because you are able to grow what you like to grow and eat what you like to eat." But once that level of choice has been made available, others highlighted the need to encourage people to choose healthy options. Stephanie Berryhill describes her line of work with the MFSI as "on the most basic level, advocating for people to choose healthier foods to eat." Similarly, Segrest names food sovereignty as a tool she uses in her role as a nutritionist and as a "great way of helping people to understand that food choices are their responsibility and it's their inherent right to choose what they want to eat ... food sovereignty is a method of making food choices ... it's about helping people, empowering people to make that choice for themselves."

At the same time that some interviewees described food sovereignty as providing choices, Don Charnon (Oneida) at Tsyunhehkwa described food sovereignty as being dependent on the choices people make:

> If you really want to exercise food sovereignty, you need to make decisions toward it. You need to make decisions toward it, and by making those decisions you support those entities that grow or make or produce the kind of food that you believe in and want in your system, for your family, for your community. So unless you choose to invest in those food places you consider worthy or acceptable or good places to get food, then they'll disappear.

The choice to eat healthy foods needs to be made available to community members, then they in turn need to take it upon themselves to make those choices to continue to support these types of initiatives.

Control

Part of the discussion around independence and access centered on notions of control—by individuals over what enters their bodies, by tribes over their own food destiny—and the political power associated with controlling a food system. Several participants recognized the power behind who or what controls your food source, both currently and historically. As Amos Hinton commented in recounting his tribe's history of relocations and rations, "If you think about your tribe's history, you have been controlled for a very long time by food.... If you don't raise your own food, someone else is controlling your destiny." Diane Wilson integrates this message into the curriculum of the Dream of Wild Health program: "Part of this cultural recovery process is the idea that you have control over your own food. One of the things we talk about here is, if you want to control people, control their food." Similarly, Christina Elias, who ran the Mashkiikii Gitigan garden in Minneapolis, leads discussions in the garden about "controlling people through controlling their food source... when you're that far away from your food source you're being completely controlled. You have no independence and no power in your life." She calls attempts at regulating seed libraries "desperate attempts at controlling us." For Milo Yellowhair, it was the realization that "food can be, and is, used as a weapon," that led him to get involved in the Slim Buttes Agricultural Development Program, in an effort to create an independent food source for his community.

Indigenous food projects are seeking to shift the locus of control over food toward individuals and tribal communities. This shift in control was reflected in participants' definitions of food sovereignty. On an individual level, for example, Melissa Nelson described the need for individuals to take responsibility for the control they do have over what goes into their bodies: "It's one of the few things as human beings that we actually have absolute control over. We don't always have control over what we have access to, but what we put in our mouths we really do. So to me, food sovereignty at the individual level is how we treat our bodies and our landscapes and what we put in our bodies, controlling the foods and waters and beverages that we intake." The sentiments expressed here around being able to "feed oneself" also speak to the desire for individual control over food sources.

On a tribal level, participants described food sovereignty as becoming possible only if the tribe as both a government and a community takes control of the food system. Lilian Hill, from the Hopi Tutskwa Permaculture Institute, declares "what food sovereignty means is for a tribal community to have more local ownership or local control over the food system." Chuck Hoskins defined food sovereignty as his tribe taking back control from the corporate agricultural system: "controlling our own food destiny. It doesn't have to be charted by Monsanto, it doesn't have to be charted by big agriculture. It can be charted by the same folks that did it generations ago, and that's the Cherokee people."

Similarly, Julie Garreau asserted, "Native people have to say 'We're going to control this.' Tribal governments need to create policies and legislation that encourage this sort of thing."

Part of taking back control of a tribal community's food system is having jurisdiction over the habitat that supports the food system. As Bob Shimek explained, "Food sovereignty to me is first of all control over where your food comes from. It's control of the type of food that's grown and produced there or grows naturally there. It's control of that habitat that's on that particular piece of land." Grace Ann Byrd and Romajean Thomas both described successes of their respective tribes in getting land back under tribal jurisdiction, which gave them potential to bolster tribal food sovereignty.

In addition to "controlling" land through political jurisdiction, protecting and sustaining land was seen as integral to food sovereignty. According to Michael Dahl (Ojibway) from White Earth, the fight against pipelines that he had been participating in with Honor the Earth was not about resisting just the pipelines themselves, but about maintaining a healthy way of living. "Right now, our rice and our sugar bushes and our berries, our gathering rights, are the main thing that we still have to our self-sustainability and our healthy living. So we need to protect that with our lives." Mike Wiggins (Ojibway), who was chairman of the Bad River Tribe at the time of the interview, described food sovereignty as being "rooted in sustainability and the caring for Mother Earth." Part of this for him was not only supporting sustainable gardening projects for the tribe, but also fighting to protect wild rice beds from a proposed taconite mine. Pati Martinson states that the goal of NAFSA is to help tribes protect everything related to traditional foods and that people feel that part of NAFSA's mission should address "that food needs protection . . . that is part of sovereignty, a real protection for the seeds, a protection for the land, a protection for the foods, that's a common goal that should be able to impact those policies."

Caitlin Krenn, who directs the Nisqually Community Garden project, described food sovereignty as not just catching and eating traditional foods like fish, "it means actually trying to sustain the rivers again, the sound, the ocean," the environment that supports those fish. This philosophy extended to farmland as well. Gayley Morgan (Pueblo) of the Tesuque Pueblo Farm defined food sovereignty in part as "just making sure your lands are worked on and taken care of. And also the water. And it goes hand in hand with the environment too, just the surroundings with the environment, the birds and the bugs and the bees, everything plays a part in it. Just as long as we have a nice, healthy environment here, it's also a piece of sustainable farms." The importance of sustainable farming and maintaining healthy soil came up in a number of conversations about how to achieve food sovereignty.

Seeds

Working to restore heritage seed varieties to Indigenous garden projects is seen as a primary goal of many of the projects I visited. Woodrow White commented, "We are trying to restore and recover our Indigenous seeds now. Everybody's in a scurry. Well, how much do we have left? Who has them? Let's get them. Do they still have germinating power? If not, we can share. I mean your tribe lives one hundred miles away but hey, it grows good, so we will share what we have left. We need to bring these seeds back. It's saving the seed."

For Amos Hinton, part of establishing a food sovereignty project for the Ponca Tribe was reclaiming traditional corn varieties, gathering these varieties from seed keepers in other states and "bringing them home." The foundation of gardening is seeds, and as such, for many communities having adequate access to their traditional seed varieties is imperative for food sovereignty. Gardeners felt a connection and obligation to these seeds and to the elders and ancestors with whom these seeds connected them. Roberto Nutlouis stated that "our work is not because of federal policies or tribal policies, it's because of our deeper connections to our lands, to the seeds that we have, that our elders passed on to us, that those have to continue."

The concerns about "control" over aspects of the food system expressed thus far also extend to seed sources and a community's ability to protect heritage seeds from multinational corporations as well as ensure a constant supply of seed for gardens. Regaining "control of our food system" begins with control over seed sources. Stephanie Berryhill describes how her Mvskoke Creek community has "definitely lost control of traditional plants," such as the corn to make *sofkey*, which is now provided to the community only by a non-Native company. She spoke about the importance of "regaining control of our food system, and specifically seed sources," so as not to be reliant on the vagaries of the market for one of their traditional foods. Similarly, Angelo McHorse asserted that if you can keep your seed from one year to the next, "well then, you have a sovereign source. Food sovereignty, you control it."

Sociologist Jack Kloppenburg has declared, "If there is to be food sovereignty, surely it will be facilitated and enabled by a struggle for seed sovereignty,"[15] a term that arose during several of the interviews I conducted. Indigenous control over seeds was an integral part of the definition of seed sovereignty provided by two of the seed keepers I interviewed. Rowen White, of the Sierra Seeds Cooperative, chairs the board of Seed Savers Exchange, and is heading up an initiative to further develop Indigenous seed-keeping networks among Haudenosaunee communities, as well as across the Upper Midwest. Seeds are her life. As we sat in the shade by one of her fields in July 2014, I asked her to define "seed sovereignty," a term she used frequently to describe the foundation of her own work and of food sovereignty more generally. She replied,

> Seed sovereignty is to me when you have an understanding of your inherent right to save seed and pass it on to future generations, and that you are exercising it at that same time. It also means that you as a person or as a community are self-informed and dictate your relationship to seed; that says that these are seeds that really do not belong to anyone. They belong to us as a community in the commons but that we can define our relationship to that seed based upon our own values and not the values of anyone else outside of our community ... seed sovereignty, at the heart, is really just taking back the action of saving seed and keeping it again year after year, generation after generation, so that we can have the security of knowing that we have seeds that will feed our children and our grandchildren. That we have the means by which to feed our people instead of relying on external sources ... [that] we can take care of ourselves, that we can sort of get back to the way it was before colonization, that we can have some sort of control or say over what foods we are able to put on our table and what foods are available for people to have access to in our communities.

Participants working with heritage seeds expressed major concern over how to protect what they saw as both living relatives and communal intellectual property from tampering with or patenting by multinational corporations. Like Rowen White, Clayton Brascoupe has not only been working with heritage seeds, but also promoting ideas of seed sovereignty across Indian Country. As we sat at his kitchen table, surrounded by ears of corn and piles of beans from his gardens, I asked him to define what seed sovereignty meant to him. His definition also centered on relationships with seed, as well as control over seed:

> We refer to these as our living relatives. So, we have to have control and ability to protect our living relatives. That's what seed sovereignty means to me. So they can't be molested, contaminated, or imprisoned. When I say

"imprisoned," I mean perhaps someone will say "this is some interesting stuff," and they grow it up for a few years and all of the sudden they say they own it. [Seed sovereignty means] the protection of our living relative, if not, then somebody else may say they own it. They're imprisoned and you can't go visit and plant your relative. Also, if you have the ability to interact with your relatives through these seeds, you also have the ability to feed yourself well.

The need to protect seeds from multinational corporations was framed in two ways: as cultural intellectual property that belonged to the entire community and all of the ancestors whose gardens had contributed to the current seed stock, and as living relatives who needed to be treated as such and protected. As Kloppenburg notes, the very nature of property is called into question when Indigenous people reject the notion of "owning" seeds, which they may see as "antagonistic towards social relations founded on cooperative, collective, multigenerational forms of knowledge production."[16] Reflecting on her interviews with both the staff of *ex situ* seed banks and participants in *in situ* Indigenous seed-saving projects, Sheryl D. Breen describes the difference between the perception of seeds as discrete material objects—"active storage containers of genetic material"—as opposed to viewing "seeds as responsive beings that are inherently embedded within ecological and spiritual webs of kinship," which highlights an important epistemological difference between the two parties in negotiating the political problems of seeds as property.[17] As seen in Clayton's commentary, seeds are "living relatives" not property, and relatives should not be "molested, contaminated, or imprisoned." Seeds are described almost as intergenerational relatives—both as children that need nurturing and protecting, and as grandparents who contain cultural wisdom that needs guarding.

Even though genetic modification and patenting were opposed for slightly different reasons, they were traced to the same common enemy—multinational agricultural corporations, particularly Monsanto. Lilian Hill insisted that tribal governments need to "take up these issues of food sovereignty" and work to "protect our crops against genetically modified organisms or other corporations that want to come in and patent our food crops and heirloom types of corn and things like that." Rowen White reported that when she does seed workshops in Native communities "it is the one thing that people want to talk about, 'Well, how are you going to protect our seeds from Monsanto?' or 'How are we going to protect our seeds from patenting?' There's no clear answer." In short, protecting seeds from modification and patenting, and ensuring access to them for community members interested in farming, were seen as integral to seed sovereignty, and thus food sovereignty.

Education and Youth

Although the role of education in these programs is important—both in promoting the movement and in improving health-related statistics—education has not been extensively highlighted in the broader food sovereignty literature. In reflecting on the work of the Cheyenne River Youth Project, Julie Garreau focused on the need to educate young people in order to address health issues: "we just need to change our diet. We just do. And we need to teach our kids now." She went on to detail how part of that effort will entail teaching people how to garden and preserve food.

People in this movement are hungry not only for the food being produced, but also for education and knowledge. As Valerie Segrest described to an audience at the Native American Culinary Association conference in the fall of 2013, "what this food sovereignty movement is hungry for is to remember the plants, our foods, the teachings. To share those memories with people, to be active in our food systems, to get your hands in the dirt. Get your head out of a book and focus on the lessons and blessings you're receiving."[18] At a subsequent Food Sovereignty Summit hosted by the Oneida Nation of Wisconsin, Segrest reiterated, "Food is our greatest teacher, without a spoken word." She highlighted that "loss of land, loss of rights, environmental toxins, and cultural repression impede our ability to access millions of fishing teachings," as well as those from cedar trees and other natural elements.[19]

Education is seen as not merely a goal, but also a responsibility. Roberto Nutlouis, of the Blackwater Mesa Coalition, explains that "we have the responsibility to share that information with the community, and that's part of our community outreach." Jeff Metoxen, from Tsyunhehkwa, described that the purpose of food sovereignty is "to share that information with your community. It's to ensure that the generations to come know this and are learning it. . . . I feel I've never stopped learning. I don't know everything there is to do *with* the white corn; I don't know everything there is to do *for* the white corn, but we're learning. And we're trying to make sure we share that knowledge with our community members, especially with the youth, and hopefully you can instill in them some pursuit of the knowledge."

Kenny Perkins noted that the apprentices he was teaching through the cultural restoration program are now able to go on to teach others. This is how food sovereignty and cultural restoration will be linked for the community: "the apprentices that we have now are able to go out and teach in the [Mohawk] language especially. They can go to the immersion school, the Freedom School, and teach everything there is to know about horticulture traditionally, culturally, and the new techniques and the modern ways of gardening, and be able to do it all in the [Mohawk] language."

Leaders of several projects recognized that the purpose of, and main contribution of, their project to the community was not necessarily an ability to feed everyone, but rather their ability to educate people about how to eat well and grow their own food. For example, on a tour of the Nisqually Community Farm, Grace Ann Byrd described the farm's mission as "providing education, providing food, as well as nutrition." Similarly, Don Charnon described one of the purposes of Tsyunhehkwa as "to be an example or a resource for people who want to grow things." Cassius Spears Sr., who is working with his family to establish a community farm for the Narragansett Tribe, lamented that "a lot of the youth don't even realize where the food comes from anymore, and the elders are getting separated from working in the soil." His goal is to bring these two groups together so that the elders can "start to teach and work with the youth again and bring out some of them old ways and old reasons."

People often spoke of "the future generations" that would benefit from healthy eating and saved seeds, as well as the youth they were working with who were natural audiences for this information, and who were going to be responsible for carrying it forward to those future generations. After hosting a boisterous group of about a dozen students from the tribal summer program, who had grazed their way through the bean patch and the apple trees, as well as through the wild grapes covering the fencerow, Woodrow White, from Whirling Thunder, noted that kids seem naturally inclined to eat well and work outside. "They naturally would like to do it anyway; they just need the place and the teachers, and they will take off."

Similarly, Romajean Thomas highlighted the natural inclination of youth to be involved, and the importance of including them in food-gathering programs. "The youth just pick it up naturally, they're ready to get out there, they're not afraid of hard work, and they're not afraid to eat right out of the environment! . . . So any way that we can continue to involve the youth and have them teaching

Hands-on corn-washing workshop at the Intertribal Food Sovereignty Summit hosted at the Meskwaki Settlement in May 2018. This Bear Island flint corn was boiled with hardwood ash, rinsed, then made into soup. *Photo by Elizabeth Hoover.*

the younger generations and get it back in the classroom" will contribute to community goals of food sovereignty.

Education is an important part of the sustainability of these projects. Amos Hinton, former director of the agricultural program for the Ponca Tribe, described to the Food Sovereignty Summit at Oneida in 2013: "If we don't educate our children, we're not going to have anything. There'll be no one to carry it on." With this in mind, many of these Indigenous food sovereignty projects have targeted youth as the focus of their work.[20]

Food Sovereignty as a Movement

To *become* food sovereign was seen as a goal for many projects and communities, but several participants also described food sovereignty as a broader movement, one that was both far-reaching and gaining traction in their own communities. Melissa Nelson described the "movement-building" aspect of food sovereignty, which has "many dimensions from the very person to the expansive political legislation." On a local level, participants described a gradual readiness for this movement in their communities. In reflecting on how work around different aspects of food and health have coalesced for him recently, Ken Parker (Seneca), former director of Food Is Our Medicine, asserted, "I think people are ready for this now. It seems like it's always been there, but now it's a bigger movement." Roberto Nutlouis observed that "the Native food movement is penetrating into the communities. People are more aware about it now." He went on to describe campaigns that the Black Mesa Water Coalition had sponsored around genetically modified organisms and traditional foods, and the way their organization focused on "continued community outreach. And just pushed the knowledge out there into the communities." In contrast to "the food movement," which Segrest labeled as being "a little bit elitist," for her own community and others she had worked with, "what this food sovereignty movement is hungry for is to remember the plants, our foods, the teachings."[21]

Rowen White described her experiences "working in the last over-a-decade in the food sovereignty movement within Indian Country," including with the "seed sovereignty movement" that "rose within our communities." This type of work was also encapsulated in Zach Paige's description of Indigenous seed alliances and the "network of growers" that is currently coming together to share seed and information, and to support each other. Rowen White expounds on this topic in chapter 7 of this book.

People who promote these projects also find community in the movement, in coming together for food sovereignty conferences and events. Julie Garreau proclaimed excitedly, "The movement is growing." Describing the interconnected tribal projects, the like-minded people who meet up at food sovereignty summits, she exclaimed, "the movement nationally . . . it's a small community but it's all over the nation." In addition to supporting the groundswell in their own communities, many of these participants recognized themselves as belonging to a broader Indigenous movement.

Food Sovereignty as a Process and a Method

In addition to constituting a movement, many of the participants I spoke with described food sovereignty as a goal with an extensive timeline, a process or a method, rather than a solidly defined destination. Chuck Hoskins, in reflecting on food issues for the Cherokee Nation, proclaimed, "Look, we didn't get here in a generation. It took many. We didn't get here overnight, so it won't be fixed overnight." Similarly, Alan Bacock, from the Big Pine Paiute Tribe in California, described food sovereignty as "a goal worth striving for. It's a vision that I would like to see develop, but it's not going to happen overnight, it's not going to happen in a year, in two years. It's going to take a long time to develop."

Jeff Metoxen detailed how at Tsyunhehkwa working toward food sovereignty is a never-ending learning process focused on Indigenous language, practices, and culture, in addition to basic knowledge about growing plants. "With food sovereignty pursuits, you're learning more about your own culture. . . . It's to share that information with your community. It's to ensure that the generations to come know this and are learning it as well. That's a big part. I look at that never-ending process you are going through. I feel I've never stopped learning . . . we're pursuing our food security. We're always pursuing our food sovereignty."

On the other hand, rather than describing food sovereignty as a goal or a process, Valerie Segrest reasoned to an audience at the 2013 Native American Culinary Association conference that "food sovereignty is a method, getting to a place of decolonizing our diets, revitalizing our traditional food culture."[22] In my 2014 interview with her she further pondered, "What does food sovereignty look like? I don't think it looks like anything. I think it's just a way of living and making food choices." She described food sovereignty as another tool, another lens through which she could work as an educator, nutritionist, and community member.

LIMITATIONS AND CONCLUSIONS

In this chapter I have discussed how participants in community-based farming and gardening projects that serve Native American communities perceive Indigenous food sovereignty. Two limitations of this particular sample for application to the broader Indigenous food sovereignty movement should be noted. The first is that although some of the study participants also engage in other types of food-procurement activities (for example, foraging, hunting, and fishing), this project is focused on farming and gardening projects. The Indigenous food sovereignty movement is also very concerned with ensuring access to treaty-guaranteed fishing, gathering, and hunting sites, as well as protection and utilization of traditional knowledge related to these activities; and some voices have been critical of an agriculture-centric version of food sovereignty.[23] This is due in part to the fact that while some tribal communities—such as the Mohawk, Seneca, Navajo, Ponca, and Pueblo communities featured here—view horticulture as a traditional activity, for other tribal nations, agriculture was an activity introduced as part of colonial oppression, deliberately intended to remove those Native nations even farther from the land and their traditional food-procurement activities.[24] And even for some of the communities that recognize horticulture as a traditional food-procurement activity, their culturally specific horticulture was forcibly replaced by federal government programs aimed at promoting a Westernized form of irrigated, monocropped agriculture.[25] That said, farming and gardening are tools some communities have used to celebrate their horticultural heritage, and others have employed to more efficiently use available tribal land to work toward food sovereignty.

A second limitation of this project is that the participants worked only in community-based projects, most of which have official nonprofit status or operate as such. I did not meet with individual farmers and ranchers who are producing food for individual profit. Dan Cornelius, of the Intertribal Agriculture Council, works to connect Native farmers and ranchers with USDA government programs in an effort to bolster overall food production for tribal communities. He asserts,

> The individual producer is so overlooked in the current food sovereignty movement. These [community] programs are critical to helping provide support but we need to get more individuals and families back into production. Look at the number of producers (nearly 72,000) and sales ($3.2 billion) and casino food service—about $4.5 billion. Sure, a $40,000 grant can help run a community garden for a year and can make an impact in how communities think about food, but true rebuilding of our food systems requires thinking about supporting individual and family production on landscape levels.[26]

A more comprehensive examination of food sovereignty among Native American farmers will need to include the voices of individual for-profit farmers as well.

DEFINING SOVEREIGNTY: A PROCESS THAT REQUIRES FOOD

Anishinaabe legal scholar Heidi Kiiwetinepinesiik Stark has asserted that

> because sovereignty is "intangible" and an inherent "dynamic cultural force," it is crucial that Indigenous peoples define for themselves a vision of their own nationhood and sovereignty, as well as the practical implications that come with this term. By looking to their own epistemologies and practices, Native peoples can put forward definitions of sovereignty that

are distinct from United States legal and political definitions of Native nations status that have operated to diminish Native sovereignty and self-government.[27]

Stark describes sovereignty as "deeply intertwined with a nation's sense of self," and as constantly undergoing transformation to meet the needs of the people of these nations. Rather than limiting sovereignty to "its restrictive legal-political context," she calls on us to see sovereignty "as a process, or a journey." Citing Standing Rock Sioux Vine Deloria Jr's proclamation: "Sovereignty is a useful word to describe the process of growth and awareness that characterizes a group of people working toward and achieving maturity," she concludes that it is through this connection to identity and self that "sovereignty becomes a process rather than a stagnant notion."[28] Amanda Cobb similarly describes the importance of thinking about sovereignty as a process rather than a final achievement: "By casting sovereignty not only in terms of process, but more particularly in narrative terms, sovereignty becomes the ongoing story of ourselves—our own continuance. Sovereignty is both the story or journey itself and what we journey towards, which is our own flourishing as self-determining peoples."[29]

As I described earlier, many study participants recognized food sovereignty not only as a movement, but also as a process, a method, and a goal. But for many of them, working toward food sovereignty was an important part of becoming truly sovereign nations in a broader sense as well. For example, despite Winona LaDuke's criticisms of the term discussed earlier in this article, at almost every food-related event where I have heard her speak, she has reiterated that "you can't say you're sovereign if you can't feed yourself." Several of the participants repeated this aphorism to me, and some elaborated upon it as a stipulation. Food sovereignty was seen either as a marker of achieved tribal sovereignty (as Jeremy McClain from the Bad River Gitiganing project described, "an ability to feed yourself is a marker of true sovereignty"), or as a necessary prerequisite toward which tribes should work before they can claim to be sovereign. As Alan Bacock (Paiute) explained, "I would say that you can't be sovereign if you can't feed yourself, . . . When we begin to use it [traditional food knowledge] again, we begin to then develop sovereignty once more. Because I don't believe that you can have that sense of sovereignty without that food connection." And as Clayton Harvey (White Mountain Apache), from the White Mountain Apache Ndée Bikíyaa project, described, food production is central to the identity of Native nations; "I think about sovereignty and I think about being Native

American, and being who you are, and that's growing your own food." Rowen White concludes, "when we are able to control our food sources and really able to dictate our seed and our food, we have a greater sense of sovereignty as a whole."

When asked to define or describe the concept of food sovereignty, participants from Native American community and farming projects across the United States highlighted a number of features common to the broader food sovereignty movement: the importance of access to healthy, culturally relevant food, land, and information; independence for individuals to make choices about their own consumption and for communities to define their own food systems; and desire to keep food dollars within the community. Some aspects of participants' definitions grew specifically out of their own histories with the land and the colonial entities that had impacted their communities. Participants described the importance of tribes having the independence and control to provide the foods they see appropriate, grown in a manner that is deemed acceptable for their constituents. Relationships to the environment, food sources, and other people were highlighted, including trade relationships. The ability to sustain both land and cultural lifestyles was emphasized, as was the ability to protect seeds as the living relatives necessary for the continuation of the food sovereignty movement. The importance of education and working with Native youth was also mentioned as a specific antidote to culture loss and the ensuing health problems that have made Indigenous communities the subject of so many public health studies.

As Kyle Powys Whyte has pointed out, "concepts of food sovereignty can come across as so many impossible ideals of community food self-sufficiency and cultural autonomy."[30] Importantly, however, this study's participants viewed food sovereignty not as an absolute that can be achieved or lost, but rather as a movement that is carrying these projects toward their goals; a process that participants expect to be undertaking for a great deal of time; and a framework through which they are working toward improved physical, cultural, and economic health.

ACKNOWLEDGMENTS

I was accompanied during a majority of this trip by filmmaker Angelo Baca (Navajo/Hopi), who is in the process of creating a documentary about the Indigenous food movement from footage collected on this trip. I want especially to acknowledge Jeff Metoxen (Oneida), and Stephanie Berryhill (Mvskoke Creek), who are no longer with us but whose words have been important to this project. This project was sponsored by a Salomon Grant from Brown University in 2013–15.

INTERVIEWS CONDUCTED

Alan Bacock (Paiute), Big Pine, California, July 14, 2014
Lannesse Baker (Ojibway), Minneapolis, Minnesota, August 13, 2014
Stephanie Berryhill (Mvskoke Creek), Okmulgee, Oklahoma, June 4, 2014
Clayton Brascoupe (Mohawk), Tesuque Pueblo, New Mexico, June 11, 2014
Grace Ann Byrd (Nisqually), Dupont, Washington, July 22, 2014
Don Charnon (Oneida), Tsyunhehkwa, Oneida Nation of Wisconsin, August 20, 2014
Tom Cook (Mohawk), Akwesasne Mohawk Reservation, New York, September 8, 2014
Michael Dahl (Ojibway), Minneapolis, Minnesota, August 7, 2014
Christina Elias, Minneapolis, Minnesota, August 13, 2014
Darlene Fairbanks (Ojibway), Minneapolis, Minnesota, August 12, 2014
Julie Garreau (Lakota), Cheyenne River Indian Reservation, South Dakota, August 1, 2014
Keith Glidewell (Paiute), Bishop, California, July 14, 2014
Richard Grounds (Yuchi/Seminole), Sapulpa, Oklahoma, June 4, 2014
Clayton Harvey (White Mountain Apache), Fort Apache, Arizona, July 3, 2014
Lilian Hill (Hopi), Kykotsmovi Village, Arizona, July 9, 2014
Amos Hinton (Ponca), Ponca City, Oklahoma, June 6, 2014
Chuck Hoskins (Cherokee), Tahlequah, Oklahoma, June 7, 2014
Ben Jacobs (Osage), Tocabe restaurant, Denver, Colorado, November 21, 2015
Terrol Dew Johnson (Tohono O'odham), Sells, Arizona, June 30, 2014
Caitlin Krenn, Dupont, Washington, July 22, 2014
Winona LaDuke (Ojibway), White Earth Reservation, Minnesota, August 29, 2014
Pati Martinson (Lakota/French), Taos, New Mexico, June 12, 2014
Jeremy McClain (Ojibway), Bad River Reservation, Wisconsin, August 18, 2014
Angelo McHorse (Taos Pueblo), Taos, New Mexico, June 12, 2014
Jeff Metoxen (Oneida), Tsyunhehkwa, Oneida Nation of Wisconsin, August 20, 2014
Gayley Morgan (Tesuque Pueblo), phone interview, July 25, 2014
Melissa Nelson (Turtle Mountain Chippewa), Novato, California, June 24, 2014

Roberto Nutlouis (Navajo), Pinon, Arizona, July 10, 2014

Zach Paige, White Earth Reservation, Minnesota, August 5, 2014

Ken Parker (Seneca), Cattaraugus Reservation, New York, September 2, 2014

Kenny Perkins (Mohawk), Akwesasne Mohawk Reservation, New York, September 5, 2014

Dan Powless (Ojibway), Bad River Gitiganing Community Garden, Wisconsin, August 19, 2014

Jayson Romero (Cochiti Pueblo), Cochiti Pueblo, New Mexico, June 11, 2014

Josh Sargent (Mohawk), Akwesasne Mohawk Reservation, New York, September 7, 2014

Valerie Segrest (Muckleshoot), Muckleshoot Tribal College, Washington, July 21, 2014

Sean Sherman (Lakota), Minneapolis, Minnesota, August 29, 2014

Bob Shimek (Ojibway), White Earth Reservation, Minnesota, August 5, 2014

Scott Shoemaker (Miami), Science Museum of Minnesota, St. Paul, August 29, 2014

Cassius Spears Sr. (Narragansett), Westerly, Rhode Island, April 30, 2016

Romajean Thomas (Muckleshoot), Muckleshoot Tribal College, Washington, July 21, 2014

George Toya (Jemez Pueblo), Nambé Pueblo, New Mexico, June 10, 2014

Lori Watso (Shakopee Mdewakanton Sioux), Shakopee Mdewakanton Sioux Tribe, Minnesota, August 7, 2014

Rowen White (Mohawk), Nevada City, California, July 16, 2014

Woodrow White (Ho-Chunk), Tomah, Wisconsin, August 21, 2014

Mike Wiggins (Ojibway), Bad River Reservation, Wisconsin, August 19, 2014

Diane Wilson (Dakota), Dream of Wild Health farm, Hugo, Minnesota, August 14, 2014

Milo Yellowhair (Lakota), Pine Ridge Reservation, South Dakota, August 3, 2014

NOTES

1. For more information, see Elizabeth Hoover, Tsyunhehkw^, Oneida Nation, Wisconsin, https://gardenwarriorsgoodseeds.com/2015/01/19/tsyunhehkw-oneida-nation-wisconsin/ or Tsyunhehkwa's Facebook site at https://www.facebook.com/tsyunhehkwa/.

2. Hoover, *The River Is in Us: Fighting Toxics in a Mohawk Community* (Minneapolis: University of Minnesota Press, 2017).

3. I am labeling as "interviews" the sessions during which I sat down with a participant and used an interview protocol to ask a standard series of questions, although these sessions were often guided by the participants and their interests as well. In contrast, during what I label "conversations" and "farm tours," with the participants' permission, I turned on my audio recorder while walking around their farms and gardens. These conversations frequently included multiple people, and the topics were primarily guided by the landscape and the project at hand. All participants in interviews as well as conversations signed informed consent forms asking if they wanted to be named or have their information remain confidential.

4. NVivo software helps analyze qualitative data by providing the framework to organize and sort interview information into different coded categories established by the researcher.

5. See gardenwarriorsgoodseeds.com.

6. Valerie Segrest, presentation at the Native American Culinary Association Conference, Arizona-Sonora Desert Museum, Tucson, AZ, November 2013. For conference information see Apaches in the Kitchen, November 23, 2013, http://apachesinthekitchen.blogspot.com/2013/11/nacas-2013-indigenous-food-symposium.html.

7. Shoemaker interview and personal communications with members of the Indigenous Seed Keepers Network, 2014–2017, especially at the Indigenous Farming Conference hosted by the White Earth Land Recovery Project in Minnesota, which includes panels and sessions by Indigenous Seed Keepers Network members every year.

8. Valerie Segrest, presentation, Food Sovereignty Summit, Green Bay, WI, April 16, 2014. For full agenda, see https://firstnations.org/sites/default/files/conferences/2014/documents/2014_Ag_Summit_4_7_2014.pdf.

9. Johnson interview; see also Tristan Reader and Terrol Dew Johnson, "Tohono O'odham *Himdag* and Agri/culture," in *Religion and Sustainable Agriculture: World Spiritual Traditions and Food Ethics*, edited by Todd LeVasseur, Pramod Parajuli, and Norman Wirzba (Lexington: University of Kentucky Press, 2016), 315–36.

10. Mariaelena Huambachano, "Food Security and Indigenous Peoples Knowledge: El Buen Vivir–Sumaq Kawsay in Peru and Tē Atānoho, New Zealand, Māori–New Zealand," *Food Studies* 5, no. 3 (2015): 33–47, 40, 42, https://doi.org/10.18848/2160-1933/CGP/v05i03/40505.

11. Winona LaDuke has described this survey during a number of her presentations at events like the Indigenous Farming Conference, held every year in March at White Earth, as well as during a TED talk discussed in John Platt, "Why Winona LaDuke Is Fighting for Food Sovereignty," mnn.com Leaderboard, March 4, 2013, https://www.mnn.com/leaderboard/storieswhy-winona-laduke-is-fighting-for-food-sovereignty.

12. Hopi Tutskwa Permaculture, "Hopi Farmers Market & Exchange," https://www.hopitutskwapermaculture.com/job-announcements.

13. Nancy J. Turner and Dawn C. Loewen, "The Original 'Free Trade': Exchange of Botanical Products and Associated Plant Knowledge in Northwestern North America," *Anthropologica* 40, no. 1 (1998): 49–70, https://doi.org/10.2307/25605872.

14. Intertribal Agriculture Council, "Mobile Farmers Market, Trade Routes Roadtrip," https://nativefoodnetwork.com/trade-routes/.

15. Jack Kloppenburg, "Seed Sovereignty: The Promise of Open Source Biology," in *Food Sovereignty: Reconnecting Food, Nature and Community*, edited by Hannah Wittman, Annette Aurélie Desmarais, and Nettie Wiehe (Halifax, NS: Fernwood, 2010), 165.

16. Kloppenburg, "Seed Sovereignty," 157.

17. Sheryl D. Breen, "Saving Seeds: The Svalbard Global Seed Vault, Native American Seed Savers, and Problems of Property," *Journal of Agriculture, Food Systems, and Community Development* 5, no. 2 (2015): 46–47, https://doi.org/10.5304/jafscd.2015.052.016.

18. Valerie Segrest, presentation at the Native American Culinary Association conference, Arizona-Sonora Desert Museum, Tucson, AZ, November 2013.

19. Valerie Segrest, presentation at the Food Sovereignty Summit, Green Bay, WI, April 16, 2014.

20. Amos Hinton, presentation at the Food Sovereignty Summit, Green Bay WI, April 2013.

21. Valerie Segrest, presentation at the Native American Culinary Association conference, Arizona-Sonora Desert Museum, Tucson, AZ, November 2013.

22. Valerie Segrest, presentation at the Native American Culinary Association conference.

23. Annette Aurélie Desmarais and Hannah Wittman, "Farmers, Foodies, and First Nations: Getting to Food Sovereignty in Canada," *Journal of Peasant Studies* 41, no. 6 (2014): 1153–73, esp. 1155, https://doi.org/10.1080/03066150.2013.876623.

A number of studies from Manitoba, Canada, focus on Indigenous food sovereignty in the context of wild foods. For example, see Karlah Rae Rudolph and Stéphane M. McLachlan, "Seeking Indigenous Food Sovereignty: Origins of and Responses to the Food Crisis in Northern Manitoba, Canada," *Local Environment* 18, no. 9 (2013): 1079–98; Asfia Gulrukh Kamal, Rene Linklater, Shirley Thompson, Joseph Dipple, and Ithinto Mechisowin Committee, "A Recipe for Change: Reclamation of Indigenous Food Sovereignty in O-Pipon-Na-Piwin Cree Nation for Decolonization, Resource Sharing, and Cultural Restoration," *Globalizations* 12, no. 4 (2015): 559–75; Shirley Thompson, Asfia Gulrukh Kamal, Mohammad Ashraful Alam, and Jacinta Wiebe, "Community Development to Feed the Family in Northern Manitoba Communities: Evaluating Food Activities Based on Their Food Sovereignty, Food Security, and Sustainable Livelihood Outcomes," *Canadian Journal of Nonprofit and Social Economy Research* 3, no. 2 (2012): 43–66; Tabitha Martens, Jaime Cidro, Michael Anthony Hart, and Stephane McLachlan, "Understanding Indigenous Food Sovereignty through an Indigenous Research Paradigm," *Journal of Indigenous Social Development* 5, no. 1 (2016): 18–37.

24. See, for example, Rudolph and McLachlan "Seeking Indigenous Food Sovereignty."

25. Reader and Johnson, "Tohono O'odham *Himdag* and Agri/culture"; David A. Cleveland, Fred Bowannie Jr., Donald F. Eriacho, Andrew Laahty, and Eric Perramond,

"Zuni Farming and United States Government Policy: The Politics of Biological and Cultural Diversity in Agriculture," *Agriculture and Human Values* 12, no. 3 (1995): 2–18, https://doi.org/10.1007/BF02217150.

26. Dan Cornelius, email correspondence with the author, July 9, 2017.

27. David Wilkins, *American Indian Politics and the American Political System* (2002): 48, quoted in Heidi Kiiwetinepinesiik Stark, "Nenabozho's Smart Berries: Rethinking Tribal Sovereignty and Accountability," *Michigan State Law Review* 339: 343.

28. Both quotations in Stark, "Nenabozho's Smart Berries," 352; Vine Deloria Jr. quoted from "Intellectual Self-Determination and Sovereignty: Looking at the Windmills in Our Minds," *Wicazo Sa Review* 13, no. 1 (1998): 27.

29. Amanda Cobb, "Understanding Tribal Sovereignty: Definitions, Conceptualizations, and Interpretations," *American Studies* 46, nos. 3–4 (2005): 125.

30. Kyle Powys Whyte, "Indigenous Food Sovereignty, Renewal, and Settler Colonialism," in *The Routledge Handbook of Food Ethics*, edited by Mary C. Rawlinson and Caleb Ward (New York: Routledge, 2017), 354–65.

DEVON A. MIHESUAH

3

SEARCHING FOR *HAKNIP ACHUKMA* (GOOD HEALTH)

Challenges to Food Sovereignty Initiatives in Oklahoma

In the past three decades, tribes and grassroots organizations comprising determined tribal members have initiated numerous food projects, including seed distribution, food summits, farmers' markets, cattle and bison ranches, and community and school gardens.[1] These enterprises are steps toward achieving what many food activists refer to as food sovereignty, though there are various ideas about what food sovereignty is, or can be. Further, there is no universal solution to achieving food sovereignty. Tribal food self-sufficiency involves the coordination of complex social, political, religious, economic, and environmental concerns. Those efforts vary by tribe because tribes adapted to colonization differently and do not have the same access to the resources; consequently, their food sovereignty goals—if they even have any—also differ. Many do not. British food activist Raj Patel, writing about the idea of food sovereignty in general, comments that "there are so many versions of the concept, it is hard to know exactly what it means," a statement that certainly applies to Indigenous people in the United States.[2]

I begin with the most complete vision of Indigenous food sovereignty as the concept is defined in the 2007 Declaration of Nyéléni: "The right of peoples to healthy and culturally appropriate food produced through ecologically sound and sustainable methods, and their right to define their own food and agricultural systems." The declaration also asserts that food sovereignty "ensures that the rights to use and manage lands, territories, waters, seeds, livestock, and biodiversity are in the hands of those of us who produce food."[3]

To be a "food sovereign" tribe ideally would mean, then, that the tribe has the right to control its food production, food quality, and food distribution. It would support tribal farmers and ranchers by supplying machinery and technology needed to plant and harvest. The tribe would not be answerable to state regulatory control, and would follow its own edicts, regulations, and ways of governance. Its members would have educational and job opportunities. The tribe collectively would decide if it wanted to purchase foods produced outside its boundaries or to trade with other groups. The tribe would have renewable energy infrastructure, such as solar and wind power.[4] Elders would be honored for their ability to teach language and impart traditional Indigenous knowledge about planting, harvesting, seed saving, hunting, basket and tool making, and ceremonies associated with sustenance. They would remind us about traditions among many tribes regarding female deities who originally brought them sustenance. Rather than viewing environmental resources as commodities for monetary gain, tribal members would show reverence for the land that sustains them. They would protect and respect the natural world because thriving relationships between healthy ecosystems and Indigenous peoples underlie tribal political, social, and religious systems. For Indigenous activists concerned with food injustice, that would be the ideal scenario. But is food sovereignty, defined in this way, possible?

OKLAHOMA

Oklahoma presents opportunities for discussion about food sovereignty because of its multifaceted history, environmental issues, and current politics, which include uneven food quality, poor Indigenous health, intratribal factionalism, trenchant racism, and the glaring dichotomy between those tribal members who are affluent and those who suffer from extreme poverty.

Oklahoma became a state in 1907. Prior to that time, it was Indian Territory. The Indian Removal Act of 1830, signed by President Andrew Jackson, was a cruel and devastating policy that forced thousands of Indians to move to Indian Territory in order to make way for white settlement in the Southeast and elsewhere.[5] Sixty-seven tribes were moved to Indian Territory, but eventually some were allowed to return to their homelands or were moved to other states. Today there are thirty-eight tribal nations in Oklahoma.[6] For Oklahoma tribes wishing to follow their foodway traditions, there is no one-size-fits-all model. Some tribes have an agricultural legacy, but also depended on wild game and plants. Plains tribes, which arrived in the 1870s, hunted bison and other animals, but they expanded their resource base by gathering wild flora and trading with other tribes and non-Indians. Some, such as the Comanches ("Lords of the Plains"), had no agricultural tradition but had access to a variety of foods and were adept at taking what they wanted.

Among the Native peoples who, under great duress and with devastating loss of life, came to Indian Territory were the "Five Tribes" (Cherokees, Chickasaws, Choctaws, Mvskoke-Creeks, and Seminoles). They found that much of their new lands in the eastern parts of Indian Territory resembled their southeastern homelands. There was plentiful game in the forests and grasslands, and fertile soils that allowed them to farm. They had ample water and a variety of nut trees and wild fruits.[7] Yet, despite the assortment of foods that many Natives grew, gathered, and hunted, by the time of the Civil War different foods were becoming available, and many Natives began suffering the consequences of altering their diets of vegetables, fruits, and game meats to rely instead on sugary, fatty, and starchy foods.[8]

The reasons for these health changes are complex. Boarding schools, missionaries, and intermarriage with whites contributed to the disassociation from tribal language, religion, and foodways. Particular population groups that were affected include tribal members who were affluent—a group that included full-blooded Native persons, though most were racially and culturally mixed—and Native students at the boarding schools, who were fed white flour and sugar three times a day. These populations developed digestive disorders, with some becoming diabetic or prediabetic. In addition, as non-Native intruders surged into Indian Territory throughout the 1800s, the plants and animals Indians once used for sustenance and medicine diminished. The ecosystems were transformed by dams, mines, deforestation, invasive species, and ranching. Those Native people who could not afford to purchase store goods survived on what they could grow; some suffered from malnutrition. The bison herds diminished, drastically altering the lifeways of Plains tribes. After being placed in Indian Territory in the late 1800s, Comanches, Cheyennes, Kiowas, and other hunters were forced to depend on inadequate rations provided by the federal government (see chapter 11).

Today, at least 16 percent of Native people in the United States suffer from diabetes and 33 percent are obese.[9] Among the Oklahoma population, Indians have the highest rates of heart disease, "unintentional injury deaths," diabetes, and asthma. They eat fewer fruits than whites, blacks, and Hispanics. The Oklahoma Department of Health assigns Native Americans a grade of D for low physical activity and incidence of obesity, and an F for "poor mental health" and "poor physical health" days.[10] Of the Cherokees who seek treatment at Cherokee clinics, 34 percent are overweight or obese.[11] The rate of diabetes on the Osage Reservation is 20.7 percent, double that of the overall rate in the United States. The rate of heart disease among reservation Osages is double that of those off-reservation. Twenty-one percent of reservation Osages live in poverty, compared to 10.3 percent of the US population.[12] Children spend less time playing outdoors and adults are increasingly separated from the land, resulting in waning interest in the natural world. Smoking and depression exacerbate Oklahoma Natives' health issues.

It is not only people in Indian Country who feel effects of environmental degradation, climate change, foodborne illnesses, industrial chemicals, and soil erosion.[13] Worldwide, water and air are polluted, seafood is overharvested, and the cost of animal feed has risen. All consumers now face prices that have risen 26.8 percent in the last ten years, affecting bread, baked goods, canned vegetables, fruit, eggs, beef, pork, and chicken.[14] The avian flu, porcine epidemic diarrhea virus, and excessively dry and wet seasons have resulted in sick animals and failed wheat, lettuce, and corn crops. Food sovereignty activists are situated in an economy in which four seed companies, Dow AgroSciences, DuPont/Pioneer, Monsanto, and Syngenta, control 80 percent of the corn market, 70 percent of the soybean market, and 50% of the world's seed supply.[15] Ten companies own almost every brand of food and beverage.[16] Environmental activist Wendell Berry sums up what we all want: "Food that is nutritionally whole and uncontaminated by pesticides and other toxic chemical residues."[17]

Many tribal members who do not qualify for government commodities find that stores are not conveniently located and the products stocked are inadequate. For example, the 2,251 square miles of the Osage reservation in Osage County has only four grocery stores, making it a "super food desert." Most of the land is devoted to livestock ranching, not agriculture, and there is no public transportation.[18] Most stores on Indian land sell produce that comes from farms that use genetically modified seeds. These pest-resistant crops grow bigger and more quickly, but they are less nutritious and leave behind eroded and depleted soils.[19]

Among Oklahoma Indians, the lack of food, or lack of nutritious food, results from a combination of their having no money to purchase it; being dependent on the government; having no control over resources; and being unable to produce food. Many Indians have eaten their traditional foods, hunted, and gardened their entire lives. Most, however, have not. Natives who take advantage of the commodities offered under the USDA's FDPIR food distribution program have about one hundred food-buying choices, but often they opt for white flour, lard, cheese, and sugary and salty items. Notably, food distribution commodities are available to low-income tribal members who live within a tribal nation's boundaries.[20] Yet not all residents on tribal lands are members of that tribe and, given the differing food histories of various tribes, their food choices are likely both to be different and to affect the diets of those around them. For example, the majority of the 233,126 persons residing within the Choctaw Nation are not Choctaws.[21] In addition, in Oklahoma's program, as long as one member of a federally recognized tribe lives in the residence, non-Indians also in the home can receive commodities under the FDPIR. Non-Indians' food choices may not include items that are culturally connected to the tribe or healthy, and their preferences might influence others in the house.

"TRADITIONAL" FOODS

Health and traditionalism intertwine. Tribal members can consume non-Indigenous foods and be healthy, but food sovereignty activists are hopeful that a return to traditional foodways will provide something more: empowering links to their cultures and histories. Of course, it must be determined what those traditional foodways are, and not everyone agrees on what is "traditional."[22] The Choctaw Nation's website and its 2017 calendar, for example, feature some reasons why many tribal members are obese and diabetic: a "traditional" recipes section heavy on unhealthy food items, including sugar, white flour, cheese, and butter, used for making grape dumplings, cheddar and corn chowder, crisp salt pork, cobblers, fried corn, fry bread, Indian tacos, and creamed Indian corn (sugar, flour, milk, canned corn, and pork). A sweet potato dish that would be flavorful without any seasoning calls for adding two cups of sugar and one cup of flour.[23] The Chickasaw Nation does the same, with "Chickasaw" appearing in the names of some dishes that lack Indigenous ingredients.[24] Similarly, Osage cooking classes teach young tribal members how to make "Indian food" such as wheat flour rolled out in the "Osage custom"—that is, fried in hot grease—as well as to how to cook chicken and dumplings, and meat with wheat gravy.[25]

Defining traditional food is tricky because some Oklahoma tribes adopted European and African foods and material goods centuries ago, and some define traditional foods as what their grandparents ate. Those traditionalists who advocate for precontact foods may clash with tribal members who argue that fry bread, mutton, and grape dumplings made with wheat flour and sugar are traditional, as are dishes made with dairy, eggs, beef, pork, and chicken. Ancestors of tribespeople in Oklahoma started growing European-introduced crops on a large scale for profit by the 1840s. After the Civil War, Choctaws cultivated sixty thousand acres of corn and potatoes, but also non-Indigenous wheat and oats. By this time, members of the Five Tribes used metal axes, plows, hoes, harrows, scrapers, shovels, spades, threshers, mowers, and reapers. Backyard gardens featured more European-introduced foods: lettuce, turnips, peas, and mustard. Many Indians cultivated European-introduced apple, peach, and plum trees. Some of the farmers along the North Fork and Arkansas Rivers grew cotton and tobacco, and ranchers raised non-Indigenous horses, cattle, mules, sheep, goats, and hogs.[26]

If one's goal is to eat only precontact foods, then the menu might include alligator, elk, waterfowl, deer, antelope, wild turkey, and bison. Those animals would have to be hunted or raised, both of which options would require financial planning. Tribes historically gathered flora such as mulberries, wild plums, grapes, onions, and nuts (such as pecan, hickory, walnut, and acorn), but today those foods may grow only on private property. Traditional foods are not consistently available, so some food projects and families might use only a few Indigenous foods as symbols of culture. For example, the Delawares were originally hunters and coastal people who were removed several times before settling in Indian Territory in 1867. They no longer have access to marine life, so they stock their ponds with fish, hold annual fishing tournaments, and teach their children to hunt.[27]

Some tribes have vested economic interests in producing foods such as cattle, wheat, hogs, and sorghum, none of which is Indigenous. However, supplying traditional foods to their members may not be their goal. Hunting tribes followed bison herds for hundreds of miles each year and, obviously, they cannot do that today. The Quapaw Tribe of Indians moved to Indian Territory in 1834 and settled on ninety-six thousand acres in what is now Quapaw County.[28] Traditionally, Quapaws hunted, gathered, and farmed, but like other tribes, they were not ranchers.[29] In June 2016, the Quapaw Tribe opened the Quapaw Mercantile, the distribution center for the Quapaw Cattle Company. The store sells beef and bison ribeye steaks, beef bacon, and bratwurst from the tribe's herds in Miami and Quapaw. They provide meat to the tribe's elder center, daycare centers, and the Quapaw and Downstream Casino restaurants. The tribe is in the process of designing its own meat-processing plant and has plans to grow feed for the animals.[30] The Iowa, Modoc, and Cheyenne and Arapaho Tribes and the Cherokee Nation raise bison. Some ranchers crossbreed bison with cattle to create beefalo.

POVERTY

"Food security" has been defined as all members of a household having, at all times, "physical and economic access to sufficient, safe, and nutritious food that meets their dietary needs and food preferences for an active and healthy life."[31] Eric Holt-Giménez writes, "Where one stands on hunger depends on where one sits."[32] Some Natives in Oklahoma are quite affluent and can buy whatever they want; others are poverty-stricken and have little opportunity for economic or social advancement.

The Choctaw Nation of Oklahoma lands consist of 10,613 square miles of rural area in 10.5 counties in southern Oklahoma. In 2016 the Choctaw Nation had nine thousand workers on a payroll of $300 million. The tribe operates seven casinos, thirteen travel plazas, twelve smoke shops, two Chili's franchises, a resort in Durant, and document-archiving companies. Along with manufacturing operations, it manages seven Black Angus cattle ranches and provides other management services. In the 2018 State of the Nation Address, Chief Batton stated that the total tribal assets were $2.4 billion.[33] In 2015, moreover, a tribal trust court settlement awarded the Choctaw and Chickasaw Nations $139.5 million and $46.5 million, respectively, for the federal government's failure to protect the tribes' interests when it sold more than one million acres of timberlands in the decades between 1908 and 1940.[34]

My tribe should not be impoverished, yet despite the millions of dollars produced each year, some of the poorest counties in the country are within the Choctaw Nation. High proportions of residents in Atoka, Coal, Haskell, Latimer, LeFlore, McCurtain, Pittsburg, and Pushmataha Counties all have high-risk factors such as smoking, obesity, physical inactivity, and low consumption of fruit and vegetables. One census tract has a poverty rate of more than 52.8 percent, with leading causes of death being heart disease and cancer.[35] Those who do find work receive low wages. Recognizing the dire situation, in 2014 then-president Barack Obama named the Choctaw Nation one of five "Promise Zones." The award entails tax incentives for businesses that invest in the community and promises them "competitive advantage" when applying for federal grants.[36]

After the Promise Zone award was received, the tribe's chief business and economic development officer enlisted several Choctaws with expertise in traditional foods, medicine, and gardening to brainstorm strategies for a farm-to-table agriculture initiative. The focus was to be on nutrition and natural medicines, in addition to Native foods, backyard gardens, "agri-art," farmers' markets, and other endeavors. The new tribal chief elected that year, however, dismissed the business development officer, thereby severing ties with those of us who had contributed a plethora of ideas to the Promise Zone initiative. The initiative's leadership then created the Choctaw Small Business Development Services (CSBDS), which currently offers advice, planning, and counseling for tribal entrepreneurs, but not financial support.[37]

One of its stated goals is that "natural, historic, and cultural resources" serve as the foundation for initiatives, including "technology-enhanced traditional farming and ranching," large greenhouses, and training for women business owners.[38] Choctaws did not traditionally ranch, so it is not clear what is meant by "traditional farming and ranching." Indeed, backyard gardens had been among the suggestions initially submitted to the Choctaw Promise Zone initiative. Families desirous of cultivating gardens would have been given seeds, basic tools, soil, and water. The tribe would finance the plowing of land, and would provide basic gardening lessons. That idea apparently has been discarded. The monthly tribal newspaper, the *BISKINIK*, includes columns about traditional foods, but there are no Indigenous gardens or classes to teach tribal members how to grow or gather them.[39]

Part of the Choctaw Nation's plan is to create an educated workforce that can succeed in the business world. This is a crucial initiative considering that Oklahoma is ranked forty-ninth in the nation in educational services and performance.[40] If that workforce education strategy is to include implementing "traditional" farming methods, that workforce must know how to cultivate traditional foods and how to save seeds. The plan calls for partnerships with Oklahoma State University, Eastern Oklahoma State College, and the Kiamichi Technology Center.[41] However, none of those schools offers courses dealing with Choctaw history and culture.

For low-income seniors residing in the nation's 10.5-county area, the Choctaw Nation has instituted the Senior Farmers' Market Nutrition Program. Qualified seniors receive $50, and an additional 3,800 participants receive $30, to purchase locally produced foods. Non-Natives over sixty years of age living in a household that includes one enrolled Choctaw are eligible for checks. Funded by both the USDA and the Choctaw Nation, the program is "designed to encourage participants to make better food choices and raise awareness of farmers and farmers markets." Only about half of the ninety-five farmers who sell produce to the market are tribal members, however, which does not advance the purpose of supporting tribal farmers.[42]

An additional concern is lack of data regarding what consumers do with the produce. The Choctaw Nation has a number of health initiatives, but there is little research regarding their successes beyond the number of people using the vouchers. The tribal newspaper includes articles about diabetes, obesity, and exercise, and the Diabetes Multi-resource Task Force travels across the Choctaw Nation to educate tribal members about healthy lifestyles. However, as seen on Choctaw Nation calendars, the nation's website, and at tribal celebrations, the tribe also provides and promotes unhealthy food.

IMPACT OF DIMINISHED HEALTH-CARE FUNDING

As of March 2018, the Trump administration has weakened the Affordable Care Act (ACA) and continues to seek its repeal. Through the Indian Healthcare Improvement Act (IHCIA), which is part of the ACA, Indian health centers can bill third-party insurers, Medicare, and Medicaid. Almost 2.2 million people who use the Indian Health Service (IHS) will face difficulty accessing health care if the ACA is repealed.[43] The IHS could potentially lose more than $800 million in funding from Medicaid programs.[44] The Choctaw Nation recently completed a 143,000-square-foot regional medical clinic, the first tribal clinic in the United States with an outpatient ambulatory surgery center. The ambitious project provides dental services, podiatry, endoscopy, pediatrics, respiratory therapy, cardiology, diabetic, and pulmonology care, in addition to behavioral health services and an on-site laboratory.[45] The tribe paid for the construction of the facility, and the IHS works with Congress to provide funding for staff. Considering that President Trump has called for a 16.2 percent cut in funding for the Department of Health and Human Services, this is cause for alarm.[46] Everyone needs medical and dental care. Nonetheless, to improve physical and mental health and avoid hospital visits to treat maladies caused by poor diet and inactivity, it is key to adopt an exercise regime and a diet of unprocessed and fresh foods, and to quit smoking.

ACCESS TO TRADITIONAL FOODS BY TREATY: HUNTING, FISHING, AND GATHERING RIGHTS

Treaties between tribes and the federal government are legally binding contracts that contain assurances of self-determination, health-care and educational services, religious freedom, and rights to hunt and fish. The federal government has a responsibility to protect tribal treaty rights, lands, and resources. Those who were forcibly sent to Indian Territory were understandably suspicious of government promises, as are their descendants. Removal treaties guaranteed that tribes would retain their lands, but Oklahoma has a long history of racism and dispossessing tribes of their property—27 million acres during the allotment period. Portraits of men such as Governor Charles N. Haskell, who stole land from tribes during the allotment period, hang in the statehouse.[47] The discovery in 1897 of oil under Osage lands not only resulted in the murders of dozens of tribal citizens at the hands of unscrupulous whites intent on taking their resources, but also caused socioeconomic rifts within the tribe.[48] University of Oklahoma students are nicknamed "Boomer Sooners," after the intrepid pioneers who illegally jumped the gun on the Appropriations Act of 1889 in order to claim land belonging to tribal peoples.

Tribes must know how to negotiate the various challenges from outside forces (such as racism, climate change, pollution) as they relate to the powers of their tribe, the states, and the federal government, as well as abrogation of treaty agreements that guarantee water, hunting, fishing, and gathering rights. Several treaties in the 1830s guaranteed to Cherokees "free and unmolested use" of lands not within the bounds of the Cherokee Nation.[49] It was not until 2015 that the Cherokee Nation became the first tribe to sign a compact giving their members hunting and fishing rights in all seventy-seven counties in Oklahoma. Cherokees over the age of sixteen can receive one "dual license" (Cherokee Nation and Oklahoma) and one free turkey and deer tag per year. Beginning in January 2017, Choctaw Nation citizens in Oklahoma did not have to pay for licenses either; the tribes pay a fee for each tag received and Oklahoma, in turn, receives federal monies for wildlife conservation.[50]

These are indeed important compacts, but one cannot (or should not) just pick up a gun and go hunting. Procuring a deer or turkey requires skill, patience, and knowledge of hunting safety and protocol. Proper equipment and clothing is expensive. Moreover, physical fitness is essential for those who stalk birds all day or who must drag a heavy animal back to camp. Then it must be dressed and butchered. Although some Natives are adept at using traditional blowguns, rabbit sticks, and bows to hunt small animals, it should be recognized that not everyone has the wherewithal to hunt game.

TRADITIONAL FOODS AND ECOSYSTEM CHANGES

A return to traditional ways of eating requires access to healthy ecosystems and their resources. After the 1830s removal to Indian Territory, serious environmental changes and resource depletion resulted from human activities, including building fences, dams, and railroads; harvesting timber, mining, and digging reservoirs; and overgrazing rangeland, which increased the likelihood of drought. In the 1830s much of the Choctaw Nation's fertile lands, for example, were crossed by streams of clear water and were lush with edible plant foods. After removal, my family settled in Atoka County, then moved to the Kully Chaha (High Spring) township in the shadow of Nvnih Chufvk (Sugar Loaf mountain), once deemed by both Choctaws and newspaper reporters as an "oasis" of springs, bountiful game, nuts, and berries.[51] Despite the relative isolation of this area, nearby cattle ranching, diversion of waterways, and deforestation caused the disappearance of many wild fruit plants, turkeys, deer, and pollinators. Many Natives stated that they were careful not to overhunt, and throughout the Choctaw Nation the complaints were the same: when white intruders arrived on their lands, the herds and flocks declined—some said to the point of "extinction"—mainly because whites engaged in unchecked sport hunting.[52] To address severe wildlife depletion, in 1895, the Oklahoma Territorial Legislature created the first game laws.[53] Shortages of fish and game and environmental problems remain, however.

BLUNDERING INTRUDERS AND ENVIRONMENTAL DAMAGE

Nancy Turner and colleagues use the term "blundering intruders" to describe policies and external projects that impede Indigenous peoples' efforts to protect their cultures, resources, and independence.[54] A major blunderer is Oklahoma's fracking industry, which opens fissures into the earth in order to extract oil and gas with high-pressure forcing of sand, liquid, and sometimes chemicals. The waste liquid from fracking often flows into underground aquifers and pollutes water and soil.[55] The rocks fracture because of the force of the injection. Disposal wells holding millions of gallons of liquid cause faults to slip, resulting in earthquakes; the state of Oklahoma has the highest number of induced earthquakes in the country. There were 889 earthquakes in 2015, and 1,055 from March 2017 to March 2018.⁵⁶ A September 2016 earthquake damaged the Pawnee Nation's administrative buildings and tribal members' homes. The nation responded by filing suit against the Bureau of Indian Affairs and Bureau of Land Management in an effort to rid their nation of drilling permits and oil and gas leases on their land, which the agencies approved without consulting the tribe or adhering to natural resource protection laws.⁵⁷

Fracking is not the only problem. In June 2017, the Oklahoma Department of Environmental Quality warned that fish in fifty-four Oklahoma lakes have high levels of mercury and that consumers should limit their intake.⁵⁸ The Poncas, who were removed to Indian Territory from Nebraska in the 1870s, report that fish in the nearby Arkansas River are contaminated by raw sewage and a ConocoPhillips refinery and other factories in Ponca City. They are also battling air pollution from carbon-black emissions. The Poncas suffer from what Mekasi Horinek, the coordinator of Bold Oklahoma, a former environmental activism group, calls a "tirade of cancer" because of "environmental racism."⁵⁹ The problems are so severe that the Ponca Nation will be the first tribal nation to add a statute enacting the "Rights of Nature."⁶⁰ Osage oil still causes serious environmental problems because the Bureau of Indian Affairs will not enforce oil and gas drilling regulations.⁶¹ The Cherokee Nation established the Inter-tribal Environmental Council (ITEC) in 1992 to protect tribal national resources and their environments. The consortium consists of thirty-nine tribes in Oklahoma, Texas, and New Mexico. Recently, the Cherokee Nation filed a restraining order against Sequoyah Fields Fuels Corporation to prevent it from dumping radioactive waste into the Arkansas and Illinois Rivers.⁶²

Conservationists will continue to resist actors who emphasize economic development over a healthy environment. In February 2017, President Trump appointed Oklahoma Attorney General Scott Pruitt to head the EPA. Pruitt's office previously sued the EPA at least a dozen times in efforts to curb environmental protection regulations, including pollution policies.⁶³ A few months later, Oklahoma Governor Mary Fallin signed into law House Bill 1123, which makes it illegal for anyone to trespass on property containing a "critical infrastructure facility," which includes pipeline interconnections for oil, gas, and chemicals. Trespassers could receive a $1,000 fine, six months in jail, or both. Those who damage or destroy property might face a $100,000 fine, ten years in prison or both.⁶⁴ The Diamond Pipeline is set to transport almost 200,000 barrels of crude oil each day from Cushing, Oklahoma, to Memphis, Tennessee, crossing

491 waterways. Peaceful protesters have camped at the Oklahoma Coalition to Defeat the Diamond Pipeline's Oka Lawa Camp (Choctaw for "many waters") since March 2017. The camp is located on private, allotted land east of McCurtain, Oklahoma, and from that safe spot protesters can educate the country about the pipeline without being harassed.[65]

POACHING, INVASIVE SPECIES, AND LOSS OF POLLINATORS

Poachers in Oklahoma illegally take many deer, elk, fish, and other animals every year and trespass onto private land. For example, the Mihesuah family allotment on Little Beaver Creek in southern Oklahoma consists of 180 acres of forest and grassland that the family has hunted and fished since 1902. Multiple times a year my husband hunts for deer, turkeys, and quail, and every time he removes illegally placed deer stands and cameras and contends with poachers, who invariably argue that they were "lost." There is also a problem with runoff from the multitude of cows that graze on ranchland surrounding the allotment. Cows destroy vegetation, contaminate groundwater, and emit nitrogen into the atmosphere. Ranchers often cut trees to provide the pastureland cattle need. In fall 2016, one neighboring white rancher clear-cut an entire swath of cottonwoods to make way for more pasture, thus producing more contaminated drainage. The bass, carp, catfish, crappie, perch, and turtles that used to inhabit Little Beaver Creek are almost gone now.

Nonindigenous flora and fauna such as poison hemlock, Dutch elm disease (a fungus), eastern red cedar (out of control because of fire suppression), tamarisk, Chinese bush clover, musk thistle, and Bradford pear have spread throughout Oklahoma.[66] *Sericea lespedeza*, a perennial legume that was introduced in Kansas in 1900 to control erosion, has spread far beyond that area and is considered a hard-to-eradicate noxious weed. It has, for example, overgrown the bison-grazing area in the Seneca-Cayuga Nation, and the animals will not eat it.[67] Many of the more than two hundred lakes in Oklahoma (all but sixty-two of which were created by dams) now contain nonindigenous zebra mussels, bighead carp, golden algae, and hydrilla, among other invasive species. Wild boar (also known as wild pigs and wild hogs) can weigh hundreds of pounds. The aggressive and intelligent animals now inhabit all seventy-seven Oklahoma counties. They reproduce quickly and destroy agriculture, livestock, and ecosystems. Rush Springs, the "Watermelon Capital of the World," is my family's favorite place to acquire watermelons, but feral hogs now destroy multiple acres when the fruit is ripe.[68]

Pollinators—butterflies, moths, flies, beetles, wasps, and hummingbirds—collect nectar from flowering plants and in the process spread pollen. Their activity is crucial to the survival of fruit, vegetable, and nut plants. Residents of Indian Territory and Oklahoma observed healthy populations of pollinators until habitat loss and pesticides reduced their numbers.[69] In recent years, Oklahoma has lost more bees than any other state to drought, pesticides, undernutrition, and varroa mites.[70] Natives stated that during the late 1800s they had access to plenty of bee trees and hives in caves and under cliffs, and many men and women kept apiaries. One man recalled finding a hive so big that he collected a "washtub" of honey.[71] In an effort to increase the pollinator population, the Monarch Watch Program at the University of Kansas and the Euchee Butterfly Farm in Bixby, Oklahoma, were awarded a $250,000 grant from the National Fish and Wildlife Foundation for planting milkweed and other plants for monarchs and pollinators. Tribal Environmental Action for Monarchs is a coalition of Chickasaw, Citizen Band Potawatomi, Miami, Mvskoke-Creek, Osage, Seminole Nations, and Eastern Shawnee tribes that have pledged to plant 35,000 milkweed plants and 28,000 native wildflowers in the next two years.[72]

THE FAMILY AND COMMUNITY GARDENING MODEL

A challenge facing those tribes desiring to provide healthier food for all their members is how to produce it on a large scale in a safe and sustainable manner. Tribes historically did raise crops in just that way, but it required a community effort. Today, larger farms invest in machinery and other technologies that make production easier, maximize profits, and minimize costs.[73] Many of those large-scale agricultural endeavors, however, use technological innovations such as fertilizers and pesticides that result in multiple environmental consequences.[74]

Anthony "Chako" Ciocco, national program coordinator of the Ancestral Lands Program on the Navajo Nation and former communications coordinator for the Mvskoke Food Sovereignty Initiative in Okmulgee, believes, "Our agricultural practices are a major part of who we are. If we were really sovereign we'd be living in the Mvskoke way."[75] Prior to the removal of the Mvskoke-Creeks in the 1830s, each town worked a large garden divided into family parcels. Everyone worked there: women cared for the small family gardens and in summer, when men did not hunt, men helped women tend the larger community gardens. Other times, women did the bulk of the labor with the assistance of older men who could no longer hunt.[76] One man blew a conch to call the men to work. They arrived at the garden with their hoes and axes while the women arrived with food for the day. William Bartram, who observed them farming in the eighteenth century, described them as "marching in order to the field as if they were going to battle." Those who did not work were fined. The farmers sang as they worked, usually through early afternoon, when they sometimes broke to play games. Children sat in small shelters that were interspersed in the fields in order to scare away pests such as birds and raccoons. Men patrolled the fields at night to deter deer. When it was time to harvest, each family gathered plants from their parcel and donated a portion of their corn crop to the "king's crib," a cache of corn for use in hard times, for guests, and for war parties.[77]

Their main foods were corn, sweet potatoes, rice, squashes, and pumpkins, as well as the nonindigenous watermelons. Creeks pounded, boiled, and then strained hickory nuts to extract the oily, sweet liquid to use in corn dishes. In addition to produce, they consumed waterfowl, rabbits, turkey, venison, alligator, bear, deer, trout, catfish, sunfish, bream, and softshell turtle, as well as European-introduced beef, goat, and pork. Creeks had festivals every month, almost all of which were dedicated to hunting or agriculture; their principal festival, called the "feast of first fruits," took place when the corn crops matured in August.[78]

Choctaws also used a plethora of flora and fauna, including acorns, alligators, blackberries, chestnuts, chinquapin, several varieties of corn (dent, flint, flour, and pop), deer, fish, geese, wild grapes, hickory nuts and oil, mulberries, mushrooms, pecans, persimmons, wild plums, potatoes, pumpkins, strawberries, sunflowers, squirrels, sweet potatoes, turkeys, walnuts, and wild onions.[79] Each Choctaw family was responsible for its own sustenance, and families cultivated backyard gardens. Men and women both procured game. Families often lived far from each other, but feasts and religious ceremonies necessarily brought families and clans together.[80]

After removal to Indian Territory, many families continued to maintain small gardens around their houses, what nineteenth-century residents of Indian Territory referred to as "patches" or "roasting-ear patches."[81] Some family gardens were large. One Atoka family that moved to Indian Territory from Mississippi in 1889 maintained an ambitious garden of corn, potatoes, pumpkins, beans, peas, and peanuts, together with an orchard of apple, peach, plum, pear, and cherry trees, as well as berry bushes and grapevines. They managed cattle, hogs, and horses, along with chickens, turkeys, and bees. Another resident cultivated five acres of corn, peas, beans, and pumpkins. When planting corn he dropped a minnow into each hole to fertilize the corn kernel.[82] The variety of cultivated plants allowed farmers to recycle nutrients and organic matter. Choctaw seed savers took great pride in saving the best kernels and stringing cobs in a dry place.[83] If people lost kernels or seeds or had a poor growing season, they could trade something of equal value with a neighbor for more seeds.

I recount the importance of family gardens as a lifeline to cultural, emotional, and physical survival through multiple generations in my first novel, *Roads of My Relations* (2000). After removal, my ancestors, like many other Choctaws and Chickasaws, cultivated backyard gardens that supplied a good portion of their diet. Understanding the seasons and knowing when to plant and harvest were crucial to survival. My parents had a variety of plant foods growing around their home, but I have duplicated the large garden my grandparents cultivated in Muskogee, which was a copy of what their ancestors had cultivated.[84] As I write this in July 2017, there remains in one of our freezers frozen peppers, okra, dried tomatoes, and squash soup from plants grown last summer, as well as approximately one-quarter of a white-tailed deer, a wild turkey, two pheasants, numerous quail, and catfish from our pond. Our modest greenhouse and inexpensive cold frames allow me to start planting in early spring and to keep plants going into the cool fall and cold winter. Since spring we have harvested potatoes, herbs, carrots, beets, spinach, bok choy, kale, broccoli, raspberries, mulberries, and strawberries. Corn, peppers, green beans, okra, squashes, and another round of potatoes are yet to come. We save seeds, make compost, use rain barrels, and maintain four large pollinator gardens around the property. Not all the foods are Indigenous and the gardens do not supply us with everything we need. Still, this kind of gardening provides quite a bounty and is realistic for families willing to spend time outside and to exert themselves. If tribal members are physically unable to garden, the tribe should provide a workforce to do it for them.

THE COMANCHE NEED FOR FOOD INITIATIVES

The Kiowas, Cheyennes, Arapahos, Plains Apaches, and Comanches now in Oklahoma did not farm historically and therefore have no agricultural tradition to revive. They face a dilemma when looking for cultural connections to traditional foods. Comanches, for example, once roamed over "Comanchería," a vast area of various ecosystems with myriad resources.[85] They historically ate mainly game meats, but they also relied on a variety of wild fruits and trade items (as well as food they raided, notably corn, squash, and sheep).[86] The Comanche Nation has founded a diabetes awareness program and an environmental program that monitors hazardous materials in eight Oklahoma counties, but as of November 2017 it has no food sustainability plan. The monthly publication *Comanche Nation News* includes recipes for such foods as patty melts made with one stick of butter and eight slices of cheese; cabbage casserole with butter, Cheez Whiz, and grated cheese; pecan pie with butter, sugar, and dark Karo syrup; cottage pudding with flour, sugar, milk, and shortening; and a host of other recipes that include overabundances of fat, lard, sugar, and salt.[87] Cultural disconnection and the lack of both resources and food initiatives are among the reasons why Comanches suffer from high rates of diabetes and obesity.

SUSTAINING ENTHUSIASM AND INSTITUTING BIOSAFETY

Producing food in the backyard may sound enticing, but the reality of the work involved deters many. To illustrate, in 2006 the then-provost of the University of Kansas allotted me an acre of land on campus to establish an Indigenous demonstration garden that would operate out of the Indigenous Studies Department. Funding provided fencing, equipment, soil, a water line, benches, birdbaths and bird feeders, gloves, plant labels, a kiosk, composting equipment, water barrels, and an information center. Colleagues from areas of the country with similar weather promised to donate heirloom seeds. The idea was to give students a hands-on experience in cultivating plants their tribes used. The planning was to be done by students so that they could research their tribes' agricultural techniques, ceremonies associated with food, names of foods and animals in their tribes' language, and how to save the foods for cold months. They would take that knowledge home to their families and tribes. Indigenous students expressed excitement at the idea of a garden featuring foods of their ancestors, but what ultimately killed the initiative was that only one student was willing to get his hands dirty.

Maybe that attitude is changing. Members often feel they have no say in their tribe's decisions. However, many understand that their collective actions can go a long way toward creating tribal cohesion and supplying food for their households and communities. In 2013, the Cherokee Nation began its "Learn to Grow" project, teaching children how to plant, cultivate, and harvest. The next year, more than 3,500 children cultivated a variety of garden produce.[88] In 2014, the Cherokee Nation distributed heirloom corn, bean, squash, and gourd seeds to more than 1,500 Cherokees.[89] In August 2015, AmeriCorps awarded the Osage Nation a two-year $1.1 million grant to create the twelve-acre Wah-Zha-Zhi "ecological park" and develop the Bird Creek Farm located near Pawhuska to include walking trails, gardens, and a farmers' market. Plans include classes on cooking and "traditional Osage dishes."[90] Vann Bighorse, who directs the Wah-Zha-Zhi Cultural Center, is attempting to collect heirloom seeds and grow crops, then distribute seeds to tribal members.[91]

It will be interesting to learn the fate of tribal heirloom seeds. Do any individuals who receive seeds sell theirs to non-Indians? Do any of them work for biotech companies? Many tribes have instituted strict research guidelines in order to protect their intellectual and cultural property, but enforcing a ban on nontribal use of heirloom seeds will be challenging, especially when those seeds leave the tribal nation.[92] In addition, potential blunders include allowing genetically modified plants to cross-pollinate with fields of heirloom plants. Without biosafety policies, tribal plants will become endangered.

SOVEREIGNTY AND FOODWAYS SYSTEMS: NOW WHAT?

A common goal among activists is to achieve tribal autonomy and the ability to supply nutritious and affordable foods to tribal members.[93] At the very least, this goal requires clean air, uncontaminated water, fertile soil, regular weather patterns, adequate pollinators, clean energy, farm equipment, and a recycling and composting system. There also must be laws to protect the environment and resources.[94] A solid health-care system must be in place. Indeed, significant hurdles must be overcome in order to return at least partially to traditional ways of eating (or at least to have nutritious food), to maintain a healthy environment, and to inspire tribal pride through recovering cultural knowledge.

The *Merriam-Webster* online dictionary defines "sovereign" as "autonomous, free from external control." The federal government declares that tribes are sovereign entities with the right to govern themselves, but also deems them "domestic dependent nations" and ultimately holds power over every tribe. As Jeff Corntassel and Cheryl Bryce wrote in 2012, "The indigenous rights discourse has limits and can only take struggles for land reclamation and justice so far."[95] Tribes can attempt revitalization of traditional foodways and will succeed in many endeavors, but until they have control over their lands and resources and are independent from neoliberal food policies, they cannot achieve food sovereignty. Mohawk activist Taiaiake Alfred reminds us that "sovereignty" is a European concept and does not adequately describe Indigenous peoples' traditional philosophies.[96] Indeed, if the goal is to revert to traditionalism, then the quest for food sovereignty is further complicated because not only are many tribal governments patterned after the US government, many tribal members have vested interests in keeping them that way.

Highly motivated individuals instigate food initiatives but the tribe as a whole does not always support them. Not every Indian has an emotional investment in eating traditional foods, and not everyone is concerned about the environment. Many avoid political activism because it can be emotionally exhausting. There are vast socioeconomic differences among members of some tribes and the internal politics among some are volatile. Some community-based food autonomy and health endeavors are hampered by inadequate management, shortage of finances, lack of nutritional knowledge, absence of long-range planning, and intratribal factionalism. As Hope Radford discovered after an investigation of food sovereignty efforts among seven tribes in Montana, "Tribes are making progress, but many people are still hungry, many people are still unhealthy, and many people are still left without a voice in deciding what their community eats and where it comes from."[97] Anyone familiar with tribal politics knows that one tribal council might approve a project requiring tribal funds—such as a food initiative—but future councils can deny that venture.

Tribes cannot overhaul their foodways without assistance from outside entities and without adhering to governmental laws and regulations. Many business owners need loans, and food project organizers seek aid from Indigenous and non-Indigenous foundations. Institutes such as the Intertribal Agriculture Council, Native Food Systems Resource Center, Seeds of Native Health, and Indigenous Food and Agriculture Initiative at the University of Arkansas School of Law have assisted tribes with heirloom seed distribution, community and school gardens, businesses, and cattle ranching initiatives.[98] However, these organizations are in turn funded by, or partnered with, foundations such as W. K. Kellogg Foundation, American Association of Retired Persons, and Walmart Foundation, among other non-Indigenous entities. The grants and advice offered by these institutes are crucial in helping certain projects flourish, but it takes much more than a few projects to make tribes truly food sovereign.

In order for food initiatives to prosper (that is, to be self-sustaining), there must be long-range plans that take into account available finances and resources, and identify people committed to furthering the goals. Tribal and community discussions are crucial in order to determine what is already being attempted, identify the most critical concerns, pinpoint policies that have negative consequences, ascertain what resources are needed and which endeavors are successful, and decide how best to proceed. The decision-making entity should consist of tribal members with knowledge about traditional plants, seed saving, cultivating, harvesting, and animal processing, as well as members with political, economic, and scientific expertise. These knowledgeable and culturally connected tribal members (not just friends and political cronies of the current leadership) should have major roles in food education and Native-owned farms.

There are numerous Indigenous food success stories. Schoolchildren cultivate garden plots, more conference papers about traditional foodways are presented each year, Indigenous haute cuisine is a new trend, and more grants are forthcoming. Many Native people are just now rediscovering their traditional foods, and any news story about an Indigenous chef or a successful garden harvest is felt to be a unique and exciting step toward a vision of food sovereignty. It remains to be seen, however, whether schoolchildren will be sufficiently inspired to continue gardening; whether recently formed pan-Indian Indigenous food organizations will benefit communities; and whether Indigenous foodie gatherings and summits will serve only those who can afford to attend them. Moreover, research is needed to determine if the tribal food initiatives that have emerged across the United States improve health. Indeed, gatherings, chefs' cooking demonstrations, food tastings, and philosophizing are easy compared to the work of confronting the political, economic, and social realities of building food sovereignty. That is why the Indigenous food sovereignty movement might stay in a state of "sovereignization"—that is, continual planning and constructing, including negotiation, protest, and debate—until these questions are answered.[99]

In January 2017, Michael Wise wrote a blog entry, "Native Foods and the Colonial Gaze," asserting that "if there are important lessons to be learned by the food movement that are buried in the Native American past, they aren't embodied by ancient vegetables or archaic fishing techniques, but by stories of Indigenous resistance and accommodation to forces of colonialism and capitalism that have refashioned the lives, livelihoods, and dinner plates of us all over the last few centuries."[100] Wise is partially correct. A common way for the Five Tribes to catch fish, for example, was to daze them by dragging the mashed perennial herb white snakeroot (*Ageratina altissima*) through the water, then netting or shooting the paralyzed fish with arrows. This "Devil's Shoestring" is toxic, and today the practice is illegal in Oklahoma.[101] Some, however, still catch fish in an "archaic" way by spearing or shooting them with a bow (my son catches catfish in our pond in this manner). Yet Wise overlooks key facts: those "ancient vegetables" in large measure accounted for the good health of Native peoples, and their cultures are founded in their relationship to the Earth that produced those plants. Many of us will continue what Wise refers to as our "quixotic quests for authenticity." We know that past diets, activities, and reverence for the natural world can help us avoid many modern health problems. Regardless of the challenges, we will endeavor to accomplish what is realistic for our tribes and communities.

ACKNOWLEDGMENTS

Thank you to Peggy Carlton, Anthony "Chako" Ciocco, Jeff Corntassel, Pat Gwin, Nicky Michael, and Chip Taylor.

NOTES

1. Indigenous Food and Agriculture Initiative, *A National Intertribal Survey and Report: Intertribal Food Systems* (Fayetteville, AR: Indigenous Food and Agriculture Initiative, 2015).

2. For a discussion of the history of the term "food sovereignty," see Raj Patel, "What Does Food Sovereignty Look Like?" *Journal of Peasant Studies* 36, no. 3 (July 2009): 663, https://doi.org/10.1080/03066150903143079.

3. "Declaration of Nyéléni: Declaration of the Forum for Food Sovereignty," February 27, 2007, https://nyeleni.org/spip.php?article290.

4. See, for example, "Winning the Future: Navajo-Hopi Land Commission Leverages DOE Grant to Advance Solar Ranch Project," October 22, 2015, https://www.energy.gov/indianenergy/articles/winning-future-navajo-hopi-land-commission-leverages-doe-grant-advance-solar; Katherine Saltzstein, "Hopi Woman Brings Power of the Sun to the People," *Native Sun News*, October 9, 2014, https://www.indianz.com/News/2014/10/09/native-sun-news-hopi-woman-bri.asp.

5. Act of June 30, 1834, Pub. L. No. 23-161, §12, 4 Stat. 729, 730 (codified as amended at 25 U.S.C. §177 (2006).

6. See Blue Clark, *Indian Tribes of Oklahoma: A Guide* (Norman: University of Oklahoma Press, 2009).

7. See Devon A. Mihesuah, "Sustenance and Health among the Five Tribes in Indian Territory, Post-Removal to Statehood," *Ethnohistory* 62, no. 2 (April 2015): 263–84.

8. See Mihesuah, "Historical Research and Diabetes in Indian Territory: Revisiting Kelly M. West's Theory of 1940," *American Indian Culture and Research Journal* 40, no. 4 (2016): 1–21, https://doi.org/10.17953/aicrj.40.4.mihesuah.

9. Anne Gordon and Vanessa Oddo, "Addressing Child Hunger and Obesity in Indian Country: Report to Congress," January 12, 2012, Mathematica Policy Research, Princeton, NJ, 5–7.

10. Oklahoma State Department of Health, *2014 State of the State's Health*, 13, 15, 24, 26, 28, 30, 35, 36, https://ok.gov/health2/documents/SOSH 2014.pdf.

11. Sarah McColl, "With Heirloom Seeds, Cherokee Nurture Cultural History and Future Health," *takepart*, January 29, 2016, http://www.takepart.com/article/2016/01/29/cherokee-seeds/.

12. Benny Polacca, "Health Survey: Reservation Osages Report 'Poorer Health' than Osages Living Elsewhere," *Osage News*, August 30, 2010, http://www.osagenews.org

/en/article/2010/08/30/health-survey-reservation-osages-report-poorer-health-osages-living-elsewhere/.

13. "Oklahoma Academy of Science Statement on Global Climate Change," November 8, 2013, http://www.oklahomaacademyofscience.org/uploads/4/6/0/5/46053599/oas_statement_of_global_climate_change__2013_.pdf.

14. Evan Comen and Thomas C. Frohlich, "The Price of Food: What Groceries Are Driving up Food Bills?" *USA Today*, July 18, 2018, https://www.usatoday.com/story/money/personalfinance/2018/07/16/what-groceries-driving-up-food-bill-look-top-20/776106002/.

15. Ken Roseboro, "The GMO Seed Monopoly: Fewer Choices, Higher Prices," *Food Democracy Now*, October 4, 2013, https://www.fooddemocracynow.org/blog/2013/oct/4/the_gmo_seed_monopoly_fewer_choices_higher_prices.

16. Kate Taylor, "These Ten Companies Control Everything You Buy," *Business Insider*, September 28, 2016.

17. Wendell Berry, *The Unsettling of America: Culture and Agriculture* (San Francisco: Sierra Club Books, 1977), 218.

18. See National Service Blog, http://www.nationalservice.gov/blogs/2014-03-07/gardening-osage.

19. See Mark Shepard, *Restoration Agriculture* (Austin, TX: Acres USA, 2013); Akihiko Michimi and Michael C. Winmerly, "Associations of Supermarket Availability with Obesity and Fruit and Vegetable Consumption in the Conterminous United States," *International Journal of Health Geographics* 9, no. 1 (2010): 49, https://doi.org/10.1186/1476-072X-9-49.

20. Food Distribution Program on Indian Reservations (FDPIR), https://www.fns.usda.gov/fdpir/food-ditribution-program-indian-reservations-fdpir.

21. Choctaw Nation, "When Catastrophe Strikes: Responses to Natural Disasters in Indian Country," https://www.choctawnation.com/news-events/press-media/when-catastrophe-strikes-responses-natural-disasters-indian-country.

22. This dynamic is discussed in detail in Mihesuah, "Indigenous Health Initiatives, Frybread, and the Marketing of Non-Traditional "Traditional" American Indian Foods," *Native American and Indigenous Studies* 3, no. 2 (2016): 45–69, https://muse.jhu.edu/article/641379.

23. Choctaw Nation, "Food," https://www.choctawnation.com/history-culture/heritage-traditions/food.

24. Chickasaw Nation, "Foods," https://www.chickasaw.net/our-nation/culture/foods.aspx. Chickasaws did not grow corn, squash, and beans together in the manner of the "three sisters."

25. Shannon Shaw Duty, "Osage Cooking Classes Begin with Young Crop of Students," *Osage News* (Pawhuska, OK), August 20, 2010.

26. Muriel H. Wright, "A Report to the General Council of the Indian Territory Meeting at Okmulgee in 1873," *Chronicles of Oklahoma* 34, no. 1 (1956): 9–10.

27. Delaware Tribal Council member Nicky Michael, personal communication. See also C. A. Weslager, *The Delaware Indians: A History* (New Brunswick, NJ: Rutgers University Press, 1990).

28. Barbara Harper, "Quapaw Traditional Lifeways Scenario," Superfund Research, Oregon State University (2008), https://superfund.oregonstate.edu/sites/superfund.oregonstate.edu/files/harper_2008_quapaw_scenario_final.pdf.

29. W. David Baird, *The Quapaw Indians: A History of the Downstream People* (Norman: University of Oklahoma Press, 1980).

30. Kimberly Barker, "Quapaw Tribe Opens New Meat Distribution Center," June 7, 2016, *Miami News-Record*; "Bumpers College, School of Law Help Quapaw Tribe with Processing Plant," December 7, 2016, University of Arkansas News, http://news.uark.edu/articles/37330/bumpers-college-school-of-law-help-quapaw-tribe-with-processing-plant. Pima and Maricopa tribal members in Arizona also are attempting to revitalize their food traditions by cultivating as many traditional foods as they can. They face challenges from federal food safety laws that restrict their food production and processing, so now they are writing their own laws, which will still ensure that foods will be properly refrigerated and free of contaminants such as salmonella and *E-coli*. One challenge is that bison is considered "exotic," so each animal must be inspected (for a fee), and the animals have to be processed in facilities approved by the Food and Drug Administration (FDA). Tristan Ahtone, "Tribes Create Their Own Food Laws to Stop USDA from Killing Native Food Economies," *Yes!* May 24, 2016, https://www.yesmagazine.org/people-power/tribes-create-their-own-food-laws-to-stop-usda-from-killing-native-food-economies-20160524.

31. Food and Agriculture Organization of the United Nations (FAO), World Food Summit, "Rome Declaration on World Food Security," November 13–17, 1996, http://www.fao.org/docrep/003/w3613e/w3613e00.htm.

32. Eric Holt-Giménez, "Food Security, Food Justice, or Food Sovereignty?" in *Cultivating Food Justice: Race, Class, and Sustainability*, edited by Alison Hope Alkon and Julian Agyeman (Cambridge, MA: MIT Press, 2011), 319.

33. 2018 State of the Nation Address,: https://www.choctawnation.com/news-events/press-media/2018-state-nation-address-caps-choctaw-labor-day-festival.

34. "U.S. Government, Chickasaw, Choctaw Tribes Announce Historic Settlement Worth Millions," *Times Record* (Fort Smith, AR), October 7, 2015.

35. Oklahoma State Department of Health, "2014 State of the State's Health," https://ok.gov/health2/documents/SOSH%202014.pdf.

36. Amy Pereira and Trymaine Lee, "Hope on the Horizon for Choctaw Nation," March 19, 2014, MSNBC, http://www.msnbc.com/msnbc/choctaw-nation-hope-on-horizon slide1.

37. Choctaw Nation Small Business Development Services, https://choctawsmallbusiness.com/.

38. American Presidency Project, "Fact Sheet: President Obama's Promise Zones Initiative," January 9, 2014, http://www.presidency.ucsb.edu/ws/index.php?pid=108123.

39. See Biskinik Newspaper Archive, https://www.choctawnation.com/biskinik-newspaper-archive.

40. Jessica McBride, "The Cost of Education," MvskokeMedia.com, June 13, 2017.

41. American Presidency Project, "Fact Sheet," 3.

42. Program director Peggy Carlton, personal communication; "2016 State of the [Choctaw] Nation," 17.

43. Dana Hertneky, "Oklahoma Native Americans Concerned about Future of Indian Healthcare," *Newson6.com*, January 31, 2017, http://www.newson6.com/story/34394277/oklahoma-native-americans-concerned-about-future-of-indian-healthcare.

44. Amanda Michelle Gomez, "Native Americans and Alaska Natives Will Disproportionately Suffer under the GOP Health Care Plan," *ThinkProgress*, June 7, 2017, https://thinkprogress.org/native-americans-and-alaska-natives-will-disproportionately-suffer-under-the-gop-health-care-plan-d695283153c2/.

45. Ronni Pierce, "A Healthy Outlook: New Regional Clinic to Open Its Doors," *BISKINIK* (Talihina, OK), February 2017.

46. Mark Trahant, "How Bad Could It Be? Don't Get Sick if Senate (or House) Bill Becomes Law," June 23, 2017, TrahantReports.com, https://trahantreports.com/.

47. See, in particular, Angie Debo, *And Still the Waters Run: The Betrayal of the Five Civilized Tribes* (Princeton, NJ: Princeton University Press, 1940).

48. Terry Wilson, *The Underground Reservation: Osage Oil* (Norman: University of Oklahoma Press, 1985); David Grann, *Killers of the Flower Moon: The Osage Murders and the Birth of the FBI* (New York: Doubleday, 2017).

49. Ralph Keen II, "Tribal Hunting and Fishing Regulatory Authority within Oklahoma," *Oklahoma Bar Journal* 86, no. 24 (September 12, 2015).

50. *2016 State of the [Choctaw] Nation*, 21.

51. Muriel H. Wright, "Notes and Documents: Sugar Loaf Mountain Resort," *Chronicles of Oklahoma* 36 (1960), 202–3; *South McAlester Capital*, July 12, 1894; interview with Elijah Conger, Indian and Pioneer Papers, vol. 2: 196–97, Western History Collections, University of Oklahoma, Norman (hereafter IPP).

52. Interview with Limon Pusley, December 28, 1937, IPP 73: 346; J. T. Poston, September 16, 1937, IPP 72: 286; Elijah W. Culberson, November 4, 1937, IPP 72: 215–16; Sarah Noah and Robert Noah, April 12 1937, IPP 67: 254 Jim Spaniard, June 25, 1937, IPP 86: 7.

53. Oklahoma Department of Wildlife Conservation, *Fishing in the Schools Manual* (Oklahoma Department of Wildlife Conservation: Oklahoma City, 2014), 4.

54. Nancy J. Turner, Fikret Berkes, Janet Stephenson, and Jonathan Dick, "Blundering Intruders: Extraneous Impacts on Two Indigenous Food Systems," *Human Ecology* 41, no. 4 (2013): 563–74, https://doi.org/10.1007/s10745-013-9591-y.

55. Darryl Fears, "This Mystery Was Solved: Scientists Say Chemicals from Fracking Wastewater Can Taint Freshwater Nearby," *Washington Post*, May 11, 2016; Jim Kelly, "On Oklahoma, Earthquakes, and Contaminated Water: The Fracking Connection," *A New Domain*, December 8, 2015, http://anewdomain.net/oklahoma-earthquakes-contaminated-water-fracking-connection/.

56. See "Recent Earthquakes Near Oklahoma," *Earthquake Track*, https://earthquaketrack.com/p/united-states/oklahoma/recent; Katie M. Keranen, Matthew

Weingarten, Geoffrey A. Abers, Barbara A. Bekins, and Shemin Ge, "Sharp Increase in Central Oklahoma Seismicity since 2008 Induced by Massive Wastewater Injection," *Science* 345, no. 6195 (July 25, 2014): 448–51. See also "Oklahoma and Fracking," http://www.sourcewatch.org/index.php/Oklahoma_and_fracking; and Jessica Fitzpatrick, "Induced Earthquakes Raise Chances of Damaging Shaking in 2016," March 28, 2016, *USGS Science Features*, https://www.usgs.gov/news/induced-earthquakes-raise-chances-damaging-shaking-2016.

57. Matthew L. M. Fletcher, "Pawnee Nation and Walter Echo-Hawk Sue over Fracking," *Turtle Talk*, November 21, 2016. See also Liz Blood, "Fracking in Bad Faith," *The Tulsa Voice*, January-B, 2017, http://www.thetulsavoice.com/January-B-2017/Fracking-in-bad-faith/.

58. "14 More Oklahoma Lakes Have Elevated Mercury Levels in Fish," *Oklahoma's News 4*, June 22, 2017, http://kfor.com/2017/06/22/14-more-oklahoma-lakes-have-elevated-mercury-levels-in-fish/.

59. "Battling Pollution on Our Lands: Mekasi Horinek," *Cultural Survival Quarterly Magazine*, September 2016.

60. Movement Rights, "Ponca Nation of Oklahoma to Recognize the Rights of Nature to Stop Fracking," *Intercontinental Cry*, October 31, 2017, https://intercontinentalcry.org/ponca-nation-oklahoma-recognize-rights-nature-stop-fracking/.

61. Chalene Toehay-Tartsah, "Osage County Landowners Speak Out against Bad Drilling Practices," *Osage News*, August 18, 2014.

62. Inter-tribal Environmental Council, http://itec.cherokee.org/; "Cherokee Nation Files, Is Granted Emergency Restraining Order," *Anadisgoi*, February 9, 2017, http://www.anadisgoi.com/archive/1519-cherokee-nation-files-is-granted-emergency-restraining-order-halting-disposal-ofradioactive-waste-near-the-arkansas-and-illinois-rivers.

63. Kristin Hugo, "Native Americans Brace for Impact as EPA Undergoes Changes," *PBS Newshour*: The Rundown, February 17, 2017, https://www.pbs.org/newshour/science/native-americans-brace-impact-epa-undergoes-changes.

64. See Bill Information for HB 1123, State of Oklahoma, 1st Session of the 56th Legislature (2017), http://www.oklegislature.gov/BillInfo.aspx?Bill=hb1123; Alleen Brown, "Oklahoma Governor Signs Anti-Protest Law Imposing Huge Fines on 'Conspirator' Organizations," *The Intercept*, May 6, 2017, https://theintercept.com/2017/05/06oklahoma-governor-signs-anti-protest-law-imposing-huge-fines-on-conspirator-organizations/.

65. Casey Smith, "The Diamond Pipeline," *Tulsa World*, February 3, 2017; Mark Hefflinger, "Fight Against Diamond Pipeline Spans Three States," *Bold Oklahoma*, January 30, 2017; Oka Lawa Camp Facebook page, https://www.facebook.com/OkaLawaCamp/.

66. Oklahoma Invasive Plant Council, "The Dirty Dozen Poster," https://okipc.wordpress.com/the-dirty-dozen/; Brianna Bailey, "The Bradford Pear: Oklahoma's Worst Tree or Just Misunderstood?" *NewsOK*, March 5, 2017.

67. Chip Taylor, personal communication.

68. Oklahoma Department of Wildlife Conservation, "Feral Hogs in Oklahoma," https://www.wildlifedepartment.com/feral-hogs-in-oklahoma "There Was Nothing I Could Do: Feral Hogs Wreaking Havoc on Oklahoma Watermelon Farmers," *Oklahoma's News 4*, September 5, 2017, http://kfor.com/2017/09/05/there-was-nothing-i-could-do-feral-hogs-wreaking-havoc-on-oklahoma-watermelon-farmers/.

69. For an overview of pollinators in Oklahoma in 1917, see Sister M. Agnes, "Biological Field Work," *Oklahoma Academy of Science* 1 (1917): 35–38, http://digital.library.okstate.edu/OAS/oas_pdf/v01/p35_38.pdf.

70. Logan Layden, "Why Oklahoma Had the Nation's Highest Percentage of Bee Deaths Last Year," *National Public Radio*, StateImpact-Oklahoma, June 25, 2015, https://stateimpact.npr.org/oklahoma/2015/06/25/why-oklahoma-had-the-nations-highest-percentage-of-bee-deaths-last-year/.

71. Edmund Flint interview, April 23, 1937, IPP 3: 527; Ben Cartarby, June 29, 1937, IPP 19: 203; Josephine Usray Lattimer, September 23, 1937, IPP 33: 84; T. P. Wilson, n.d., IPP 11: 498; Elijah W. Culberson, November 4, 1937, IPP 22: 216; W. C. Mead interview, January 17, 1938, IPP 62: 17; Johnnie Gipson interview, April 21, 1927, IPP 34: 175.

72. Chip Taylor, personal communication. See also Tribal Environmental Action for Monarchs, http://www.nativebutterflies.org/saving-the-monarch; Trilateral Committee for Wildlife and Ecosystem Conservation and Management, "Native American Tribes Pledge to Save the Monarch," http://www.trilat.org/index.php?option=com_content&view=article&id=1197:native-american-tribes-pledge-to-save-the-monarch&catid=17&Itemid=256.

73. Tamar Haspel, "Small vs. Large: Which Size Farm Is Better for the Planet?" *Washington Post*, September 2, 2014.

74. Rajeev Bhat, "Food Sustainability Challenges in the Developing World," in *Sustainability Challenges in the Agrofood Sector* (New York: Wiley, 2017), 2, 4.

75. National Family Farm Coalition and Grassroots International, *Food Sovereignty* (Washington, DC: Grassroots International, 2010), 11.

76. "Observations on the Creek and Cherokee Indians, by William Bartram, 1789," *Transactions of the American Ethnological Society* III (1853): 39–40. See also the series of "Mvskoke Country" articles authored by James Treat, https://mvskokecountry.wordpress.com/category/mvskoke-country/.

77. William Bartram, *Bartram: Travels and Other Writings* (New York: Literary Classics of the United States, 1996), 506–7. See also James Adair, *History of the American Indians* (London: Edward and Charles Dilly, 1775), 405–10.

78. Bartram, *Travels*, 56, 319, 557–60, 404–5.

79. T. N. Campbell, "Choctaw Subsistence: Ethnographic Notes from the Lincecum Manuscript," *Florida Anthropologists* 12, no. 1 (1959): 9–24; H. B. Cushman, *History of the Choctaw, Chickasaw, and Natchez Indians* (Greenville, TX: Headlight, 1899), 74, 168, 231–32, 250, 272.

80. Campbell, "Choctaw Subsistence," 10–11; John R. Swanton, "Aboriginal Culture of the Southeast," in *Forty-Second Annual Report of the Bureau of American Ethnology* (Washington, DC: GPO, 1924–25), 695. There is no evidence that Choctaws planted corn, squash, and beans together in the manner of the Three Sisters.

81. See Mihesuah, "Sustenance and Health among the Five Tribes in Indian Territory."

82. J. C. Moncrief interview, November 1, 1933, IPP 64: 57.

83. Meton Ludlow interview, April 26, 1934, IPP 56: 182.

84. I expound on this in "The Garden Meal," in Linda Murray Berzok, ed., *Storied Dishes: What Our Family Recipes Tell Us about Who We Are and Where We've Been* (Santa Barbara, CA: ABC-CLIO, 2010), 57–60.

85. For general works on the Comanches, see John Frances Bannon, *The Spanish Borderlands Frontier, 1513–1821* (New York: Holt, Rinehart and Winston, 1970); Pekka Hämäläinen, *The Comanche Empire* (New Haven, CT: Yale University Press, 2009); Ernest Wallace and E. Adamson Hoebel, *The Comanches: Lords of the South Plains* (Norman: University of Oklahoma Press, 1952); Elizabeth A. H. John, *Storms Brewed in Other Men's Worlds: The Confrontation of Indians, Spanish, and French in the Southwest, 1540–1795* (College Station: Texas A&M University Press, 1975); Thomas W. Kavanagh, *The Comanches: A History, 1706–1875* (Lincoln: University of Nebraska Press, 1999); Stanley Noyes, *Los Comanches: The Horse People, 1751–1845* (Albuquerque: University of New Mexico Press, 1993); W. W. Newcomb Jr., "Comanches: Terror of the Southern Plains," in *The Indians of Texas: From Prehistoric to Modern Times* (Austin: University of Texas Press, 1961), 155–91.

86. Gustav G. Carlson and Volney H. Jones, "Some Notes on Uses of Plants by the Comanche Indians," *Papers of the Michigan Academy of Science* 25 (1940): 517–42. See also Mihesuah, "Comanche Traditional Foodways and the Decline of Health," *Great Plains Journal* (forthcoming).

87. *Comanche Nation News*, https://comanchenation.com/tcnn.

88. Sheila Stogsdill, "Cherokee Nation Garden Project Seeks to Teach Nutrition in Oklahoma," *The Oklahoman*, June 15, 2014, https://newsok.com/article/4913312/cherokee-nation-garden-project-seeks-to-teach-nutrition-in-oklahoma.

89. Rick Wells, "Cherokee Seed Bank Program Provides Connection to Past," February 3, 2017, *News9.com*, http://www.news9.com/story/34423103/cherokee-seed-bank-program-provides-connection-to-past.

90. Lenzy Krehbiel-Burton, "Osage Nation Awarded AmeriCorps Grant for Park, Gardens to Address Diabetes, Obesity," *Tulsa World*, August 20, 2015, https://www.tulsaworld.com/news/state/osage-nation-awarded-americorps-grant-for-park-gardens-to-address/article_577b7c3a-2a23-5fa8-afae-23a4ae71d1e5.html.

91. Tara Madden, "Community Gardens Being Grown by Osage Nation TA–WA AmeriCorps," *Osage News*, August, 14, 2014, http://osagenews.org/en/article/2014/08/14/community-gardens-being-grown-osage-nation-tawa-americorps/.

92. Pat Gwin, personal communication.

93. Kyle Powys Whyte, "Indigenous Food Sovereignty, Renewal, and US Settler Colonialism," in *The Routledge Handbook of Food Ethics*, edited by Mary Rawlinson and Caleb Ward (London: Routledge, 2016), 354–65.

94. Bhat, "Food Sustainability Challenges in the Developing World," 3–4.

95. Jeff Corntassel and Cheryl Bryce, "Practicing Sustainable Self-Determination: Indigenous Approaches to Cultural Restoration and Revitalization," *Brown Journal of World Affairs* 18, no. 2 (2012): 152.

96. For a discussion about the implications of using the term "sovereignty," see Taiaiake Alfred, "Sovereignty," in *A Companion to American Indian History*, edited by Philip J. Deloria and Neal Salisbury (Malden, MA: Blackwell, 2004), 460–74.

97. Hope Radford, "Native American Food Sovereignty in Montana," August 2016, 6, http://aeromt.org/wp-content/uploads/2016/10/Native-American-Food-Sovereignty-in-Montana-2016-1-1.pdf.

98. See Intertribal Agriculture Council, http://www.indianaglink.com/our-programs/technical-assistance-program/; Native Food Systems Resource Center, http://www.nativefoodsystems.org/about; Seeds of Native Health, http://seedsofnativehealth.org/partners/; "Smokehouses, Farmers' Markets and More," *Indian Country Today*, June 20, 2017, https://newsmaven.io/indiancountrytoday/archive/smokehouses-farmers-markets-and-more-growing-food-sovereignty-_Kx8qz4fboy8K-PWRq2IqWw/; University of Arkansas School of Law Indigenous Food and Agriculture Initiative, http://indigenousfoodandag.com/about-us/. The initiative offers strategic planning and technical support for tribal governance infrastructure in the areas of business and economic development, financial markets and asset management, health and nutrition policies, and intellectual property rights. It also supports increased admission of students into land grant universities, and creation of academic programs in food and agriculture.

99. Food sovereignty construction is discussed in Christina M. Schiavoni, "The Contested Terrain of Food Sovereignty Construction: Toward a Historical, Relational and Interactive Approach," *Journal of Peasant Studies* 44, no. 1 (2017): 1–32.

100. Michael Wise, "Native Foods and the Colonial Gaze," *Process: A Blog for American History*, January 10, 2017, http://www.processhistory.org/wise-native-foods/.

101. Elizabeth Ross, June 10, 1937, IPP 109: 190–92; Emiziah Bohanan, May 10, 1937, IPP 9: 139; T. J. Johnson, July 16, 1937, IPP 48: 402; Elizabeth Witcher, April 18, 1939, IPP 99: 390. 29 OK Stat §29-6-301a (2016), "Prohibited Means of Taking Game or Nongame Fish: Poison, Explosive, or Electrical Shock Devices," http://law.justia.com/codes/oklahoma/2016/title-29/section-29-6-301a/.

KEVIN K. J. CHANG
CHARLES K. H. YOUNG
BRENDA F. ASUNCION
WALLACE K. ITO
KAWIKA B. WINTER
WAYNE C. TANAKA

4

KUAʻĀINA ULU ʻAUAMO

Grassroots Growing through Shared Responsibility

OLI HŌʻALA PUALU
A chant to call forth the *pualu* fish

E ala ē, e alu pū
   Arise, move forward together!

Ke ala liʻu o Hoakalani
   Here on the long path of Hoakalani (an ancient name for the Hawaiian Archipelago),

Kīʻei aku i ka hālāwai ē. Hū
   Gaze out to the horizon, whoa!

Kau mai i ka ʻauamo
   Place the carrying stick upon the shoulders,

ʻAʻaliʻi kūmakua
    It's made of the ʻaʻaliʻi of the upland forest,

Maka kaʻapuni i ka ʻumeke pohepohe
    The eyes [of the net] surround the round carrying gourd,

Kāʻeo ʻana! Kāʻeo lā
    It's overflowing!

Kōkō weluwelu pāhono ʻia
    The old/tattered net has been mended,

Ua paʻa ē! Ua paʻa lā
    It's once again solid.

Aweawe pualu, E Alu Pū
    Here come the traces of the *pualu fish* in the calm water.
    Let us gather together,

Kīʻililī? Kūʻolokū
    [In the language of the birds] Are you ready? We stand ready!

In this essay we set forth who we are through the lens of E Alu Pū, a network focused on empowered community-based biocultural resource management in Hawaiʻi and the activities we and many other communities and networks engage in. We explain our vision, collective history, work and faith that we can all together care for Hawaiʻi.

On an early morning in September 2017, a fisherman and his eldest son launch their canoe in search of ʻōpelu (mackerel, or scad, *Decapterus macarellus*). They head to the ʻōpelu koʻa (mackerel fishing grounds) to feed the ʻōpelu, a longtime ʻohana (family) tradition. The *palu* (vegetable chum) consists of pumpkin, avocado, and papaya collected from the family farm, grated into bite-sized pieces and cooked. Upon reaching the koʻa the son fills a weighted cloth bag with a handful of palu and tosses it over the side. The fisherman looks through his glass box to see if the fish are attracted. It may take one or several tries before the fish arrive or he decides to move on to another koʻa. They are not there today to catch the fish; that will come later in the season when the fish are larger and sufficiently trained to stay around the koʻa by repeated feedings from the fisherman. On this occasion the fish aggregate in good numbers, and the fisherman feeds them all the palu then leaves.

Within a half hour another fisherman, who has watched the feeding, arrives on the koʻa, chums using animal bait—a new practice—and drops his net. The first fisherman was there to care for the ʻōpelu, to observe and feed them. He will return when the fish and the catch are large enough because the ʻōpelu have been accustomed to aggregate for feeding. The second fisherman thinks only of his immediate need to catch and sell the fish. He can only hope that the fish will be there when he returns again.

No one is present to determine which is the best practice, a situation that needs to be addressed.

True story.

E ALU PŪ: MOVING FORWARD TOGETHER TO MĀLAMA HAWAIʻI

E ala ē, e alu pū!
Rise, move forward together!
Ke ala liʻu o Hoakalani.
Here on the long path of Hoakalani (the Hawaiian Archipelago).

We are a part of E Alu Pū. Our *oli* (chant) opens with these lines. Our oli frames our purpose. It is a *kāhea* (call) we use to enter most spaces together. To ground ourselves. To ask for permission. To reclaim an old space. Or to shift the energy in a space when the voice and intentions of our communities should be recognized.

Our oli recognizes the long and complex journey of healing Hawaiʻi's social-ecological systems and the food systems born from them. It recognizes this healing as part of a deeper sociocultural and historical process to rebuild and reinvigorate civic relationships, and to revive governance and institutional frameworks that suit an island community. Neither E Alu Pū nor this chapter seeks to right all injustices nor claims a monopoly on the right answers to correct the wrongs of the past. Our focus is part of a path toward reconciliation and a more just system of governance; our contention is that communities, more specifically rural Native Hawaiian and *kamaʻāina* (local people; generational residents) communities who believe in values and practices like those set forth here, are a part of the solution.

We work from the inside out and from the bottom up. Our journey begins with our relationship with ourselves, each other, and our ʻāina (land and the connecting bodies of water; literally, a "source which feeds"). We reclaim our community agency and traditions of *konohiki* (a resource management approach that invites community stewardship), *laulima* (working together;

sharing responsibility), and *mālama ʻāina* and *aloha ʻāina* (care and love for the resources that provide life), among other values, to affirm and embrace the best of our island communities and their roles in the care of Hawaiʻi. We are also informed and guided by current scientific knowledge and ongoing research. All can help lead us to a new era of biocultural resource management and governance in Hawaiʻi.[1]

We step out together. According to the ancestral traditions of Pacific Islanders, our understanding of the past sets out in front of us a road map and lessons for our future. We continue to discover, interact with, and embrace our history and sociocultural reality. We confront the complexities of systemic change, and the *hihia* (negative entanglements) and *hukihuki* (tensions) that accompany it, with an active, open heart and mind. For us, our foundation is "place" and the relevant cultural values, wisdom, and practices shaped by the island communities that have been here since time immemorial.

E Alu Pū means "move forward together." We are a network of more than thirty rural and Native Hawaiian community-based mālama ʻāina groups from throughout Nā Kai ʻEwalu (the main Hawaiian islands, literally "the eight seas"[2]). Our members are Native Hawaiians, *keiki o ka ʻāina* (children of the land), and others who gather to build a movement to empower one another to care for the places that have sustained our communities, the places we love and call home. The root of our creation came from our *kupuna* (elder) founder and *lawaiʻa loea* (expert fisherman) Uncle Kelson "Mac" Poepoe,[3] who called for opportunities for communities to share knowledge, practices, and values for the management of their particular places. The organization Kuaʻāina Ulu ʻAuamo (KUA) was created to help us gather, connect, mobilize, and tell our collective story, and in KUA, E Alu Pū found a home.

E Alu Pū is a play on words described by one of our community leaders, Hiʻilei Kawelo, who recalled watching the *pualu*, a type of surgeonfish: "Every summer in the fishpond, pualu school up and are easily seen moving through the shallow waters as a single unit." We work to empower each other in our places, and school together when the season is right to collectively navigate the tides of change.

E Alu Pū members gathered and grew together over the past fifteen years. We committed to gather because of a shared belief that empowered community stewardship leads to an abundant, productive ecological system that supports community well-being—ʻāina momona (abundance; literally, "the fat lands"). Today, through KUA's facilitation and support, we are joined by two other networks, Hui Mālama Loko Iʻa (the groups that care for fishponds, a collective of almost forty traditional Hawaiian fishpond restoration groups) and the Limu Hui (seaweed *loea*; individual experts and gatherers of native seaweed for diet, spiritual, and health reasons).

When we say "place" the Hawaiian term and metaphor of *kīpuka* feels most appropriate, as defined by ethnic studies scholar and history professor Davianna McGregor in her book *Nā Kuaʻāina: Living Hawaiian Culture*:

> Even as Pele (goddess of the volcano) claims and reconstructs the forest landscape, she leaves intact whole sections of the forest, with tall old-growth, ʻōhiʻa trees, tree ferns, creeping vines, and mosses. These oases are called *kīpuka*. The beauty of these natural *kīpuka* is not only their ability to resist and withstand destructive forces of change, but also their ability to regenerate life on the barren lava that surrounds them.... Like the dynamic life forces in a natural *kīpuka*, cultural *kīpuka* are communities from which Native Hawaiian culture can be regenerated and revitalized in the setting of contemporary Hawaiʻi.[4]

In many cases, our places maintain their traditional names and *palena* (boundaries) as communities continue to or are newly inspired to maintain traditional relationships with them. As a matter of context, we are a geographically isolated island community in the middle of the Pacific Ocean, more than 2,500 miles from anywhere. There are significant points at the foundation of our treatise here: (1) Hawaiʻi was once largely food self-sufficient with a population close to the present-day one;[5] (2) seafood comprised the primary source of protein for our population;[6] (3) our Native Hawaiian population has recovered from a low of about twenty-four thousand to approximately half a million, but more than half of these Native Hawaiians no longer live on the islands;[7] (4) of all ethnic groups in Hawaiʻi, Native Hawaiians are disproportionately represented in negative social statistics (chronic disease and mortality rates, incarceration, housing instability, etc.);[8] (5) a traditional Hawaiian diet has been shown to improve health, especially among Native Hawaiians;[9] (6) in conservation circles Hawaiʻi is referred to as the "endangered species capital of the world" because we have more than 44 percent of the endangered species in the United States;[10] (7) the pillars of our economy are tourism and military defense;[11] (8) as of this writing visitor forecasts for 2018 are more than 9.8 million;[12] (9) we import almost 90 percent of our food;[13] and (10) state funding to protect the environment and agriculture is approximately 1.4–1.5 percent of the annual state budget.[14]

Under these daunting conditions we pursue a vision of 'āina momona—an abundant, productive ecological system that supports community well-being—a recurring theme in this chapter. We call upon a community development tradition and spirit of laulima, or shared power, responsibility, and governance for our places.

Our efforts follow the basic stages of movement building, well articulated by education advocate Parker Palmer, who said, "The genius of movements is paradoxical; they abandon the logic of organizations in order to gather the power necessary to rewrite the logic of organizations." If we wish to truly change our situation, our systems, and the way we govern, we must do so outside of the boundaries of convention and convenience, from the bottom up and the top down. In simplified terms, we evolve and cycle through the stages Palmer identifies: (1) isolated individuals decide to stop leading "divided lives"; (2) these people discover one another and form groups of mutual support; (3) empowered by community, they learn to translate "private problems" into public issues; and (4) alternative rewards emerge to sustain the movement's vision, which may force the conventional reward system to change.[15]

These stages have demonstrated relevance to our work. When we decide to gather, to share lessons we have learned, we recognize that we are not alone; our private concerns and passions and feelings of being alone become the basis for a broader sense of community. We begin to unify and speak together when we can. We make our issues "public," and we take action to achieve what we could not do alone, revealing "alternative rewards" and outcomes that can ultimately benefit us all. We realize the alternative rewards we seek will lead us to return to a state of 'āina momona, including a community and government institutional shift toward a konohiki mindset, economy, and infrastructure, as we summarize here.

We won't always agree on the nuances of the ways ahead, but we can commit to a focus on community- and place-based mālama 'āina; on having *aloha* (love, compassion, and respect) for each other; on shaping and developing an active, open mindset and process guided by shared values; on gathering, breaking bread, and sharing knowledge and experience across generations; and on speaking from our hearts and our love for Hawai'i and our places. Further, we can communicate and share values when we put hands to soil, stream, or reef and co-create a context where we find ways to move forward together to attain 'āina momona.[16] We don't just talk, we commit to do, and to continuously reflect on and seek new ways to accelerate our work together.

UNITY IN INTEGRITY: NO LONGER LIVING DIVIDED LIVES

Hawai'i's people, as well as its natural and biocultural resources—storied landscapes, sacred forests, agroforests, streams and rivers, agricultural fields, fisheries, and reefs—have been deeply damaged by more than two hundred years of political, economic, and social upheaval and change.[17] Rarely do people, even those who live here, understand Hawai'i's historic and ongoing role in the US campaign of influence in the Pacific and Asia, not to mention the consequences of colonization on Native Hawaiians. Few know about the historical erosion of Native Hawaiian governance; the overthrow of the internationally recognized Hawaiian Kingdom; the post-overthrow policies to wipe out Native Hawaiian language, culture, and identity; or the historical and ongoing severance of relationships and familial connections to ancestral lands and the natural world.[18]

But we choose not to belabor this typical narrative of colonialism and occupation. Not all was lost, nor were Native Hawaiians idle during the last two centuries of change. Pre-overthrow Native Hawaiians developed customs and traditions around food and environmental stewardship based on intimate knowledge and expertise derived from long-term, multigenerational observations of the natural world. They developed a governance system, including management practices, for 'āina (referred to in the contemporary period as the *ahupua'a* system); became one of the most literate populations in the world; entered into treaties with other nation-states; and published books on history, tradition, and culture. And because their sense of cultural and national identity persevered, they planted the seeds that future generations would use to reconnect with each other, their culture, their places, and even the world.

The overthrow of the Hawaiian Kingdom in 1893 by Western plantation interests set a foundation for even greater sociocultural, socioeconomic, and environmental impacts to the islands. Nevertheless, Native Hawaiians demonstrated resilience, using every tool available to reverse the setbacks that had befallen them.[19] As noted, their population has now recovered. They have remained steadfast in civic and political engagement, while rural kīpuka communities have served as repositories for traditional knowledge and values. More recently, cultural and civic protest and activism in the 1960s and 1970s garnered broad community support and unity, sparking a cultural renaissance that has progressed markedly over the past forty-plus years. Notably, the Hawai'i State Constitution was amended in 1978 to, among other things, recognize the Native Hawaiian language as an official language of the state, establish a semi-autonomous Office of Hawaiian Affairs (OHA) charged with "bettering the conditions" of the Native Hawaiian community, and provide constitutional protection of Native Hawaiian traditional and customary rights.[20] Since then, ʻŌlelo Hawaiʻi (the Hawaiian language) has continued to recover relatively rapidly, with Hawaiian language immersion programs providing public education in the language for all grade levels, K–12; OHA has become the state's thirteenth largest landowner and a leading advocate for Native Hawaiian and environmental interests; Native Hawaiian traditional and customary rights are now recognized to varying degrees in state statutes, county ordinances, and the courts; and the successful completion of the recent Mālama Honua (worldwide voyage) of the canoe *Hōkūleʻa* has garnered international recognition and awakened Native Hawaiians and others to the potential for Hawaiʻi's indigenous knowledge and values to affect the world.

At the foundation of the 1960–79 renaissance and the ongoing Native Hawaiian movement today are traditional ecological knowledge (TEK) and its attendant values, customs, and practices that previous generations fought to preserve and which were never completely lost. This TEK includes cultural legacies retained by ʻohana and individual kūpuna, *kumu* (knowledge sources), *kahu* (guardians), and *kahuna* and *loea* (experts and professionals) who preserved traditions through practice or oral history;[21] Hawaiian-language speakers also continually rediscover knowledge captured generations ago in Hawaiian-language books and newspapers, archival interviews, and even Hawaiian Kingdom laws and court testimonies.[22] This growing base of traditional understanding not only continues to connect and unify communities throughout the islands, but also connects Hawaiʻi to the world. TEK, whether retained, rediscovered, or redeveloped by E Alu Pū communities, has indeed been a foundation of our work and weaves our movement into the larger cultural renaissance that continues to evolve today.

Palmer's third stage of movement building occurs when isolated individuals or communities decide to stop living "divided lives"—feeling one way on the inside but living differently on the outside—and discover they are not alone. Private issues begin to reveal themselves as shared, public problems. E Alu Pū has shown that this process occurs over time when place-based community management efforts across Nā Kai ʻEwalu—especially between cultural kīpuka—are brought together. In this stage it is not enough to battle with the symptoms of systemic dysfunction in isolation. Some communities in Hawaiʻi are uniquely endowed with a cultural legacy, and when they are brought together, a new way of being starts to reveal itself.

TAKING UP THE *'AUAMO*: COLLECTIVE RESPONSIBILITY

Kī'ei aku i ka hālāwai ē. Hū
  Gaze out to the horizon. Whoa!
Kau mai i ka 'auamo
  Place the carrying stick upon the shoulders.

Our oli recognizes the long path to a state of 'āina momona. *Hū* (Whew!) It is a heavy lift, indeed! We acknowledge the *kuleana* (rights/privileges founded on responsibilities/obligations) to care for Hawai'i.[23] This responsibility extends beyond the horizon. A collective kuleana is embodied in the *kaona* (poetic symbolism or hidden metaphor) of the 'auamo, the palanquin or carrying stick borne on the shoulders of many people to bear the burden of something greater than one can carry alone.

Mindful of our kīpuka, together we take up the 'auamo, and place it on our shoulders as a statement of metaphor. Along with leveraging a heavy load, the metaphor alludes to the spot on the body where one's ancestors are carried (practitioners of ancient forms of hula aspire to dance in the foundational position of 'aiha'a, with knees bent low to the earth indicating the weight of ancestral presence and associated knowledge carried on one's shoulders). In this living generation, we, like those before, us choose to carry the burdens of our responsibilities with style and grace.

BUILDING ON A LEGACY OF BIOCULTURAL RESOURCE MANAGEMENT

As previously described, the Hawaiian renaissance arose from revealed and reinvigorated sentiments and traditions of aloha 'āina (love for 'āina) and mālama 'āina (care for 'āina) as embodied in traditional knowledge, values and ethics, and regional place-based practices. For E Alu Pū, a particular foundation of our own movement and collective vision is a shared understanding of and value for a land management tradition today called the ahupua'a system.[24] Historically, this system became uniform across the islands as populations increased and power became centralized. Under this system, islands were generally emplaced with palena that divided land and the nearshore into nuanced districts and subdistricts which included cultural uses and spiritual realms. Land districts known as ahupua'a (often themselves subdivided into smaller units, such as *'ili*) supported communities and often extended from the ridges of the mountains to and into the ocean, with the seaward boundary formally recognized during the Kingdom era as the fringing coral reefs or one mile out to sea.[25] Ahupua'a were associated with governance and direct management of resources at different levels. Most importantly, they included a strong tradition of community-based natural resource management. Management on the ground included *maka'āinana* (commoners), such as the *mahi'ai* (farmer), lawai'a (fisher), *mahi i'a* (fish farmer), and a *konohiki* (resource manager, used to refer to the landlords/landowners of ahupua'a after the introduction of Western property concepts to Hawai'i). Konohiki, appointed by the *ali'i* (ruling class) played a key role in maintaining a vision and mindset to help ensure that *pili* (an intimate relationship) existed between the people and their place. Professor Carlos Andrade's book about the community of Hā'ena, on the north shore of Kaua'i, articulates this konohiki tradition:

> In traditional society, *konohiki* were bridges connecting the governing and the governed. *Konohiki* had to gather in the fruits of the *ahupua'a*, for *ali'i* and *mō'ī* and *akua*. However, they needed to ensure that the producers of these fruits, the *maka'āinana* were well cared for and fairly treated. If not, according to traditional custom and practice, *maka'āinana* were free to move and invest their time and energy under more deserving konohiki—an option easily made possible by the extensive kinship networks enjoyed by most families extending far outside a single *ahupua'a* (Lam 1989: 240)....
> [K]onohiki also had to have respect from the people and enough charisma to draw in and make *maka'āinana* feel confidence about investing their lives and energy in the long term success of the *ahupua'a*.[26]

Although the post-annexation territorial administration sought to, and in many cases succeeded in, eliminating formal recognition of the konohiki "rights" codified under the Kingdom of Hawai'i, konohiki practices were maintained, particularly in kīpuka communities that had largely escaped the growing Western influence and control. Many such communities today are represented in E Alu Pū, Hui Mālama Loko I'a, and the Limu Hui, who seek to perpetuate our islands' legacy of biocultural resource management through practice as well as outreach and education. Some community members within these groups even continue to be recognized by their own communities as konohiki, based on their knowledge, skills, and roles as caretakers of the pili between people and place.

In general, however, contemporary citizens look to the state of Hawai'i to assume the resource management functions of the former ahupua'a and konohiki system. We contend that the state cannot and should not be fully responsible for this. For example, whereas decision-making power and management were once shared with place-based leaders, decisions are now made in a top-down and centralized manner by the Department of Land and Natural Resources (DLNR), headquartered in the state capital of Honolulu on the island of Oʻahu. This responsibility is then subdivided into divisions with siloed and sometimes competing purposes and budgets, divisions that can claim only tenuous community connections and limited enforcement capacity at best. Further, laws, policies, and changes in societal relationship orientations between and among resource users exacerbate these dysfunctions. To compound this, Hawaiʻi's political leadership continues to fund DLNR at around 1 percent of the entire state budget, and there is no indication that this funding level will change any time soon.[27]

This reality presents an opportunity and motivation for our communities to reconnect with a tradition of biocultural resource management and to reestablish the associated practices in our kīpuka (places). With the foundation of lessons from our kūpuna and the legacy of abundance in these islands, our values affirmed and constitution strong, we move from feeling divided onto a pathway toward becoming whole.

TAKING UP THE ʻAUAMO FROM KĪPUKA TO ALL OF HAWAIʻI

Aʻaliʻi kūmakua
   It is made of ʻaʻaliʻi of the upland forest,

Maka kaʻapuni ka ʻumeke pohepohe
   The eyes (of the net) surround the round carrying gourd

The ʻauamo is a tool for leverage made of ʻaʻaliʻi, a strong and durable wood born up *mauka* (in the mountains). The ʻaʻaliʻi is a small shrub that grows in the windiest places in Hawaiʻi, yet it never seems to fall over. Because of its nature, it is cited in proverbial sayings as, *"He ʻaʻaliʻi kū makani"* (an ʻaʻaliʻi standing in the wind) in reference to individuals who continue to stand in the face of adversity.

The ʻaʻaliʻi of our movement and our work are the leaders, visionaries, teachers, and learners who stand and grow taller in tradition and knowledge, despite the prevailing winds of our modern-day challenges. One who most certainly fits this description is the late Uncle Henry Chang Wo Jr., a founder of both E Alu Pū and the Limu Hui, as well as a staunch defender of traditional gathering practices and the resources they rely upon. His tenacity and steadfast dedication to maintain the integrity of subsistence resources and practices, as well as his humble leadership and accomplishments discussed further herein, made him a beloved figure and role model for many in our communities. Today, Wo's words continue to remind us to stand strong as the ʻaʻaliʻi, rooted in our traditions and values, endeavoring to fulfill our kuleana notwithstanding the historical upheavals and modern-day dysfunctions we endeavor to transform: "We watch that first raindrop that hits the island. We follow the raindrop all the way to the ocean, we don't let that raindrop get dirty. Because when that mountain water and that ocean water meet, when they come together that's when the ocean *hānau*, that's when the ocean gives birth. Our fishes depend on that, the estuary, they all need that water from the mountain. If it was only the ocean itself, too salty, she *make* (dies) and that is what is happening now."

When we gather, share our knowledge and experiences, our challenges and dreams, we grow the ʻaʻaliʻi of our ʻauamo. At the core of this growth are the shared examples of individuals and communities that have led community-based mālama ʻāina efforts. Many of our kūpuna, like Uncle Henry, also retain memories of ʻohana or loea practice; these memories not only remind us of what once was, but inspire our resolve to continue reviving traditional biocultural community-based management paradigms. E Alu Pū founding member Uncle Kelson "Mac" Poepoe has proven to be a particularly critical source of wisdom, guidance, and inspiration for our communities.[28] Community groups that are also functionally filling the konohiki role, such as the Hui Makaʻāinana o Makana in Hāʻena on the island of Kauaʻi, similarly inspire and provide lessons in perseverance for others who have joined our efforts.[29] As our gatherings have shown us, throughout Hawaiʻi, in their own various ways, communities and community leaders can collectively shape a contemporary konohiki mindset through a focus on and understanding of their places and through living or reactivating the practices of their area—and with this understanding, the ʻaʻaliʻi of our ʻauamo grows strong.

Held by the ʻauamo of our oli is a tattered net wrapped around an old gourd. These items connect us to the waters that flow down the valley and into the ocean. That old net is our community that we need to care for and re-mend. That gourd holds the *wai* (water), our greatest asset, the source of life for our lands and our oceans. When we gather, we not only grow the ʻaʻaliʻi of our ʻauamo, but we also begin to mend the net which will carry the gourd of life that it is our kuleana to protect. For Hawaiʻi, a key challenge is the fact that the time-tested and community-based konohiki management systems and the resources they managed "have suffered from more than two centuries of erosion associated with the history of Hawaiʻi and influences of foreign cultures and governments ... [TEK] systems are often incomplete or fragmented, and the multicultural demography of Hawaiian communities means that cultural knowledge systems and practices can vary widely."[30] Mending our tattered net requires a deep understanding of how knowledge, resource, and management systems fragmented, and how present-day efforts can restore or replace the connections that are lost or eroded. In our gatherings, through the stories of kūpuna and the research of those yearning to regain knowledge lost, we piece together an ever-more-detailed understanding of the systems of the past and how they worked; this, in turn, gives us key insights into the gaps in information, understanding, or practices we can no longer account for. Shared similarities, differences, and lessons learned between communities also provide crucial hints as to how to fill these gaps and mend the broken eyes of our net.

The acts of strengthening our resolve and mending our net, our community, our systems of care, and our resources are synchronous. Through our gatherings, we join our experiences, knowledge and visions; we grow the 'a'ali'i of our 'auamo; we lash together the tattered eyes of our net and, through our subsequent efforts, we begin to fill the gourd that is our kuleana to carry.

HAWAIIAN LAND TENURE PHILOSOPHIES AND 'ĀINA MOMONA

Kā'eo 'ana! Kā'eo lā
    It's overflowing!

Traditional food systems in Hawai'i resulted in production levels that exceeded the needs of the population.[31] This is true for productivity both on the land and in the sea, and is the state referred to as 'āina momona throughout this chapter. The work of our communities is based on a vision of restoring 'āina momona throughout Hawai'i, based on traditional practices, knowledge, and values; this vision, the overflowing gourd of our oli, brings us together and drives our efforts, representing the ultimate outcome of the "alternative rewards" that we collectively pursue. In this section, we will briefly explore some of the philosophies, practices, and contemporary understandings of the systems that served as the foundation of the traditional food system of Hawai'i, and that are critical to informing a return to 'āina momona in the twenty-first century.

Hawaiians designated social-ecological zones from the mountains to the shore and into the sea as a means to holistically and synergistically manage biocultural resources and ensure their abundance. These social-ecological zones, or *wao*, tended to run horizontally across the landscape as they intersected with the palena that divided ahupuaʻa and larger land divisions. Social-ecological zone designations differed in name and diversity across islands in the archipelago and in accordance with many factors, but they were implemented with the same ultimate goal of ʻāina momona.[32]

Traditional Hawaiian knowledge teaches that health of the uplands determines the health of everything below; productive food systems thus depend on healthy forests. The layout of the various wao reflected this wisdom and dictated acceptable and unacceptable activities and practices from the top of the mountains to and beyond the shoreline. Imbued with spiritual and social beliefs, wao also reflected key practical functions of the ecosystem. For example, the *wao akua* (region of god[s]), or sacred forest, captured fog, retained rainwater, maintained the islands' core watersheds, and served as refugia for endemic and culturally significant biodiversity. Vast groves of koa trees (*Acacia koa*) in this and other wao, along with the many other leguminous species that once filled our forests, sequestered atmospheric nitrogen and transported it into the soil, which ultimately leached down to the fields below. In addition, formerly immense colonies of ground-nesting seabirds—which once blackened the skies but are now critically endangered—cycled phosphorus from the sea to the uplands, ensuring fertility in the forests as well as the agricultural fields below. Certain zones in the forests were also once tended as a source of wild foods such as birds, eggs, fruits, and ferns; the streams flowing out of the forests once teemed with various species of fish, shrimp, and mollusks—all important sources of protein for humans and wildlife alike, and all managed to ensure their perpetual abundance.

Collectively, most of the food in the ancient Hawaiian system came out of the social-ecological zone known as the *wao kanaka* (region of mankind). The wao kanaka was designated for field agriculture, aquaculture, habitation, recreation, and temple worship. Native Hawaiian historians and scholars of the nineteenth and twentieth centuries documented various kinds of field systems within the wao kanaka that hosted a diverse suite of crops, fitted into a multitude of conditions ranging from lava flows with no topsoil to depleted soils on islands five million years old, from the uplands to sea level, from some of the rainiest spots on earth to desert conditions and from wetlands to arid plains.[33] Hundreds of cultivars of *kalo* (taro)[34] and *ʻuala* (sweet potato), developed to meet the variety of available conditions, served as staple crops that facilitated the sustainable existence of nearly a million people in Hawaiʻi.[35] A large variety of other crops, such as *ulu* (breadfruit), *maia* (banana), *uhi* (yam), *pia* (arrowroot), *hoi* (bitter yam), *hala* (pandanus), and even *ape* (elephant ear) also filled niches in the landscape and in the diet, helping ensure the resilience of traditional food systems while guarding against times of famine. Such diversity and ingenuity in food cultivation reflect both the intimate knowledge of and sense of connection between people and place embodied in the konohiki system. Spiritual and religious associations of various crops, most notably kalo, also reinforced communal understandings and values such as respect for elders, the importance of ancestral connections, and the functions of the family system.

A notable connection between traditional food systems, the cultural identity of Native Hawaiian people, and the structure of the traditional family system is poignantly made in the story of Hāloa, a genealogical epic on the origin of our islands, the taro plant, and its younger sibling—the first Hawaiian. An abridged version is shared here as a means to convey its overarching themes.

In the cosmogonic genealogy of Wākea (Sky Father) and Papa (Earth Mother), many things were born out of their relationship with each other and others. Among the first were the islands themselves. In a following sequence, Wākea and Hoʻohōkū-ka-lani (The one who made the stars in the heavens) gave birth to two children. The first was a stillborn boy buried beside the house. Up sprouted a plant, never before seen. It had a long stem upon which rested a leaf that quivered when the wind blew. This was the first kalo plant, named Hā-loa-naka-lau-kapalili. In time Hoʻohōkū-ka-lani bore another son: a strong and healthy boy, the first Hawaiian, named Hāloa, in honor of his elder sibling the kalo plant. Kalo, as the elder, provides food for the younger generation; and the Hawaiian, as the junior, bestows the utmost level of respect upon and cares for his elder. Among other themes and kaona (hidden meanings) gleaned from the study of this story, the following resonate:

Respect for elders: Hawaiian family systems are founded on the concept of respect for elders. This origin story says the most immediate elder to the Hawaiian people is kalo, also known as "the source of life." Beyond that, the islands themselves—along with all the life found here before us and, ultimately, the elements of nature which created all things through their various unions—all of these things are ancestors of the Hawaiian people, and should thus be afforded due respect.

Connection to the ancestors: Kalo has an edible corm used to make poi (the Hawaiian staple food) and other foods. It also has edible leaves and flowers.

Once harvested, the corm is cut and cooked and leaves are used in separate preparations. What remains is a planting stalk, which is replanted. That stalk matures, is harvested, and is planted again. The cycle continues as the stalk is planted again and again and again through the generations. This is one of the aspects of the hidden meaning. "Hāloa" is both the first part of the elder brother's name and the full name of the younger sibling. It is something they share in common. "Hā" can refer to the stalk of the plant; "loa" can mean everlasting. As this everlasting stalk is replanted generation after generation, we are actually touching the same plant as our ancestors before us. Beyond that, "Hā" can also refer to one's "breath of life." As time continues, the everlasting stalk is paired with the everlasting breath of life. Both are ultimately connected with the generations before.

Function of the family system: A successfully functioning traditional family system mirrors successful functions observed in nature. Those familiar with kalo know that once a planting stalk is in the ground it is called the *makua*, which means "parent." In time that makua will pop out baby plants on the side called 'ōhā. The word "'ōhā" is the root of the Hawaiian word for family, 'ohana. In one of the ancient methods of planting kalo, two makua would be planted together. This method is now rarely practiced, if at all. The abandonment of such planting methodologies correlates with a historical period associated with a disintegration of the family system. Perhaps this is a coincidence, perhaps not.

The lessons and understandings in Native Hawaiian land-tenure philosophies and practices inform the work of our communities in countless and nuanced ways. For example, the loss of traditional land concepts; the destruction of resources; and the disruption of Native Hawaiian lifestyles, land use, and governance all inform our communities' recognition of present-day challenges and barriers to achieving 'āina momona. Meanwhile, the embedded values of respecting elders and honoring ancestral connections, including that with kalo, potentially provide key hints to restoring appropriate governance and decision-making protocols. Perhaps most importantly, however, is the underlying theme of interconnectedness and interdependence among people, place, food, environment, past, present, and future. Such a theme also underlies the konohiki management approach, and its revival in our—that is, all Hawai'i residents'—collective understanding may prove indispensable to our progress toward the overflowing gourd of 'āina momona, from mauka to makai.

RETURN TO ĀINA MOMONA IN THE MAKAI: HĀNAI A 'AI

As revealed in the interconnection and interdependence recognized in traditional Hawaiian land tenure, including in the konohiki approach to resource management, 'āina momona requires at its foundation the practical application of the concept of giving before taking in every facet of life. This approach should be reflected in protocol, ceremony, and everyday practice, not only in our interpersonal relationships, but our relationships with all that is around us. It is most commonly expressed in the 'Ōlelo No'eau (proverb), *aloha aku, aloha mai*,[36] which expresses the notion that you must give (love) first in order to receive. In relation to biocultural resource management specifically, there is a saying from the Kona Hema district of Hawai'i Island: Hānai a 'ai. It literally means "feed and then eat," but the real meaning is revealed in the story behind it.

The following interview with Kupuna Smith Kaleohano of Ho'okena, Hawai'i Island, illustrates Hānai a 'ai through one community-based management approach practiced by one of our communities for a key ocean resource, the 'ōpelu fish. It exemplifies ways in which the traditions of the past can—potentially—lead us to 'āina momona in the future:

> Kaleohano: You couldn't catch 'ōpelu if it wasn't the season for catching 'ōpelu because there was a season for only feeding the 'ōpelu and there was a season for only catching the 'ōpelu.
>
> The head fisherman decided when to set the restriction for just feeding the 'ōpelu. Yes, the head fisherman would decide this, and this restriction was based for the 'ōpelu fishing, not for all fishes—just for the 'ōpelu fish. No one could take their nets out during this restriction of four to five months. If you're ever caught with your net out there during the restricted time they would take your net away and burn it.
>
> . . .
>
> Larry Kimura (interviewer): And how was a head fisherman selected?
> Kaleohano: He was selected from among the fishermen. They chose him. And the only way you could become a head fisherman was to prove your excellence, you had to be the best of all the fishermen. There could only be one head fisherman, and he had to be the best. If there was fifteen canoes, fifteen of these fishermen would study who would be the best fisherman within five years, say, and then they would select him to be the head fisherman. You see, a lot depended on his good

luck and, of course, his knowledge. And everyone listened to him and trusted him because they knew they couldn't outbeat him. And when that happened, everything went smoothly. That's why when the head fisherman decided it was time just to feed the ʻōpelu, then it was feeding season and [ʻōpelu was] not fished.[37]

In this extract, Kupuna Kaleohano describes the customary practice where resource management decisions were made at a local level by a local resident who possessed intimate knowledge of the resource, demonstrated leadership skills, and would be respected, trusted, and followed by others. In function, this person was konohiki.

As one of the important seasonal food sources for Hawaiians, the ʻōpelu serves as a good example of how marine resources are better managed when decisions are made at a local level, consistent with a konohiki approach and mindset.[38] ʻŌpelu generally spawn between March and June. The season varies by locale. Accordingly, Hawaiians traditionally placed a *kapu* (off-season) on ʻōpelu fishing during the months of January to July, with the specific closure dates determined by a head fisherman, like Kaleohano. During kapu season, the fish were fed with excess vegetables grown in the upland farms, at koʻa identified by fishermen and assigned to specific fisher families, who were expected to care for these critical fishing grounds.[39] Placing a kapu on ʻōpelu and feeding (*hānai*) the fish when they are spawning help to safeguard against overfishing and ensure a successful spawn for the next cycle, whereas fishing for ʻōpelu during their spawning season would do the opposite and deplete the fishery. Notably, abundant ʻōpelu also support larger food fish, such as *ʻahi* and *aku*.

Unfortunately, state-driven resource management provides little protection for konohiki practices, including even the time-tested ʻōpelu koʻa practices. In many areas, ʻōpelu are legally fished year-round, primarily for the commercial market, often in manners that foreclose traditional koʻa practices. Today, the only affirmative legal protection for traditional ʻōpelu practices is a hard-fought regulation prohibiting the use of meat-based attractant in ʻōpelu net fishing in the waters off Miloliʻi, Hawaiʻi Island.[40] Nonetheless, Native Hawaiian fishermen from Miloliʻi and Hoʻokena have active projects, involving countless volunteer hours, to revive *ʻōpelu* fishing practices and establish seasonal closures and minimum size limits; it remains to be seen whether the state DLNR will facilitate or promote rule-making to enforce these limits for the general population.

E ALU PŪ AND REVIVING COMMUNITY-BASED KONOHIKI FISHERY MANAGEMENT

The foundation for a movement is a context in which "alternative rewards emerge to sustain the movement's vision, which may force the conventional reward system to change." This context includes a contemporary renewal of Hawaiian traditions of community-based natural resource management that includes the building of a konohiki mindset not just in our community members but also throughout our state, such that the focus, needs, and capacity of management leads to policies, procedures, infrastructure, jobs, and skills for future generations to mālama Hawai'i.

A major realization of the E Alu Pū communities was that the alternative rewards they sought might be embodied in the form of a Community-Based Subsistence Fishing Area (CBSFA), a regulatory opportunity inspired by the work of E Alu Pū founder Uncle Mac Poepoe. Another realization was that the principle of hānai a 'ai must also be applied in engagement with the state, which does not necessarily practice that concept itself.

The CBSFA law was born in 1994, after extensive study of Native Hawaiian subsistence fishing on the island of Moloka'i and based on the work of the Mo'omomi community. This law created a pathway for community-led efforts founded on the reaffirmation and protection of "fishing practices customarily and traditionally exercised for purposes of Native Hawaiian subsistence, culture, and religion."[41] In its promulgation of the CBSFA legislation in 1994 the state legislature affirmed its constitutional mandate to protect the "traditional and customary" gathering rights of Hawaiians by directing the DLNR to partner with community organizations and create CBSFA designations and management plans in accordance with traditional practices.

Despite the legislature's directive, however, the first permanent CBSFA rules were not adopted until 2015—more than twenty years after the passage of the CBSFA legislation—in application to Hā'ena, Kaua'i. Even these rules would not have been possible without the persistence and leadership of the Hui Maka'āinana o Makana, an organization composed of original *hoa'āina* (tenant) farmers of Hā'ena who have also restored their old family *lo'i kalo* (taro ponds). In the spirit of *hānai a ai*, members of E Alu Pū, who had witnessed and perhaps identified with the decades-long struggles of the Hui, offered their collective support to help the Hui overcome the final political and bureaucratic hurdles preventing adoption of the rules.[42] With successful passage of the rules, one

Community members in a prayer circle outside the public hearing for the Hāʻena CBSFA rules public hearing in Hanalei, Kauaʻi 2014. *Photo by Kimberly Moa.*

mechanism and pathway for realizing the alternative rewards of E Alu Pū has now become apparent.

CBSFAs and supporting programs allow for forms of co-governance and co-management more akin to the konohiki approach, a departure from the current top-down, one-size-fits-all approach to resource management in Hawaiʻi.[43] E Alu Pū communities realize that CBSFA designations and rules, developed collaboratively and based on a community's intimate familiarity with and knowledge of its unique place, can help to address present and future resource conditions, consistent with traditional community-based management approaches.

Given the demonstrated reluctance of the state to adjust its top-down paradigm, achieving CBSFA designation and rules can be an intimidating challenge, particularly for communities that may not be sufficiently resourced for a long and exhaustive public and political process. However, that challenge alone can build capacity within a community. It requires that a community address not just the biological complexities and interrelationships of resource management, but the social and economic implications as well.

Finally, communities now realize that a network like E Alu Pū can unite their unique and common strengths to develop support and strength statewide and among government agencies and partners. Realization of the power of networks has led to the formation of two other coalitions, the Hui Mālama Loko Iʻa and the Limu Hui, described in the final sections of this chapter.

LIMU: ALGAE

Although *limu* can refer to various algae and plant species, including mosses, liverworts, and lichen, the word as used today almost always refers to seaweed. At one time in the not-too-distant past, limu had a significant role in many aspects of Hawaiian culture and ʻāina momona. Nowhere else in the world has limu been used as extensively as it once was by Native Hawaiian people, who gathered and used limu for food, medicinal, and spiritual purposes. Today, reviving and maintaining traditional understandings of and connections to limu may be key to fostering and perpetuating the biocultural systems that limu depends upon, and that depend upon limu. Limu Hui seeks to fulfill this critical component of achieving ʻāina momona.

In traditional times limu was the salad, spice, and relish to an otherwise monotonous diet.[44] Along with fish and poi, limu served as the third major component of the healthy Native Hawaiian traditional diet, providing flavor as well as key vitamins and minerals, such as vitamins A and B12, riboflavin, vitamin C, calcium, magnesium, potassium, and iodine.[45]

Perhaps in testament to its traditional importance and value, limu continued to be gathered despite the social and political upheavals of the nineteenth century, including the overthrow of the Hawaiian Kingdom and subsequent annexation of the islands. A 1901 survey of the commercial harvest of limu in the Hawaiian Islands recorded a total of 42,764 pounds sold at a dollar value of $5,316.[46]

In 1904, a botany professor from the University of California at Berkeley, along with her Hawaiian interpreter, visited all the Hawaiian Islands to determine the extent of use of limu. She used two main methods of data collection. One was to visit the many markets where limu was sold and to record the names and uses of limu. A second method was to visit the shoreline at low tide and survey the women harvesting limu. This study resulted in a list of more than seventy-five different species of limu that were being sold or gathered for food.[47]

Sadly, the continual upheavals of the twentieth century caused the rapid loss of limu grounds and associated human connections to limu. In 1974, the late Dr. Isabella Abbott, a Native Hawaiian professor and ethnobotanist then recognized as the leading authority on limu, documented only eighteen species of limu in use.[48] When kūpuna are asked why there is less limu now than in the past, their responses almost always refer to "loss of freshwater." Indeed, many of our highly desirable, nearshore limu species, such as 'ele'ele (*Ulva enteromorpha*), *pakeleawa'a* (*Grateloupia filicina*), and *wawae'iole* (*Codium edule*) thrive in places where freshwater springs or streams along the shoreline create a low-salinity, high-nitrate ocean environment. These are some of the more desirable limu for people and for fish. *Kai* (seawater) is high in phosphate and low in nitrate. *Wai* (mountain water) is high in nitrate and low in phosphate. Freshwater, in the forms of surface stream flow and groundwater upwelling, provide the needed nitrogen for desirable limu.

Diversion of surface stream flow began with the introduction of sugarcane and pineapple production in Hawai'i, after the first successful sugar plantation began at Kōloa, Kaua'i in 1835.[49] Over more than 150 years, streams on the windward, or wet, side of the islands were increasingly diverted by large sugar and later pineapple plantations to irrigate their crops on the leeward, or dry, side of the islands. These diversions left many streambeds dry and many more with just a trickle of water. The result was a major decline in nearshore limu abundance.

The ongoing loss of limu undermined the larger systems of interdependence critical to the 'āina momona of the past, such as native fish and other organisms that rely upon limu. The loss of these critical food sources, limu and fish, disrupted subsistence lifestyles, knowledge, and connection to place in communities that would otherwise be able to monitor and maintain the health of their local resources.

For example, the Windward Maui community of Ke'anae is named after its former abundance of 'anae, the Hawaiian word for large broodstock-sized mullet. However, 'anae are no longer found in Ke'anae. Kūpuna from that community note the decline of mullet began with the decline in limu, such as pakeleawa'a and 'ele'ele. The decline of limu began with a plantation diversion of water from streams flowing into Ke'anae.

Urbanization has also played a major role in the loss of coastal groundwater discharge necessary for limu habitat. Freshwater aquifers sitting under all our islands feed the coastal springs that various limu depend upon. A boom in urban development, particularly after statehood, has placed ever-growing pressure on these aquifers and the springs they feed. Roads, shopping centers, housing developments, and other urban land uses create impermeable surfaces that prevent rainwater from percolating into aquifers; meanwhile, the high urban demand for fresh water further depletes groundwater and reduces or even halts coastal discharge.

Significantly, the widespread disappearance of limu has resulted in the loss of human knowledge and connections to this critical component of ʻāina momona. Without limu, there is little basis upon which *loea limu* (limu experts) can pass on knowledge of limu and limu practices that have been maintained for generations. These experts' intimate knowledge of limu ecology may be vital to understanding how best to restore, care for, and perpetuate limu and its interconnected systems, and their knowledge of limu practices may be key to a broader appreciation of why it is so critical to do so. With the passing of each loea limu, we lose more of that traditional cultural knowledge.

As one of the few remaining loea limu, Uncle Henry Chang Wo Jr. recognized the importance of his knowledge and dedicated the latter part of his life to passing it on. He helped found the ʻEwa Limu Project to educate people about limu and to organize limu planting and restoration efforts. He focused those efforts in a place where limu once thrived in great abundance but had declined at an alarming rate.[50] Inspired by his experience with the collective action of E Alu Pū, he helped give birth to the Limu Hui, which today bring together kūpuna, younger generations, and researchers to capture and pass on traditional limu knowledge, and to revive limu and the foundation of our nearshore ecosystems.

A true ʻaʻaliʻi kū makani, Uncle Henry pointed the way, and the Limu Hui continues on his path.

LOKO IʻA: AQUACULTURE AND MARICULTURE

> Fishponds, *loko iʻa*, were things that beautified the land, and a land with many fishponds was called a "fat" land (ʻāina momona).[51]

Long before Native Hawaiians discovered Captain Cook, loko iʻa contributed to ʻāina momona. Loko iʻa in ancient Hawaiʻi formed a technologically advanced, efficient, and extensive form of aquaculture found nowhere else in the world.[52] While techniques of herding or trapping adult fish with rocks in shallow tidal areas, in estuaries, and along their inland migration routes can be found around the globe, Hawaiians developed technologically unique and efficient fishponds, advancing the cultivation practice of mahi iʻa (fish farmer) as distinct from lawaiʻa (fisher). The variety of loko iʻa designs and construction methods reflected an unparalleled understanding of engineering, hydrology, ecology, biology, and agriculture, including environmental, ecological, and

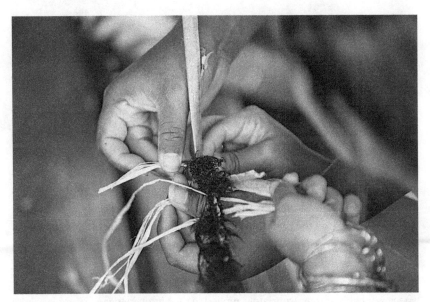
Limu lei weaving at the Global E Alu Pu in 2016. *Photo by Mark Lee, Holladay Photo.*

social processes specific to Hawaiʻi.[53] Constructed using natural coastal features and tidal cycles, they produced an estimated three hundred pounds of fish per acre per year, producing a reliable supply of fish notwithstanding adverse weather, surf, or fishing conditions.[54] They also contributed to social cohesion, as the surrounding community contributed to their construction and maintenance and shared the fish they produced.[55] Loko iʻa have their own place in *moʻolelo* (Hawaiian stories), and continue to hold spiritual significance and power as places where akua (gods) and ʻaumakua (ancestral deities) often gather. Loko iʻa are thus celebrated for their past and future potential to contribute to the economic and food self-sufficiency needs of their ahupuaʻa and Nā Kai ʻEwalu, and achieving a state of ʻāina momona.

Unfortunately, after a thousand years of generational knowledge, experimentation, adaptation, and practice that informed their creation and unparalleled success, loko iʻa saw rapid declines following Western contact and the subsequent disruption of the environmental and social systems integral to their functioning. Statewide surveys have identified an upper estimate of 488 formerly active fishponds on six islands, including several locations containing clusters or concentrations of ponds.[56] An inventory in 1901, however, identified only 99 ponds actively being used in commercial trade; these relatively few ponds nonetheless produced an estimated 680,000 pounds of fish per year, including 485,000 pounds of ʻamaʻama (striped mullet, *Mugil cephalus*) and 194,000 pounds of ʻawa (milkfish, *Chanos chanos*). The inventory also speculated that there had been more than twice as many ponds in use only thirty years prior.[57] By 1970, loko iʻa produced only 20,000 pounds of fish—less than 1 percent of what they likely produced at their peak.[58]

Today, the majority of loko iʻa sites are in highly degraded condition, some completely covered with dredged fill and unrecognizable as fishponds. Barriers to loko iʻa restoration include altered watersheds and the diversion of water (necessary to create the productive brackish water environment for coastal loko iʻa), establishment of invasive species such as mangrove, permit requirements that are not well designed to accommodate loko iʻa restoration—including misplaced and burdensome permitting processes intended to protect sensitive species or habitats inappropriately applied to the highly managed systems of loko iʻa and *loʻi kalo* (taro fields)—and loss and scattering of generational knowledge for management and care of loko iʻa. Additionally, many fishponds are owned by the state,[59] but are not maintained as loko iʻa. A number of fishponds in Hawaiʻi are also owned by the federal government or private entities, and likewise are not maintained or supported as loko iʻa.

Yet loko iʻa retain the potential to contribute to a healthy and robust food system for their ahupuaʻa. Over recent decades, Hawaiian communities and *kiaʻi loko* (fishpond guardians and caretakers) have worked to restore loko iʻa around the islands and reclaim the knowledge and practice of loko iʻa culture. In recognition of their shared challenges and in an effort to increase collaboration and accelerate restoration and food production, kiaʻi loko from around the islands formed the Hui Mālama Loko Iʻa, a network of loko iʻa and kiaʻi loko from six Hawaiian islands. Since 2004, members of the Hui Mālama Loko Iʻa have met annually and opportunistically to strengthen working relationships and share experience and expertise about their respective places and work.

Most recently, the Hui Mālama Loko Iʻa has leveraged its collective influence to streamline the permitting processes in collaboration with the state of Hawaiʻi, and has generally improved and grown co-management relationships with government and private entities.[60] Many of the loko iʻa the Hui manages serve as kīpuka for education and the renewal of traditional practices and values in contemporary ways. The restoration of loko iʻa provides opportunities for

Native Hawaiians and the larger community to renew ʻāina momona on many fronts and in balanced ways.

In April 2016, Hui Mālama Loko Iʻa adopted the following guiding statements for their continued collaboration as a network:

> Vision: Perpetuate *ʻāina momona* through *loko iʻa* culture.
>
> Mission: Empowering a network of *kiaʻi loko* whose *kuleana* is to reactivate, restore, and cultivate *loko iʻa* guided by *loko iʻa* culture in pursuit of *ʻāina momona* for *ʻohana* and communities.

BY CARING FOR HAWAIʻI, WE REBUILD AND CARE FOR OUR COMMUNITY

Kōkō weluwelu pāhono ʻia,
   he old/tattered net has been mended,

Ua paʻa ē! Ua paʻa lā!
   It's once again solid!

Our oli recognizes that although change is inevitable, our community has agency and a legacy of traditional tools and knowledge to help shape that change. We can carry forth the best of community in making this happen. We can mend the old net that connected and bonded our communities together. We can refill the gourd.

By meeting together, sharing experiences and knowledge, partnering with others, and building community, we can co-create a narrative that taps into the potential of Hawaiʻi's people to mālama Hawaiʻi and begin a return to ʻāina momona. By gathering, mobilizing, and communicating as a network, building consensus on our values, and acting and reflecting on our actions together in an iterative manner we co-create the change we wish to see in our community and the world.

We have only just begun to show our state and our greater community that we can work together to change things for the better. Most importantly, we must continue to demonstrate to ourselves that we can do so. As the konohiki would inspire a spirit of laulima and mālama āina in the community of the ahupuaʻa to open pathways for water to flow, pathways to ʻāina momona in the present day require us all to move away from isolation and back toward a stronger sense of community and stewardship as an expression of island citizenship. Indeed, this is happening not just on statewide basis but at the mokupuni (island) and moku (larger regional district which includes ahupuaʻa) levels.[61]

WE STAND READY. ARE YOU?

Aweawe pualu, E Alu Pū!
Here come the traces of the pualu in the calm water,
let us gather together!

Kīʻililī? Kūʻolokū!
Are you ready? We stand ready!

As we turn our eyes to the horizon, we see the kuleana to mālama Hawaiʻi is great. We take a deep breath as we feel the weight of the task, and together we pick up the ʻauamo. We move forward with our work to be better together and together to better care for Hawaiʻi.

As dawn breaks, ripples appear on the water, signaling a growing movement of pualu fish: they shift and adapt to changes, separate to overcome or surround danger or obstacles, then as always, come together again and reassemble in order to move forward together.

NOTES

1. For more information and an in-depth analysis of Western and Indigenous frameworks on natural resource management, see Kekuhi Kealiʻikanakaʻolehaililani and Christian P. Giardina, "Embracing the Sacred: An Indigenous Framework for Tomorrow's Sustainability Science," *Sustainability Science* 11, no. 1 (January 2016): 57–67; Fikret Berkes, *Sacred Ecology: Traditional Ecological Knowledge and Management Systems* (Philadelphia: Taylor and Francis, 1999); and Gregory Cajete, *Native Science: Natural Laws of Interdependence* (Santa Fe, NM: Clear Light, 2000).

2. The eight main Hawaiian islands are connected by these eight seas. Kanalu G. T. Young, *Rethinking the Native Hawaiian Past* (New York: Garland, 1998).

3. "Uncle" is a title of respect in local Hawaiian vernacular, one bestowed upon those in our community whose words should be heeded once uttered. This title is founded in the concept of extended family and is not limited to blood relations.

4. Davianna Pōmaikaʻi McGregor, *Nā Kuaʻāina: Living Hawaiian Culture* (Honolulu: University of Hawaiʻi Press, 2007), 7–8.

5. Population estimates suggest that the precontact Hawaiian population was as high as eight hundred thousand people. David E. Stannard, *Before the Horror: The Population of Hawaiʻi on the Eve of Western Contact*, 3rd. ed. (Honolulu: University of Hawaiʻi Press, 1994), 59. Hawaiʻi now has more than one million residents. US Census Bureau, "Annual Estimates of the Resident Population: April 1, 2010 to July 1, 2016 (NST-EST2016-01)," https://www.census.gov/data/tables/2016/demo/popest/nation-total.html.

6. Dennis Kawaharada, *Hawaiian Fishing Legends with Notes on Ancient Fishing Implements and Practices* (Honolulu: Kalamakū Press, 1992), xi.

7. Sarah Kehaulani Goo, "After 200 Years, Native Hawaiians Make a Comeback," Pew Research Center, April 6, 2015, http://www.pewresearch.org/fact-tank/2015/04/06/native-hawaiian-population/.

8. Office of Hawaiian Affairs Research Division, *Native Hawaiian Health Fact Sheet 2015*, vol. 1: *Chronic Diseases* (Honolulu: Office of Hawaiian Affairs, 2015); Office of Hawaiian Affairs (OHA), *The Disparate Treatment of Native Hawaiians in the Criminal Justice System* (Honolulu: OHA, 2010); OHA Research Division, *Income Inequality and Native Hawaiian Communities in the Wake of the Great Recession: 2005 to 2013* (Honolulu: OHA, 2014); see also the OHA Research Division website at https://www.oha.org/research.

9. Terry Shintani, Sheila Beckham, Jon Tang, Helen Kanawaliwali O'Connor, and Claire Hughes, "Waianae Diet Program: Long-Term Follow-Up," *Hawai'i Medical Journal* 58, no. 5 (1999): 117–22; Ben DiPietro, "Obese Hawaiians Learn from Thin Ancestors' Diet," *Los Angeles Times*, March 29, 1992.

10. See State of Hawai'i Division of Forestry and Wildlife Native Ecosystems Protection and Management, "Rare Plant Program," http://dlnr.hawaii.gov/ecosystems/rare-plants/.

11. Hawai'i 2050 Sustainability Task Force, *Hawai'i 2050 Sustainability Plan* (Honolulu: Office of the Auditor, 2008), www.oahumpo.org/wp-content/uploads/2013/02/Hawaii2050_Plan_FINAL.pdf.

12. State of Hawai'i Department of Business, Economic Development, and Tourism, "DBEDT Forecasts (3Q 2018): Visitor Arrivals," http://dbedt.hawaii.gov/visitor/tourism-forecast/.

13. George Kent, "Food Security in Hawai'i," in *Thinking Like an Island: Navigating a Sustainable Future in Hawai'i*, edited by Jennifer Chirico and Gregory S. Farley (Honolulu: University of Hawai'i Press, 2015), 28–45.

14. State of Hawai'i Department of Budget and Finance, *The FB 2015–17 Executive Biennium Budget: Budget in Brief* (Honolulu: Department of Budget and Finance, 2014), 17, https://budget.hawaii.gov/wp-content/uploads/2012/11/FB15-17-Budget-in-Brief-Ige-Administration.pdf.

15. Parker Palmer, "Divided No More," *Change Magazine* 24, no. 2 (1992): 10–17.

16. E Alu Pū affiliated organizations agree to work together and sign an *'aelike* (creed) that sets the foundation for their relationship and work as a network. These organizations are (from Maui) Kīpahulu 'Ohana, Hōlani Hāna; (from Wailuku) Wailuku Community Marine Managed Area, Waihe'e Limu Project; (from Lāna'i) Kūpa'a No Lāna'i, Maunalei Ahupua'a Mauka Makai Managed Area; (from Moloka'i) Hui Mālama o Mo'omomi, Ka Honua Momona; (from Hawai'i) Kama'āina United to Protect the 'Āina, Pa'a Pono Miloli'i, Ka 'Ohana o Hōnaunau, Hui Aloha Kīholo, Ka'ūpūlehu Marine Life Advisory Committee Families; (from Kaua'i) Hui Maka'āinana o Makana, Limahuli Garden and Preserve, Waipā Foundation, Hanalei Hui Watershed, Mālama Kōloa; (from O'ahu) 'Ewa Limu Project, Ho'ōla Hou Iā Kalauao, Ka'ala Farm, Loko Ea Foundation, Mālama Pūpūkea-Waimea, Hui o Hau'ula, Ho'āla 'Āina Kūpono/ Kahana Kilo Kai, Paepae o He'eia, Kāko'o 'Ōiwi, God's Country Waimānalo, Hakipu'u 'Ohana, Ho'oulu 'Āina.

17. For more information, see Queen Liliʻuokalani, *Hawaii's Story by Hawaii's Queen* (Boston: Lee & Shepard, 1898), http://digital.library.upenn.edu/women/liliuokalani/hawaii/hawaii.html; Lilikalā Kameʻeleihiwa, *Native Land and Foreign Desires: Pehea Lā E Pono Ai*? (Honolulu: Bishop Museum Press, 1992).

18. The Hawaiian Kingdom had treaties with other nation-states, including the United States, Great Britain, France, Austria-Hungary, Belgium, Germany, Japan, Russia, Spain, and Italy: "The Hawaiian Kingdom . . . entered into treaties and received formal recognition as a sovereign, independent nation from nearly every major world power." Davianna Pōmaikaʻi McGregor & Melody Kapilialoha MacKenzie, *Moʻolelo Ea O Nā Hawaiʻi: History of Native Hawaiian Governance in Hawaiʻi* (Honolulu: OHA, 2014), 30–31, https://www.doi.gov/sites/doi.opengov.ibmcloud.com/files/uploads/McGregor-and-MacKenzie-History_of_Native_Hawaiian_Governance.pdf. The Apology Resolution, Public Law 103-150, 107 Stat. 1510 (1993), "To acknowledge the 100th anniversary of the January 17, 1893, overthrow of the Kingdom of Hawaii, and to offer an apology to Native Hawaiians on behalf of the United States for the overthrow of the Kingdom of Hawaii."

19. Post-overthrow achievements included, among others, successfully electing, through the Native Hawaiian–established Home Rule Party, the overwhelming majority of both houses of the territorial legislature (which subsequently conducted its business in Hawaiian and passed laws—eventually vetoed—that would have increased access to medical care, reduced taxes on the poor, and eliminated the poll tax that prevented many from voting); electing Native Hawaiians to many county posts; influencing both the local Democratic and Republican Party platforms; electing Native Hawaiians as territorial delegates to the US Congress; successfully lobbying Congress to pass the Hawaiian Homes Commission Act; restoring the Royal Order of Kamehameha; and forming Hawaiian Civic Clubs to further advance Native Hawaiian interests. McGregor and MacKenzie, *Moʻolelo Ea O Nā Hawaiʻi*, 374–433; see also Kawika Burgess, "Follow in the Footsteps of Our Kupuna," *Honolulu Star Advertiser*, November 1, 2015, http://www.staradvertiser.com/2015/11/01/editorial/island-voices/follow-in-the-footsteps-of-our-kupuna-2/.

20. Haw. Const. art. XII, §7, art. XII, §5, and art. XV, §4.

21. These could include, notably, *konohiki* (land managers), *mahiʻai* (farmers), *lawaiʻa* (fishers), *kahuna laʻau lapaʻau* (herbal healers), *kumu hula* (hula teachers), *kahuna kakalaleo* (expert orators and chanters), cooks, musicians, weavers, and others.

22. Examples of resource databases are "Papakilo Database," OHA, http://www.papakilodatabase.com/main/main.php; Sea Grant, University of Hawaiʻi, Hawaiian Language Newspaper Translation Project, http://seagrant.soest.hawaii.edu/?q=hawaiian-language-newspaper-translation-project.

23. "Kule.ana, nvt. Right, privilege, concern, responsibility, title, business, property, estate, portion, jurisdiction, authority, liability, interest, claim, ownership, tenure, affair, province; reason, cause, function, justification; small piece of property, as within an ahupuaʻa; blood relative through whom a relationship to less close relatives is traced, as to in-laws." *Nā Puke Wehewehe ʻŌlelo Hawaiʻi*, s.v. "kuleana," http://wehewehe.org/.

24. The ahupuaʻa was the most predominant land designation; however, it was part of a larger, nuanced land management approach starting from the *mokupuni* (island), to the *moku* (district), to the ahupuaʻa (could be a valley or something greater), *ʻili*, *ʻili kupono*, *ʻili lele*, and so on. Further realms were based on human relationships with them for resource procurement. As an example, from mauka (mountain) to makai (ocean) were many realms such as *wao akua* (realm of the gods; upper watershed) to the *wao kanaka* (realm of humanity). See, for example, Kamanamaikalani Beamer, *No Mākou Ka Mana: Liberating the Nation* (Honolulu: Kamehameha Publishing, 2014); and Marion Kelly, "Ahupuaʻa: A Kanaka Maoli System of Natural Resource Enhancement, Utilization, and Preservation," Paper presented at the Save Our Seas Conference, Princeville, HI, June 12–14, 1997.

25. These divisions that included the ocean evolved under Hawaiian Kingdom law and passed to the current government, the state of Hawaiʻi, which now recognizes in statute: "(a) The fishing grounds from the reefs, and where there happens to be no reefs, from the distance of one geographical mile seaward of the beach at low watermark, in law, shall be considered the private fishery of the konohiki, whose lands by ancient regulations, belong to the same." Haw. Rev. Stat. §187A-23; see also Alan Murakami and Wayne Tanaka, "Chapter 10: Konohiki Fishing Rights," in *Native Hawaiian Law: A Treatise*, edited by Melody Kapilialoha MacKenzie, Susan K. Serrano, and D. Kapuaʻala Sproat (Honolulu: Kamehameha Publishing, 2015), 617–18, 633.

26. Carlos Andrade, *Hāʻena: Through the Eyes of the Ancestors* (Honolulu: Latitude 20, 2008).

27. See State of Hawaiʻi Department of Budget and Finance, *FB 2015–17 Executive Biennium Budget*.

28. For a brief description of Uncle Mac's work, see *Nā Loea: The Masters*, season 1, episode 3, "Mac Poepoe: Mālama Moʻomomi," produced by Keoni Lee and Matt Yamashita, released April 1, 2014, on ʻŌiwi TV, https://oiwi.tv/oiwitv/na-loea-malama-moomomi/.

29. See Alden Alayvilla, "Hui Makaʻainana o Makana Educates, Cultivates, Inspires," *The Garden Island*, December 4, 2016.

30. Alan Friedlander, Janna M. Shackeroff, and John N. Kittinger, "Customary Marine Resource Knowledge and Use in Contemporary Hawaiʻi," *Pacific Science* 67, no. 3 (2013): 448.

31. "Interview of John K. Kaʻimikaua," *A Mau A Mau: To Continue Forever*, produced and directed by Kaʻoi Kaʻimikaua and Nalani Minton (Honolulu: Nalani Minton, 2009).

32. Kawika K. B. Winter and Matthew Lucas, "Spatial Modeling of Social-Ecological Management Zones of the Aliʻi Era on the Island of Kauaʻi with Implications for Large-scale Biocultural Conservation and Forest Restoration Efforts in Hawaiʻi," in "Scaling Up Restoration Efforts in the Pacific Regions," special issue, *Pacific Science* 71, no. 4 (October 2017): 457–77.

33. E. S. Craighill Handy and Mary Kawena Pukui, *Native Planters in Old Hawaii: Their Life, Lore, and Environment* (Honolulu: Bishop Museum Press, 1972); Samuel Manaiakalani Kamakau, *The Works of the People of Old: Ka Hana a Ka Poe Kahiko* (Honolulu: Bishop Museum Press, 1976).

34. Kawika K. B. Winter, "Kalo [Hawaiian Taro: *Colocasia esculenta* (L.) Schott] Varieties: An Assessment of Nomenclatural Synonymy and Biodiversity," *Ethnobotany Research and Applications* 10 (2012): 403–22.

35. T. N. Ladefoged, P. V. Kirch, S. Gon III, O. Chadwick, A. Hartshorn, and P. M. Vitousek, "Opportunities and Constraints for Intensive Agriculture in the Hawaiian Archipelago Prior to European Contact," *Journal of Archaeological Science* 36, no. 10 (October 2009): 2374–83.

36. Mary Kawena Pukui, *'Ōlelo No'eau: Hawaiian Proverbs & Poetical Sayings* (Honolulu: Bishop Museum Press, 1993).

37. Smith Hawila Kaleohano, interview by Larry Kimura at Ke'ei, South Kona, Hawai'i Island, July 1976.

38. Reflecting the importance of this species, 'ōpelu is one of the important fish referenced in the *Kumulipo*, a celebrated Hawaiian creation oli, and in the historical accounts of renowned Hawaiian scholars Kamakau, Malo, and Handy and Handy, among others. Queen Lili'uokalani, *The Kumulipo: An Hawaiian Creation Myth* (Boston: Lee and Shepard, 1897; reprint, Honolulu: Pueo Press, 2016).

39. In Smith Kaleohano's example of fifteen canoes and fifteen fishermen, the fifteen canoes represent fifteen different fishermen, some of whom may be from the same families. One had to earn and ascend to the position of head fisherman on a canoe. Given that fishing traditions were and still are handed down from father to son within specific families, very likely the canoe was handed down as well. It was uncommon for someone from outside the family to be taught those valued and protected practices. The collective community incorporates all the professions necessary for survival. Collective but separate. Fishermen being only one. Kaleohano, interview.

40. Haw. Admin. Rules §13-95-18.

41. For more on the CBSFA law, see Haw. Rev. Stat. §188-22.6.

42. See Will Caron, "Kaua'i Overwhelmingly Supports Hā'ena Subsistence Fishing Plan," *Hawai'i Independent*, August 4, 2015, http://hawaiiindependent.net/story/kauai-overwhelmingly-supports-haaena-subsistence-fishing-plan.

43. CBSFA implementation is accompanied by management approaches that include an emphasis on community outreach and a community role in assisting enforcement through a program called Makai Watch. For the past three years, the state has granted official recognition of the Hawai'i Makai Watch program and has adopted CBSFA designation procedure guidelines that outline these and other important components of implementation. See State of Hawai'i DLNR, "Makai Watch Program Receives Official Recognition by the Board of Land and Natural Resources," news release, October 21, 2014, http://dlnr.hawaii.gov/makaiwatch/2014/11/26/makai-watch-program-receives-official-recognition-by-the-board-of-land-and-natural-resources/; and Erin Zanre, *Community-Based Subsistence Fishing Area Designation Procedures Guide: Standardized Operating Procedures for Community-Based Subsistence Fishing Area Designation under Hawai'i Revised Statutes §188-22.6* (Hilo: Research Corporation of Hawai'i, 2014), http://dlnr.hawaii.gov/coralreefs/files/2015/02/CBSFA-Designation-Procedures-Guide_v.1.pdf.

44. Heather J. Fortner, *The Limu Eater: A Cookbook of Hawaiian Seaweed* (Honolulu: University of Hawai'i Sea Grant Program, 1978).

45. Isabelle Aiona Abbott, *Limu: An Ethnobotanical Study of Some Hawaiian Seaweeds* (Lawai, HI: National Tropical Botanical Garden, 1996).

46. John N. Cobb, *Commercial Fisheries of the Hawaiian Islands* (Washington, DC: US Fish Commission, 1901), 464–65.

47. William Albert Setchell, "Limu," in *University of California Publications in Botany* 2, no. 3 (April 1905): 91–113.

48. Abbott, *Limu*.

49. On the history of Kaua'i's first plantation, visit the Grove Farm Sugar Plantation Museum website, https://grovefarm.org/kauai-history/.

50. Notable among Uncle Henry's accomplishments was the establishment of the first and only limu management area in the state. See Haw. Rev. Stat. §188-22.8; and Samson Ka'ala Reiny, "The House of Limu: Clinging on to the Past," *Hawai'i Independent*, July 3, 2009, http://hawaiiindependent.net/story/the-house-of-limu-clinging-on-to-the-past.

51. See Kepā Maly and Onaona Maly, *Ka Hana Lawai'a a me nā Ko'a o nā Kai 'Ewalu: A History of Fishing Practices and Marine Fisheries of the Hawaiian Islands* (Hilo, HI: Kumu Pono Associates, 2003), 22.

52. Patrick V. Kirch, *Feathered Gods and Fishhooks* (Honolulu: University of Hawai'i Press, 1985).

53. Reflection from practitioners of the local nonprofit Paepae o He'eia, formed in 2001 to restore and mālama He'eia fishpond on O'ahu.

54. Calculated from an estimate of 336 kg/ha/year. See Barry A. Costa-Pierce, "Aquaculture in Ancient Hawaii," *BioScience* 37, no. 5 (1987): 329.

55. Maly and Maly, *Ka Hana Lawai'a*, 22–23; Barry A. Costa-Pierce, "The *Ahupua'a* Aquaculture Ecosystem in Hawai'i," in *Ecological Aquaculture: The Evolution of the Blue Revolution* (Oxford: Blackwell Science, 2002), 31–32.

56. DHM, Inc., *Hawaiian Fishpond Study: Islands of Hawai'i, Maui, Lana'i and Kaua'i* (Honolulu: DHM, 1990).

57. John N. Cobb, *Commercial Fisheries of the Hawaiian Islands*. Bulletin of the US Fish Commission, 1901, 428–33.

58. "Fishponds," *Ahupua'a, Fishponds, and Lo'i*, directed by Nā Maka o ka 'Āina, produced by Nalani Minton (Honolulu, 1992).

59. The DLNR's jurisdiction encompasses nearly 1.3 million acres of state lands, beaches, and coastal waters, as well as 750 miles of coastline (the fourth longest in the country). Those 1.3 million acres are a large portion of the 1.8 million acres of Hawaiian Kingdom Crown and Government lands ("ceded lands") that were taken during the overthrow and "ceded" to the United States upon annexation; much of these 1.3 million acres were subsequently transferred back to the state upon statehood, to hold in trust for Native Hawaiians and the public. Also "ceded" and transferred to the state in trust were all submerged lands and resources surrounding the Hawaiian Islands, out to three nautical miles. See Melody Kapilialoha MacKenzie, "Historical Background," in *Native*

Hawaiian Law, 23–33; and Melody Kapilialoha MacKenzie and Wayne Tanaka, "Papahānaumokuākea: The Northwestern Hawaiian Islands," in *Native Hawaiian Law*, 712.

60. Trisha Kehaulani Watson and Michael Cain, *Hoʻāla Loko Iʻa: Permit Application Guidebook* (Honolulu: State of Hawaii, DLNR, and ʻĀpuakēhau, 2016), http://www.fpir.noaa.gov/Library/SFD/2016-loko-ia-handbook.pdf. For an overview of this effort from 2012 to 2015, see http://dlnr.hawaii.gov/occl/hoala-loko-ia/.

61. Beyond E Alu Pū there have been ongoing efforts among kalo farmers (such as Onipaʻa na Hui Kalo) and medicinal practitioners (such as Kahuna Lāʻau Lapaʻau ʻO Hawaiʻi). Also, regional-based marine management and fishpond networks are emerging on Hawaiʻi Island, including Kai Kuleana, Hui Loko, and the Maui Nui Makai Network. Other efforts have focused on the development of the state of Hawaiʻi Aha Moku Advisory Committee, established to advise the state DLNR on regional management decisions and practices. Indeed, to varying degrees, islands have established island- and moku-based councils to advise the committee, consistent with the traditional, bottom-up approach.

MELANIE M. LINDHOLM

5
ALASKA NATIVE PERCEPTIONS OF FOOD, HEALTH, AND COMMUNITY WELL-BEING
Challenging Nutritional Colonialism

Alaska Native populations have undergone relatively rapid changes in nearly every aspect of their lives over the past half century. Overall lifestyles have shifted from subsistence-based to wage-based, from traditional to Western, and from self-sustainability to reliance on outside sources. In this chapter, I use data gathered during my qualitative study of Alaska Native foodways; in 2014 I interviewed twenty respondents to investigate the effects of these changes on health and community well-being in Alaska. Tribal identity, community of origin, and age were all self-identified by each respondent. Although the sample size was small, the respondents reported significant diversity in both tribal affiliation and community of origin. I was also fortunate to have multiple generations represented (12 women, 8 men; age range: 25–87; see table 5.1). I used grounded theory method and NVivo qualitative data analysis software to discover patterns and themes in the interviews. Because in the literature there appears to be a lack of concern for and documentation of Native peoples' perceptions of the changes in their food systems and the effects on their communities, my research aims to identify social patterns around changes in the food that individuals and communities eat and possible effects of these changes on all aspects of health. It also documents how Alaska Native individuals and communities are adaptive and resilient; and it seeks to honor, acknowledge, and highlight the personal perspectives and lived experiences of respondents and their views regarding food, health, and community well-being.

TOWARD A BROADER DEFINITION OF NOURISHMENT

From a Western scientific perspective, hunter-gatherer diets are the most ancient and best suited for human physiology. Traditional Alaska Native subsistence foods (especially those that are marine-based) are arguably particularly healthy due to high omega-3 factors. Numerous studies confirm that diets high in omega-3s offer protective factors against cancer, heart disease, diabetes, and even psychological disorders.[1] Traditionally, Alaska Natives rarely experienced such disorders, whereas they now have some of the highest rates in the world. This is one of the consequences of what Western researchers call a "diet transition," but what Alaska Native interviewees describe as cultural genocide.

Nourishment feeds more than just the body—it also feeds the soul. Interviewees in my study state that for Alaska Native peoples, traditional foods represent cultural, spiritual, emotional, social, physical, and mental nourishment. Subsistence diets sustained their peoples and culture for thousands of years in one of the harshest environments on Earth. Their history of Arctic survival and resilience is born out of their connection to the land and sea. Consequently, their cultural identities and well-being are place-based. Alaska Natives respondents believe their health and well-being literally depend on these connections and access to culturally significant food sources. This is why separation from land and subsistence lifestyles can be so detrimental to various aspects of health. Disruption of traditional ways of life and reduced access to subsistence resources can result in mental stress, sedentary lifestyles, and loss of connections to personal identity, cultural history, land, family, and spirit.

Alaska Natives practicing subsistence lifestyles tend to have favorable health indicators on Western health assessments.[2] Still, Western assessments of health and well-being are not always compatible with Indigenous views or understandings because they reflect a very narrow definition of health, whereas traditional Alaska Native perspectives are holistic and collective. Indigenous worldviews often see all things as interconnected; individuals and communities

TABLE 5.1
RESPONDENTS' DEMOGRAPHIC INFORMATION

Age[a]	Gender	Community of Origin	Tribal Identity
69	W	Healy Lake, Tanacross	Athabascan
56	M	Beaver/Ft. Yukon, Utqiaġvik/Barrow	Nuiqsut
65	M	Fairbanks	Iñupiaq
50	M	Venetie	Gwich'in Athabascan
64	M	Utqiaġvik/Barrow, Iviksuk	Inuit, Inupiat
46	W	Bethel	Yup'ik
25	W	Fairbanks	Athabascan
25	W	Chalkyitsik, Fairbanks	Doyon
36	W	Selawik	Iñupiaq
36	W	Eklu	Athabascan
57	W	"Interior village"	Inupiat Eskimo, Koyukon Athabascan
43	W	Tanacross	Mendas Cha-ag, Athabascan
36	W	Kaltag	Koyukon Athabascan
26	W	Eagle Village, Fairbanks	Athabascan
48	M	Teller, Anchorage	Iñupiaq
27	M	"the community"	Athabascan, Navajo
69	W	Healy Lake, Dot Lake	Athabascan
59	M	Shishmaref, Teller	Kaweramiut
66	W	Tanacross	Athabascan
87	M	Old Minto	Athabascan

[a]Average age = 50 years

are considered inseparable from land and animals. Likewise, both individual and community health depends on culturally significant connections to food from the natural environment. Most importantly, loss of land or access to subsistence activities does not reduce the importance of traditional foods. In fact, a decrease in consumption of subsistence foods does not diminish their significance, but actually increases their ideological power.[3] In other words, not having traditional foods in one's diet elevates rather than diminishes their status. Because traditional foods are associated with good emotional, mental, and social health, especially when people share subsistence foods, this affect is often greatly missed when Native foods are limited; respondents report feeling deep happiness when seeing or eating Native foods after an absence of them.

In an Arctic environment, Alaska Natives describe traditional foods as helping them to feel physically stronger and warmer, as well as to enjoy longer-lasting energy. Additionally, as respondents repeatedly note, food is "much more than just getting something to eat." It involves cultural aspects related to identity, or "how to find out who you are."[4] Learning more about oneself is part of participating in subsistence activities. Additionally, traditional foods are viewed more as processes than as objects. For instance, as an Inupiat Eskimo and Koyukon Athabascan woman stated, "I have a whole relationship with food. I like to hunt. I like to fish. I like to process food. I like to preserve it. I like to feed people."[5] Food from the land is also viewed as connected to heritage because it has provided for their ancestors for thousands of years and knowledge of it should be passed to the next generation. Interviewees expressed concern that today's youth are missing the many benefits of culturally significant foods. As an Iñupiaq Elder explained, "If the youth aren't eating traditional foods, what else aren't they learning about who they are?"[6]

Respondents emphasize that control over local availability, accessibility, quality, and cultural appropriateness is imperative to Native well-being. Respondents point to the differences between Western and Indigenous definitions of what is acceptable nourishment. Imported, processed products simply cannot fully meet the needs of Native people. Reasons cited for this claim include the risk of relying on a corporate food system (designed for profits) that inherently lacks culturally appropriate, nutrient-dense, locally controlled options. Interviewees perceive Western foods as having short-term benefits, providing a false sense of well-being, contributing to more illnesses, and lacking spiritual significance. Products of the Western food system may prevent physical starvation, but they are inadequate food substitutes because they are not nourishing in any other sense of the word.

WHEN VALUES MISALIGN

In addition to recognizing that all things are connected and interdependent, respondents told me that health and healing at both the individual and community levels depend on traditional Alaska Native values of balance and harmony. However, only a limited degree of balance is possible in the wake of Western influences. In Alaska, outside forces originating from colonialist value systems have changed almost every facet of Native lifestyles, from education to language to housing, but probably none has been as influential as capitalist values centered on profits. The idea of a wage economy, driven solely by money, runs counter to Native values of reciprocity, sharing, and sustainability. Not only does capitalism negate Indigenous lifeways, but it fosters all kinds of social inequalities. This is incompatible with Native cultures that value cooperation rather than competition.

The dominant hyper-consumerist society is focused on short-term profits without regard for the long-term damage of resource extraction. Respondents say this contradicts Native values of showing respect for animals and nature, as well as preserving the environment for future generations. Considering that Indigenous cultures value and respect the Earth that sustained their ancestors and to which future generations are entitled, and that they consider all areas of health to be interconnected and community-based, they often encounter incongruences with the capitalist ideology driving the current consumerist food system. Degradation of ecosystems and communities reduces local control over the quality and appropriateness of food. The result is an increased dependence on the global food industry and vulnerability to variations in price, availability, and quality. Commercial foods eliminate traditional roles in the food chain that are fundamental to cultural health and instead make access dependent on one's ability to pay.[7] Some Natives would not be able to buy sufficient nutrition if they were forced to depend solely on store-bought foods.[8] Thus, not only is the current capitalist food system incongruent with Indigenous subsistence systems both economically and culturally, but it also creates dependency on an outside system that offers suboptimal nourishment because it is profit-based rather than wellness-based.

Changes in food systems at all stages (from production to consumption) can either undermine or support health. Traditional foodways contribute to people's responsibility in the community, strengthen social networks, and support connections to land essential for well-being. As traditional subsistence areas are cut off and roles change, however, the culture destabilizes and food uncertainty endangers all levels of health. Changes regarding involvement in production, including knowledge about how and where food originates, are of major concern. As an Athabascan woman explained, "We've lost concern about where it [food] comes from and how it gets there. I see our community having become a little lazier with wanting to know the origins or doing any work to find out what's in the food. It used to be you worked pretty hard to get the food on your plate, so you knew what was involved."[9]

Forced to participate in the wage economy, reliant on corporate-controlled food sources, and lured by convenience and marketing, many respondents say they have lost the skills to produce their own food. This is devastating for a culture that historically valued self-sufficiency and sustainability. Arctic Indigenous peoples were intimately engaged with where their food came from, how it was made, who made it, how it was processed, who processed it, how it was cooked, and who cooked it. Respondents expressed concern that current market-based food procurement offers little to no information about the process or connection to the people involved before food arrives at the store. Because traditional food harvest, preparation, and communal consumption are critical cultural components of all Native cultures, the loss of any traditional food creates a sense of loss of a magnitude that is difficult for non-Natives to understand. Interviewees say they prefer traditional diets in part because they view subsistence as about more than just food. In Alaska, some Natives have advocated for a more sustainable, culturally informed, and socially just food system where value is placed on nourishment and shared local resources.

From an Indigenous worldview, capitalism's treadmill of production and assumptions of endless resource extraction are incompatible with nature, harmony, and balance. Traditional values of preserving the land and water for future generations contradict the values of corporate interests in Arctic oil extraction and mining. Alaska Natives see this as a threat, not only to the way of life that sustained their ancestors, but also to future generations that may not have access to land and water subsistence resources essential for cultural preservation and health.

In addition to respect for self, others, and the environment, Native values are detailed and sacred when it comes to respect for animals. As an Inupiat Eskimo and Koyukon Athabascan woman explained, "We harvest in a respectful, spirit-filled way. There's a real spiritual component to going out and harvesting food from the land. It's respect for the animals and the respect for nature, and just respect for yourself. Going out and harvesting—it just fills your spirit."[10] Likewise, an Athabascan woman said, "I think you have a greater connection to your spiritual self or the higher power you have when you are actually doing traditional ways of living. But I don't think there's as much spiritual connection to Western food."[11] The personal connection to and respect for animals is vital here. As a Koyukon Athabascan woman explained, "In our culture there's real connections to animals and how you hunt and how you take care of them when you're doing it" and "the amount of respect you show."[12] Traditional rules, warnings, and practices around how to treat an animal dictate subsistence hunting and fishing. Interviewees believe that the person doing the hunting or fishing has the greatest responsibility for and the most to gain from a respectful encounter. Carefully prescribed actions and words must be completed thoughtfully if the hunter is to be successful, now and in the future. Each species embodies an animal spirit that must be honored and respected for its particular gifts and powers.[13] Animals are sacred because they make life possible in harsh Arctic environments. Harvesting is done quickly to minimize suffering and always with profound respect, thanking the animal for giving its life. Every part of the animal is used; nothing is wasted. Interviewees say wasting is disrespectful, as is overharvesting, which threatens future hunting success. Respectful and therefore successful hunters hold high social status within the community because many rely on their skills for survival.

Now contrast this with corporate capitalist values that routinely exploit animals for meat production on an industrial scale designed to generate profits. In the industrial meat production model, animals are anything but respected. Injected with antibiotics and hormones, crammed into confined spaces, fed genetically modified fodder, and slaughtered in fear and terror, animals arguably experience the most disrespectful treatment possible. Yet Native communities that can no longer hunt are forced to buy meat of unknown ethical origins, potentially running counter to their traditional value of showing respect for animals.

Respondents express often feeling torn between "being successful" in a capitalist culture that values wage employment and a Western education or "being successful" in a subsistence culture that values self-sufficiency and respect. Likewise, traditional values of reciprocity, sharing, and cooperation are often incompatible with colonial ideals favoring competition, inequality, and dependence. One could argue that capitalist values are opposite to Native values in nearly every way possible.

UNDERSTANDING NUTRITIONAL COLONIALISM

Outsiders often perceived Indigenous traditional lifestyles as backward, uncivilized, or wrong. For Natives, forced assimilation, Christianization, and Western education often resulted in confusion, depression, loss of positive identity, struggles to conform to Western standards, and in some cases complete rejection of Native culture in favor of a Western lifestyle and way of thinking.[14] I have coined the term "nutritional colonialism" to describe the values and practices of the dominant food system that affect all areas of Native well-being. In addition to its focus on profits, nutritional colonialism is characterized by negation of subsistence lifestyles, cultural suppression or marginalization, removal of control over resources, lack of food sovereignty, fostered dependence, environmental degradation, increase in chronic diseases, and negation of any dominant sense of responsibility. Products of nutritional colonialism contribute to sedentary lifestyles, require cash, contain suboptimal nourishment, and are culturally insignificant.

Indigenous populations have experienced relatively rapid dietary changes due to industrial influences and corporate monopoly of the global food system. Alaska Native peoples are suffering negative health effects due to the rapid transition from traditional diets toward non-Native foods that are typically processed, with long shelf lives, and require extensive and expensive transportation to remote Arctic regions. Native rural communities often incur high costs for supplementing or replacing subsistence foods with imported store-bought foods, whereas Natives in urban areas such as Fairbanks or Anchorage often incur high costs to take time off work and travel to hunting or fishing grounds. This combination of corporate control over what foods are available, who can afford them, and how they are produced can be termed nutritional colonization because it exploits people's labor, health, environment, and well-being. This system creates an abundance of food, but fosters disease, hunger, and poverty through its mechanisms of production and distribution and consigns those without sufficient income or time to the domain of poor-quality, nonnutritious and unethical food choices.[15] While commercial foods may satisfy hunger in the short term, in the long run the risk of institutionalizing inadequacies and health problems is too costly for already marginalized peoples.[16]

Stability and security of communities are maintained when people can provide themselves with food that is safe and culturally significant, but this situation is increasingly rare as global corporate agribusiness monopolizes the food system at all levels. Respondents expressed concern that the worldwide industrial use of chemical fertilizers, herbicides, and pesticides, along with tightening corporate control, increasingly threatens local production, food safety, food security, health, and ecology. The current industrial food system endorses systematic cruelty to animals, requires unsustainable levels of fossil fuels and water, contributes greatly to global climate change, and creates a foundation for disease—all of which conflict with Indigenous lifeways and values. This system promotes accumulation benefiting a few; prioritizes shareholder profits; destroys local economies; and devastates individuals, families, communities, cultures, and the natural environment. Yet the system intentionally maintains nutritional colonization by, among other means, using the power of legislation to control access to subsistence resources.

An example of nutritional colonization in Alaska is state and federal conservation legislation that restricts Natives' access to their subsistence lands. Hunting seasons and restrictions on the number, species, and place of game hunted have significantly disrupted the Native subsistence economy. Because the sale of subsistence foods is prohibited by Alaska state law, Natives are deprived of income they could earn to purchase Western goods, which effectively exiles them from the consumer system to which they had been encouraged to convert. Without the ability to purchase outside goods, reliance on subsistence food is necessary for survival, but ever-changing regulations mean Natives knowingly or unknowingly break the law to obtain food. Native people have never been consulted, yet they bear the burden of decisions made by people in federal institutions thousands of miles away who know nothing about Native ways of life. Bureaucratic and political struggles between people seeking the legal right to continue their way of life on ancestral lands and those seeking to exploit resources for economic gain have been ongoing for decades. The debate over land claims has created more racial divisions. For every proposal about who should own and control land, some decision makers believed Natives were getting too much that outsiders had rights to develop, whereas others thought Natives were getting too little of what they inherently owned.[17] Lobbyists, lawyers, politicians, corporations, and interested outsiders effectively have been arguing over where, when, and how Natives could live. Despite a millennium of careful stewardship, Natives find their rights and abilities in the modern world are constantly questioned. Donald Craig Mitchell, attorney and nationally recognized expert on federal Indian law, explains that the Alaska Native Claims Settlement Act (ANCSA) of 1971 required Alaska Natives to organize business corporations, effectively forcing them to participate in social values *antithetical to* those embodied in their traditional cultures, which evolved from participation in subsistence lifeways. In corporate culture, humans become shareholders, land becomes an asset, boards of directors have responsibility to corporations instead of to humans, and success becomes measured by profit, whereas Alaska Native culture values distribution rather than retention of wealth.[18] According to Yup'ik anthropologist Oscar Kawagley, ANCSA equated to cultural genocide: it forced Native people to transition from hunter-gatherers to corporate businesspeople in a very short time, and Western development demands created an uneasy tension between profit and preservation.[19] Mitchell argues that Natives' participation in outside systems and institutions has profoundly affected the evolution of their traditional cultures, as has the intrusion of the systems and institutions themselves in village life—and the price Alaska Natives have paid is to be trapped in a cycle of poverty and dependence on white institutions over which they have little control and from which they have no realistic expectation of escape.[20]

Additionally, environmental degradation related to industry, climate change, and contamination threaten traditional food sources. Climate change and industrial pollution are both byproducts of the relatively recent changes in the commercial, corporate, capitalist food system. For example, industrial agriculture produces more greenhouse gases than the entire transportation sector.[21] Because Arctic regions are often the first affected by climate change, predictions of weather and migration of Arctic animals have become less reliable. Unpredictable changes also limit access to fish and game because policies and management cannot respond to environmental change as quickly as the hunter or fisher needs. This means that subsistence-dependent members often must break the law in order to survive.[22] Outside management also means that villagers no longer need to observe or think about the signs and signals of nature because regulations about when and where to hunt or fish are now imposed on them by the Alaska Department of Fish and Game, but these regulations do not always correspond to opportune times and places for success. Thus, laws and regulations enacted to conserve resources often serve to isolate people from nature. Interviewees say this is especially concerning because traditional hunting is tied not only to physical places, but also to the emotions and cultural significance they provide. Limitations on hunting are further exacerbated by increased dependence on costly and inconsistent delivery of industrial foodstuffs from the outside. When needs cannot be met by local food sources, village residents are forced to buy market food patterned in line with what is available to other Americans living near the poverty line, except that Alaska Native communities are additionally limited by unreliable shipments.[23]

On top of these challenges, northern Indigenous peoples are now faced with industrial contaminants in their traditional foods.[24] Respondents perceive traditional foods as less contaminated by those additives found in commercial foods (preservatives, artificial flavors), but as potentially more contaminated by military and industrial pollutants. Interviewees are concerned that game animals and fish may be contaminated, that some animals look unhealthy, and that the land they lived on has been polluted. In many cases, the extent and length of contamination is unknown to the people who most depend on subsistence resources. Despite the potential for contamination, respondents feel they have no choice but to eat traditional foods because they cannot not always afford to buy market foods. Individuals and communities must now weigh the multiple nutritional and socioeconomic benefits of traditional food against the risk of contaminants in culturally important food resources.[25] Despite increased environmental pollution, many Indigenous people continue to eat traditional foods because they believe the nutritional and cultural value outweighs the risk of contamination. Enduring beliefs about the qualities received by consuming traditional foods are central to cultural expression, identity, and well-being, even when contamination threatens that source. Respondents view traditional foods as being affected by political, economic, environmental, and other changes in the world, and argue that traditional foods should be protected because they are too precious to jeopardize.

Although Western assessments of health are narrow in comparison to Native assessments, they offer one demonstration of how nutritional colonialism significantly affects health outcomes in Alaska. Decreases in subsistence activities and traditional food consumption correlate with rises in rates of diseases associated with Western society (such as obesity, diabetes, cancer, heart disease, stroke). Arctic people who adapted to survive in their environment by storing fat are no longer threatened by starvation, so their current lifestyle of overeating and inactivity may combine with the tendency to store fat to contribute to high rates of obesity and diabetes. Before Western influence, Alaska Natives were thought to have very low rates of chronic disease due to their reliance on marine foods (high in omega-3 and selenium) combined with high levels of physical exercise required for subsistence activities. As Western lifestyles and foods have become prevalent, so have Western disease rates accordingly. The most alarming increases are seen in Alaska Native youth; current diabetes prevalence rates may foretell a diabetes epidemic in this population.[26] This prediction poses a major public health challenge for affected communities because young people with diabetes will have more years of disease burden and a higher probability of developing costly and disabling diabetes-related complications relatively early in life. Diabetes prevalence rates are especially significant because diabetes is a risk factor for developing cancer. Alaska Native people have among the highest incidence and mortality rates for all cancers. This is partly due to the geographic isolation of Alaskan rural communities, which makes medical care delivery challenging and expensive. Sophisticated equipment is often unavailable in remote locations, meaning Natives are diagnosed at later stages of disease. The same remoteness and geographic isolation also influence the types of consumer foods available in Native villages: they must be products that can survive lengthy transport. This means less fresh food and more processed, shelf-stable products that correlate with increased disease rates are available. Additionally, Alaska has among the highest rates in the nation for other health risk factors, such as substance use (alcohol, tobacco, illegal drugs) and violence (rape, domestic violence, child sexual assault, homicide, suicide).[27] Most respondents see these rates as symptoms of what they view as cultural genocide.

FINDING BALANCE

The preceding examples show how resources, environments, and health have changed due to nutritional colonization. Among strategies to counteract the negative effects on Native life, some respondents suggest that Native people avoid all industrially produced profit-motivated food products. Others emphasize that food policies need to account for Indigenous people's human right to enjoy their culture and traditional foods, which could help counteract dependency and chronic diseases in their communities. Health interventions should be place-based and emphasize traditional diets, especially marine-based foods, due to their nutritional benefits and cultural significance. Subsistence activities also provide a combination of physical activity and relationship building that contributes to well-being. Identity and sense of place are essential to mental health, meaning that intervention programs must recognize the cultural significance of subsistence activities. In addition to support for subsistence diets, providing culturally appropriate community-based nutrition education will help Alaska Natives make better consumer food choices when subsistence foods are unavailable. Most importantly, too many Alaska Native groups are cut off from their traditional subsistence lands and waters by either physical barriers caused by resource extraction or climate change, or political and legal regulations. It makes little sense to emphasize the importance of subsistence diets or lifestyles when access to them is impossible.

Respondents say local control of resources is essential. However, community food sovereignty threatens corporate profit, power, and control. In the current system, government policies cannot be expected to support food sovereignty. Food corporations have many more lobbyists and attorneys, and much more political influence, than a small community could secure. When considering ethical, health, and environmental concerns, governments typically side with corporations rather than Native communities. Therefore, respondents say that changes at the community level must originate locally. Communities must regain their dignity by refusing to accept what corporations tell them about what they must want, what they must eat, how they must live, and how they must work. Interviewees say that the emphasis should be on local self-reliance and the right to healthy, safe, and culturally appropriate foods that are available, accessible, and shared. Respondents suggest a return to more subsistence-based lifestyles and diets would improve all areas of individual and community health.

Above all, care must be taken to listen to Native people. Dr. Michael Oleksa, recognized as an Elder by the Alaska Federation of Natives, warns that governments and outside professionals may provide services, but the more external, nonreciprocal help that is imported, the more dependent, depressed, confused, and frustrated the Native population becomes:

> The more others try to help, the worse the problems get.... No temporary hired professional can really change the dynamics of the dependence cycle. No one from outside the community can transform it, make it a better, happier, healthier place.... Only its residents and citizens can change the situation, and no one else.... A reawakening, a revitalization of the traditional culture, the Way of the Human Being, lies at the foundation of a new chapter that is beginning to emerge in many regions. Young people are reaffirming their belief in themselves, in their community, in their people, and rejecting the false dichotomies that have created the old either/or dilemma. They are embracing both identities and claim both as legitimately their own. We can be who we are, and we can live successfully in the modern world. We can do both. We *must* do both. That is how we become Real People. We adapt. We change, but we also hold on to all that is good, true, and beautiful in our story, in our way of life, in our culture.[28]

Even while recognizing the significance of subsistence lifestyles, Natives are still faced with demands from dominant Western institutions. They meet the challenge of having one foot in Alaska's subsistence culture and one foot in America's capitalist culture by finding ways to navigate within both. For most interviewees, this balance means not being able to participate fully in either. At best, partial participation in both cultures still necessitates juggling competing value systems. From a social justice perspective, it seems an almost impossible expectation for Natives to successfully balance life in both the wage and subsistence economies, both Western education and Native ways of knowing, both resource extraction and environmental preservation, both dependency and self-sufficiency. And yet, Indigenous peoples are finding ways to do just that. Alaska Natives are drawing upon their history as adaptive, resilient people to survive in a different harsh environment.

Respondents describe seeking a pairing of cultures in order to find balance. For instance, an Athabascan Elder suggested, "Pair up a hunter and an office worker to work together so that they can share in the benefits of the two cultures that we're living: the hunting culture and the workforce culture."[29] Natives perceive themselves to be dependent on both the cash economy and the subsistence economy, such that they "cannot do without either one." Some are fortunate enough to find wage employment that still allows for time off to hunt, fish, and connect with nature. Interviewees report that when traditional foods are not available, they are making efforts to buy healthier market options. Parents and grandparents are still teaching children traditional lifeways, including sending them to culture camp, even though these compete with the values instilled in their Western educations. A Nuiqsut Elder suggests that people need "role models: people that go out and hunt and fish and trap and show the younger people how to do subsistence activities . . . how to harvest wild food."[30] Additionally, interviewees advocate for more cultural awareness, both individually and collectively. Respondents describe their efforts to be more mindful, to choose healthier options, to exercise, and to learn about and participate in more subsistence activities. Respondents intentionally challenge dominant institutions by sharing subsistence foods with friends and family, and at traditional potlach events. The majority of respondents mention that sharing and trading of traditional foods is a key trait of subsistence activities, that many rely on this network, and that "sharing Native foods is instantly a good feeling."[31] Not only is this sharing and trading system recognized as an important component of community well-being, but it is a role expectation. As a Kaweramiut Elder explained, "The importance of being a hunter-gatherer is sharing what you have . . . a community that shares the wealth of being a hunter-gatherer society . . . brings a community together and makes them stronger."[32]

Respondents consistently advocate for greater cultural education and awareness; more participation in tribal decision-making; increased mindfulness when choosing consumer foods; greater participation in subsistence activities; and continued sharing of traditional foods, expressions of gratitude, care for Elders, and signs of respect (for self, animals, and environment)—all essentially

Wild berries gathered in Alaska, 2015. *Photo by Diane McEachern*

challenging values of the dominant society. By negotiating such balancing acts, Alaska Natives are resisting nutritional colonialism.

CONCLUSION

Many Alaska Native peoples have experienced rapid sociocultural changes that have driven a transition away from subsistence diets. Traditional foods have been replaced by Western foodstuffs at speeds that physiological adaptation cannot match. Given the speed of this nutritional transition, adequate adaptation has been impossible. Human bodies cannot physiologically adjust to drastic diet and lifestyle changes in just a few generations. These changes correlate with increased incidences of Western illnesses (diabetes, heart disease, stroke, cancer, obesity) thought to be almost nonexistent in Native populations prior to the introduction of Western foods and lifestyles. Because Indigenous foods have cultural significance, nutritional transitions have also damaged mental and social health. The process of this transition was primarily driven by nutritional colonialism.

Nutritional colonialism is characterized by (1) negation of subsistence lifestyles; (2) cultural suppression and marginalization; (3) denial of control over price, availability, accessibility, quality, and appropriateness of food choices; (4) lack of food sovereignty; (5) fostering of dependency and sedentary lifestyles; (6) the necessity for cash; (7) profit-focused/profit-based food provisioning systems; (8) negation of any dominant sense of responsibility; (9) environmental damage; and (10) increased rates of chronic diseases. These elements of nutritional colonialism are largely incompatible with Alaska Native cultural values of reciprocity, sharing, cooperation, self-sufficiency, respect, gratitude, and sustainability. The hope lies in Natives' resistance to nutritional colonialism. For instance, respondents continue sharing traditional foods regardless of current policies that restrict subsistence food acquisition.

Alaska Native respondents perceive traditional foods as healthy; culturally, emotionally, socially and spiritually significant; shared; respected; and valued. In comparison, market foods are perceived as unhealthy; tied to money and urban influences; lesser in value; and associated with junk food, fast food, grocery stores, chronic diseases, and a diminishment of knowledge about their origins. Respondents' perceptions of Western influences revolve around cultural changes, increased dependency, Christianization, and money. Respondents voiced concern about determining what was most important and what improvements can be made; most of these comments center on regulations, community involvement, education, and industrial practices.

A component of challenging nutritional colonialism is refusing to accept what corporations tell people they must want and eat, how they must live, and how they must work. Respondents prioritize local self-reliance and the right to healthy, safe, and culturally appropriate foods that are available, accessible, and shared. Reducing outside reliance on the global food system is perceived to be essential for Arctic community health. Because cultural health is considered a byproduct of physical, mental, and social health, a return to more subsistence-based lifestyles and diets is perceived as essential for all areas of individual and community health.

NOTES

1. Amanda I. Adler, Edward J. Boyko, Cynthia D. Schraer, Neil J. Murphy, "Lower Prevalence of Impaired Glucose Tolerance and Diabetes Associated with Daily Seal Oil or Salmon Consumption among Alaska Natives," *Diabetes Care* 17, no. 12 (1994): 1498–1501; Andrea Bersamin, Bret R. Luick, I. B. King, Judith S. Stern, and Sheri Zidenberg-Cherr, "Westernizing Diets Influence Fat Intake, Red Blood Cell Fatty Acid

Composition, and Health in Remote Alaskan Native Communities in the Center for Alaska Native Health Study," *Journal of the Ameican Dietetic Association* 108, no. 2 (Feb 2008): 266–73; Michael Davidson, Lisa R. Bulkow, and Bruce G. Gellin, "Cardiac Mortality in Alaska's Indigenous and Non-Native Residents," *International Journal of Epidemiology* 22, no. 1 (Feb 1993): 62–71; Joseph R. Hibbeln, "Fish Consumption and Major Depression," *Lancet* 351, no. 9110 (April 18 1998): 1213; Geneviève Pilon, Jérôme Ruzzin, Lauri-Eve Rioux, Charles Lavigne, Phillip James White, Livar Frøyland, and André Marette, "Differential Effects of Various Fish Proteins in Altering Body Weight, Adiposity, Inflammatory Status, and Insulin Sensitivity in High-Fat–Fed Rats," *Metabolism* 60, no. 8 (2011): 1122–30; Alex Richardson, "The Importance of Omega-3 Fatty Acids for Behaviour, Cognition, and Mood," *Food and Nutrition Research* 47, no. 2 (2003): 92–98; Jaakko A. Tanskanen, Joseph R. Hibbeln, Jukka Hintikka, Kaisa Haatainen, Kirsi Honkalampi, and Heimo Viinamaki, "Fish Consumption, Depression, and Suicidality in a General Population," *Archives of General Psychiatry* 58, no. 5 (2011): 512–13.

2. Andrea Bersamin, Bret R. Luick, Christopher Wolsko, Bert B. Boyer, Cecile Lardon, S. E. Hopkins, Judith S. Stern, and Sheri Zidenberg-Cherr, "Enculturation, Perceived Stress, and Physical Activity: Implications for Metabolic Risk among the Yup'ik—The Center for Alaska Native Health Research Study," *Ethnicity and Health* 19, no. 3 (2014): 255–69.

3. Kirk Dombrowski, "Subsistence Livelihood, Native Identity, and Internal Differentiation in Southeast Alaska," *Anthropologica* 49, no. 2 (2007): 211–29.

4. Melanie Lindholm, "Alaska Native Perceptions of Food, Health, and Community Well-Being: Challenging Nutritional Colonialism" (MA thesis, University of Alaska, Fairbanks, 2014), 71.

5. Lindholm, "Alaska Native Perceptions," 70. "Elder" is capitalized as a sign of respect, following the preferred style of the Northwest Arctic Native Association, http://nana-dev.com/about/the_nana_logo/writing_style_guide/

6. Lindholm, "Alaska Native Perceptions," 71.

7. Philip A. Loring and Craig S. Gerlach, "Food, Culture, and Human Health in Alaska: An Integrative Health Approach to Food Security," *Environmental Science and Policy* 12, no. 4 (2009): 466–78.

8. Kristen Borre, "Inuit Blood and Diet: A Biocultural Model of Physiology and Cultural Identity," *Medical Anthropology Quarterly* 5, no. 1 (1991): 48–62.

9. Lindholm, "Alaska Native Perceptions," 79.

10. Lindholm, "Alaska Native Perceptions," 74.

11. Lindholm, "Alaska Native Perceptions," 74.

12. Lindholm, "Alaska Native Perceptions," 75.

13. Richard K. Nelson, *Make Prayers to the Raven* (Chicago: University of Chicago Press, 1986).

14. Harold Napoleon, *Yuuyaraq: The Way of the Human Being* (Fairbanks: Alaska Native Knowledge Network, 2005).

15. Raj Patel, *Stuffed and Starved: The Hidden Battle for the World Food System* (Brooklyn, NY: Melville House, 2012), 319–20.

16. Loring and Gerlach, "Food, Culture, and Human Health in Alaska," 474.

17. Donald Craig Mitchell, *Sold American: The Story of Alaska Natives and Their Land, 1876–1959* (Fairbanks: University of Alaska Press, 2003), 399.

18. Mitchell, *Sold American*, 13.

19. Oscar Kawagley, *A Yupiaq Worldview: A Pathway to Ecology and Spirit* (Long Grove, IL: Waveland Press, 2006), 25.

20. Mitchell, *Sold American*, 8.

21. Food and Agriculture Organization of the United Nations, "Livestock's Long Shadow: Environmental Issues and Options" (New York: FAO, 2006), xxi.

22. Dombrowski, "Subsistence Livelihood," 221.

23. Loring and Gerlach, Food, Culture, and Human Health in Alaska," 470.

24. Marla Cone, *Silent Snow: The Slow Poisoning of the Arctic* (New York: Grove Press, 2005), 4, 23.

25. Harriet V. Kuhnlein and Laurie Hing Man Chan, "Environment and Contaminants in Traditional Food Systems of Northern Indigenous Peoples," *Annual Review of Nutrition* 20 (2000): 595–626.

26. Kelly J. Acton, Nilka Ríos Burrows, Kelly Moore, Linda Querec, Linda S. Geiss, and Michael M. Engelgau, "Trends in Diabetes Prevalence among American Indian and Alaska Native Children, Adolescents, and Young Adults," *American Journal of Public Health* 92, no. 9 (2002): 1485–90.

27. Alaska Department of Health and Social Services, Division of Public Health, "Healthy Alaskans 2010: Targets and Strategies for Improved Health" (Anchorage: Alaska DHSS, 2002).

28. Michael Oleska, *Another Culture/Another World* (Juneau: Association of Alaska School Boards, 2005), 144–45.

29. Lindholm, "Alaska Native Perceptions," 105.

30. Lindholm, "Alaska Native Perceptions," 107.

31. Lindholm, "Alaska Native Perceptions," 72.

32. Lindholm, "Alaska Native Perceptions," 76.

DENISA LIVINGSTON

6

HEALTHY DINÉ NATION INITIATIVES
Empowering Our Communities

Ch'iyáán nihits'íís bá yá'át'ééhi éí iiná at'é (Healthy food is life).

On April 1, 2015, the local news reported, "It is no April fool's joke, the 'junk food' tax is in effect as of today on the Navajo Nation." What had seemed an unrelenting political journey for us had become one worth traveling. But how did we get to this point? Why did we feel the urgency to impose a new tax? Why did we purposely pursue this unpopular and unfamiliar strategy? We share our story in the following pages with gratitude, rumination, and candor.

We have been called soda tax warriors, food rangers, and the FBI (for "fry bread investigators"), and the list continues to grow. These appellations could describe our ambition and determination, but they have also captured the support of our communities—though not instantaneously.

Our Navajo Nation was hijacked by the processed food industry, creating food swamps and food apartheid (we avoid the term "food desert" because a desert ecosystem is full of life). For the first time in our history, our people are dying not from starvation but from "#diabesity," the term we have coined for diabetes and obesity. Also for the first time we are experiencing the negative consequences of the culture of food swamps. One out of three Diné here is prediabetic or diabetic, and this statistic continues to grow. Every Navajo family is affected. This is our reality.

In 2011, in response to the suffering of our people, Diné community members aimed to bring awareness to our diabetes epidemic with an emphasis on the incessant inventories of unhealthy, processed foods widely available on the Navajo Nation. As a result of these meetings, the Diné Community Advocacy Alliance (DCAA) was formed in March 2012. The alliance members agreed to goals of addressing the high rates of obesity, diabetes, and complications of chronic health issues that affect all generations of Diné in our communities, from children to elders. Through DCAA, we as volunteer Diné grassroots-level community health advocates mobilized to raise awareness and provide information and education regarding these pressing concerns in our communities. Fundamentally, our group promoted the slogan *Shánah daniidlįįgo as'ah neildeehdoo* (Let's live a long life).

Our mission has been as challenging to implement as this slogan is to pronounce. As community members, we pursued legislative efforts in an endeavor to create a Healthy Diné Nation. In the summer of 2013, with our legislative sponsor, then Navajo Nation Council Delegate Danny Simpson, we introduced a bill to the twenty-second Navajo Nation Council. This controversial bill, known as the Junk Food Tax, would impose a 5 percent tax on "junk food," with the percentage increasing every year, and a corresponding elimination of 5 percent of the sales tax on healthy food on our Navajo Nation. Our proposed bill was reviewed in several legislative committees: Law and Order; Resources and Development; Budget and Finance; Health, Education, Human Services; *Naabik'íyáti'* (cordial discussion); followed by the full Navajo Nation Council. It failed. In general, some council members supported both initiatives; others supported one or the other, but not, as we had hoped, both.

This first attempt was met with overwhelming negative responses and resistance to addressing "junk food." The tribal council expressed several reasons, but the principal ones were (1) the tax was regressive; (2) we would be taxing our elders; (3) we were stepping onto sensitive ground of addiction; and (4) this proposal would tax a tribal member's breakfast, lunch, *and* dinner. Regarding the community, whenever we communicated using the words "junk food," we created an uncomfortable space of rejection. We learned these two words instigated negative implications and reactions. As we strategized to move forward, we redesigned the platform to a more positive message, changing the Junk Food Tax to the Healthy Diné Nation Act.

We searched for a scientific definition of "junk food" to include in the legislative language, but we discovered no formal definition existed. Further, there were no suitable guidelines to formulate a listing of which foods would qualify as "junk food." Therefore, we removed the two words "junk food" from the entire bill with the understanding that we would need to formulate a definition and list to be considered in the bill.

In our efforts, local health promotion specialists reminded us constantly that there is "no such thing as junk food, there is junk and there is food," which is the reason I use quotation marks around the term throughout this chapter. The message was even clearer: these two words never existed in our history until recently; it did not exist in our Diné language; and internet searches would return only pictures of soda, chips, and candy, which were not necessarily the biggest culprits on our nation. There were many memes with messages of health but nothing that conveyed these messages in our language. We needed to reach our tribal citizens, especially our elders who paved the way of survival, who are our knowledge bearers of traditional healthy foods, and who would be our advocates. How could we reach them and at the same time reach our youth? We concluded that our language had to unite us.

We elevated the conversations around the experiences of Hwéeldi, referring to the Long Walk and meaning "suffering" or "place of suffering" in Diné. In the early 1860s our Navajo people were forced to walk hundreds of miles to a US government–run concentration camp at Bosque Redondo, New Mexico. There, the US Army distributed rations of foods that were unfamiliar to our people, including salt, sugar, and flour. It became apparent that part of our frustration arose from the fact that to this day we were still trying to address the legacy of prison food, and we questioned what this would be called in our language. We needed to bridge our history with our current efforts.

Many discussions and inquiries about how to label "junk food" in our language resulted in formulating *ch'iyáán bizhool*, *ch'iyáán* meaning "food" and *bizhool* meaning "leftovers" or "scraps," the oppressive cheap food, and the nonnutritious, unhealthy articles. We used this term to create more fruitful and positive discussions around the Diné diet. Our community members were more likely to engage in a conversation about a Healthy Diné Nation and our future than about "junk food."

We continued our assessment and reexamined the language in the bill and food definitions. Health professionals and dietitians from across the country guided us in creating definitions and descriptions; together, we examined the options and settled on "minimal-to-no nutritional value food items" to replace "junk food." In addition to changing the language in the legislation, the new description encouraged us also to change the terminology we used when speaking with community members to "unhealthy foods." In the first bill, we had defined "junk food" based on milligrams of sodium, sugar, and fat, but we removed these criteria as well. We were aware that manufacturers are continuously reformulating food items in order to label them as "healthy" rather than "junk food" in subliminal marketing, so we voted to use the broad definition of "minimal-to-no nutritional value food items" that are "high in salt, saturated fat, and sugar," in order to encompass any future modified and newly engineered unhealthy food items.

Throughout this time, our advocates collected resolutions in support of the initiative from more than half of the 110 Navajo Chapters, as well as letters of support from local community groups, statewide organizations, and educational institutions across the country. With youth advocates, we also engaged in one-on-one communication with tribal members; occupied media outlets from the radio to newspapers and newsletters; increased our social media presence on Facebook, Twitter, and blogs; sponsored local events like our Healthy Diné Nation Zumba Celebrations; gave presentations; staffed booths; and offered healthy food demonstrations and grocery store tours.

On so many occasions that our leaders came to hold this expectation of our group, we provided healthy meals, a variety of fruit-infused waters, and healthy snacks for the council members and staff. We were also the first advocacy group to conduct Zumba dance sessions in front of the Navajo Nation Council Chamber. At first, delegates did not accept this very well, commenting, for example, that "club music" was blasting outside the legislative sessions, but after many months, several of our elected tribal leaders began to join our exercise sessions, dancing beside us with laughter and smiles.

Regarding the elimination of sales tax on healthy foods, on the Navajo Nation all foods and beverages—including fruit, vegetables, and water—were taxed at a rate of 5 percent, whereas surrounding towns located outside the Navajo Nation sold healthy food and water tax-free. We demanded the same for our nation. We argued that we were already living with limited access to healthy food and in some places no access to healthy, fresh foods or clean water. We placed emphasis on our healthful traditional Indigenous foods, including ethnic foods, being included in this bill. We were hungry for change and insisted that the foods which empowered and strengthen us must be tax-free. This was an overdue policy change and a complementary sister piece of legislation addressing ch'iyáán bizhool.

Thus, shortly after the summer council session of 2013, we split the original bill into two revised and improved bills. For the fall session, we reintroduced the Tax-Free Healthy Food Bill and the Junk Food Tax (renamed the Healthy Diné Nation Act), which passed with an acceptable 2 percent added friendly awareness tax on unhealthy food instead of the originally proposed 5 percent. Essentially, then, there would be no tax on healthy foods and a total of 7 percent sales tax on "junk food."

Throughout our committee presentations, we reminded our honorable council delegates of our Fundamental Law of Diné, which states that our leaders "shall enact policies and laws to address" our needs and that "every child and elder [shall] be respected, honored, and protected with healthy physical and mental environment." We continued to remind them of their responsibility and "duty to respect, preserve, and protect all" because they are "designated as stewards" for our people.[1] We created uplifting messages of gratitude and inspiration, and shared them to make the point that the council members held power to create healthy change.

Finally, with increased support for addressing our culture of ch'iyáán bizhool, the improved bills were enacted in January 2014. The following month then Navajo Nation President Ben Shelly vetoed the bills. In April, under the sponsorship of Navajo Nation Council Delegate Jonathan Hale, we introduced another bill to override the presidential vetoes. The Elimination of Sales Tax on Fresh Fruits, Fresh Vegetables, Water, Nuts, Seeds, and Nut Butters passed, but the Healthy Diné Nation Act failed.

At this point, we received much criticism, this time from well-known non-Indigenous public health professionals who argued an additional 2 percent tax would have minimal effect. We were aware of studies indicating that a 15 to 25 percent tax would be needed to influence behavior, but our Navajo Nation Council did not support such a steep tax since the ceiling across our nation was generally 7 percent, except in the Navajo townships of Kayenta and Tuba City, which imposed township taxes in addition to the Navajo Nation taxes. So, the next essential information we needed was an estimation of the revenue that would be generated from the sales tax.

The Office of the Navajo Tax Commission willingly provided the information we sought. At an 80 percent inventory of "junk food," the estimated revenue would be $2.7 million, but we knew the inventory and sales were higher, closer to 90 percent or $3.1 million, and with many stores at 99 percent "junk food" inventory, we estimated about $4 to $5 million a year or even more.

Despite the lack of confidence that the 2 percent unhealthy foods tax would be effective, we believed in our cause. In the summer of 2014, we reintroduced the Healthy Diné Nation Act of 2014. It passed that fall, and after much consultation with President Shelly, he signed the act into law on November 21, 2014.

In the midst of this third attempt to pass the act, the elimination of the 5 percent Navajo Nation sales tax on healthy foods became effective on October 1, 2014, affecting six categories of foods: fresh fruits, fresh vegetables, nuts, nut butters, seeds, and water. The law highlighted traditional Diné foods and special ethnic foods like sumac berries, yucca, juniper, blue corn, yellow corn, white corn, frozen or dry hominy, posole, dried beans, and wild rice.

The next date we looked forward to was April 1, 2015, when the radio would echo, "It is no April fool's joke, the junk food tax is in effect as of today on the Navajo Nation," an announcement that marked the effective date of the first unhealthy foods tax in the United States. Although it seemed like we had already traveled a long road, we were only getting started.

The Healthy Diné Nation Act of 2014, or in brief for tax reporting, the Unhealthy Foods 2% Sales Tax, compounded the 5 percent Navajo Nation sales tax, and township taxes where applicable, on the following five categories of products:

1. Beverages: any artificially sweetened, naturally sweetened, or sugar-sweetened drinks, including powders, gels, drops, sparkling drinks, both alcoholic and nonalcoholic, excluding unsweetened hot tea, unsweetened hot coffee, unflavored milk, and unsweetened, unflavored water
2. Sweets: candy, frozen desserts, pastries, pudding and gelatin-based desserts, and fried or baked goods
3. Chips and crisps: crispy snack foods that are fried, baked, or toasted, such as potato chips, tortilla chips, pita chips, or cheese puffs
4. Fast food: ready-to-eat, quickly available, quickly served foods, including any canned, precooked, or potted meats
5. Flavor enhancers: salt, sugar, and sweeteners.

The lists of healthy, whole, fresh foods and processed, unhealthy food items for both laws were detailed and defined. These laws applied to all retail food establishments on the Navajo Nation, including convenience stores, grocery stores, restaurants, fast-food eateries, casinos, and even pawnshops or any other places where food items were retailed.

Simultaneously, during the month of April 2015, we assisted the Navajo Nation Division of Community Development with the next policy step, the Community Wellness Development Projects Fund Management Plan, which also passed in April, under the sponsorship of former Navajo Nation Council Delegate and Vice President Jonathan Nez. This plan created a special revenue account to ensure 100 percent of the unhealthy foods tax would be directed to fund Community Wellness Projects in all 110 Navajo Chapters.

This plan sought to empower our community members to champion projects that would improve their social and physical environments and to implement projects that were needed in their area. The fund management plan included a list of allowable projects, excluding meetings, described as, "Any community-based wellness projects that are planned, implemented, directed, and reported by members of the Navajo Nation communities," in the following areas:

A. Instruction
    i. Fitness classes (for example, Zumba, aerobics, core training, indoor cycling)
    ii. Traditional, intergenerational, and contemporary wellness workshops (for example, Navajo philosophical and educational teachings, tai chi, yoga)
    iii. Health coaching (for example, healthy eating education, goal-setting, self-care)
    iv. Navajo traditional craft classes (for example, jewelry making, beading, weaving)
    v. Traditional and nontraditional healthy food preparation workshops (for example, making *chiłchin* [red sumac berry pudding] or blue corn mush, cleaning and preparing corn, preparing piñons or Navajo tea)
    vi. Workshops on processing healthy foods (for example, canning, food safety)

B. Equipment
    i. Wellness and exercise equipment
    ii. Supplies
    iii. Storage facilities
    iv. Maintenance, conservation, or improvement of any of these projects

C. Built Recreational Environment
   i. Walking trails, running trails, biking trails
   ii. Skate parks, community parks
   iii. Picnic grounds
   iv. Playgrounds
   v. Basketball and volleyball courts
   vi. Baseball and softball fields
   vii. Swimming pools
   viii. Maintenance, conservation, or improvement of any of these projects
D. Social Setting
   i. Recreational, health, youth clubs (for example, senior citizens' events, walking clubs)
   ii. Equine therapy (for example, activities and interaction with horses, trail rides, introduction to horses, saddling, training)
   iii. Maintenance, conservation, or improvement of any of these projects
E. Education
   i. Health education materials
   ii. Presentations
   iii. Libraries
F. Community Food and Water Initiatives
   i. Healthy food initiatives
   ii. Community food cooperatives
   iii. Farming and vegetable gardens
   iv. Greenhouses
   v. Farmers' markets
   vi. Clean water initiatives
   vii. Clean communities initiatives (for example, community trash pick-up day)
   viii. Recycling initiatives
   ix. Improvements to retail stores (for example, posting signs for taxable, unhealthy foods and tax-exempt, healthy foods)
   x. Agricultural projects (for example, 4-H activities)
   xi. Maintenance, conservation, or improvement of any of these projects

G. Health Emergency Preparedness
   i. First-aid, CPR, AED certification, and so on
H. Matching Funds: Any matching funds projects funded by federal, state, county, or public entities that have not been addressed by the Navajo Nation or other tribal budgets.

For the next thirteen months after this victory, we advocated for a distribution policy to be finalized so that our 110 Navajo Chapters could receive funds to pursue their Community Wellness Projects. More than half of our communities had wellness plans ready and were waiting to take advantage of this opportunity. We pressured the Division of Community Development once more, and with Mr. Jonathan Hale as our sponsor for the second time, we proposed the Navajo Nation Chapter Project Guideline and Distribution Policy. Finally, in June 2016, it passed and went into effect immediately.

For the next year, we strongly advocated for reporting of official revenue numbers from the Unhealthy Foods Tax. The first revenue report for 2015 was $334,038, for the first quarter of 2016, it was $473,850, and the figures thereafter were in similar ranges. Notwithstanding the minimal confidence from health professionals, as of the last reporting quarter of 2017, we had raised a total of about $4.1 million from the Unhealthy Foods Tax, despite partial compliance with the law.

It has been several years since the taxes went to effect, and we have advocated for further support for a successful implementation in partnership with our government. Since our efforts were focused on taxation, in earlier years, tribal leaders who opposed us instructed the programs we needed most for technical support to disengage from our efforts. We lost support in many areas, programs withdrew, and we were abandoned to complete the work alone. Now, we are trying to reengage programs and inspire our people who work for these programs to contribute to the work that will create healthier communities. Although we have faced silos, uninterested departments and programs, and tribal employees who oppose the initiatives, we continue to advocate for and set an example of working synergistically *with* our tribal government to fully implement these laws.

Now we are addressing the absence of tax regulations, and as of October 2017, together with then–Navajo Nation president Russell Begaye and Danny Simpson, we signed a memorandum of understanding (MOU) between the Navajo Nation Executive Branch, the Office of the Navajo Tax Commission, and DCAA. This MOU created a formal partnership for us, as community members, to provide assistance to the tax commission regarding implementation and enforcement of, and compliance with, the healthy and unhealthy food laws. Despite the MOU, we are still facing opposition and challenges from our tribal government against fully implementing the laws and addressing our concerns of improving healthy food access.

Many times, we have expressed to our tribal leaders and tribal employees that we are constituents championing a healthy cause together with them, but this message has not been accepted very well. We felt a lack of trust and were not welcome in the meetings addressing the laws and policies with tribal departments. We were viewed as volunteers who did not represent a formal program or department and did not have formal titles. Many times, our tribal elected leaders made decisions without participation or voices from our people. It was critical for us to be included in every aspect of our tribal government. We created a movement to enable our people to be a part of the legislative processes, implementation, and enforcement. Therefore, we consider the MOU a significant success to help our leaders understand that we community members need to be seen as assets rather than liabilities, and that we have the passion, expertise, and experience to create healthy change, not only in our communities, but also in partnership with our tribal government.

On the other hand, the biggest obstacle and threat to our work continues to be elected tribal council members who oppose our healthy initiatives, and Diné hired as lobbyists by multi-billion-dollar processed food industries to fight our work. This has caused division in our tribal government and tribal leadership. To this day, some tribal leaders still oppose the Healthy Diné Nation Act, still arguing that it is a regressive tax, even as we are seeing progressive benefits.

Although we have encountered strong opposition, we have also built bridges, opened dialogues, and gained support in ways that seemed very unlikely at the beginning. As regards our community members who are not eager to see healthy change, their behavior and choices do not influence our efforts or resolve. Our dedication has stemmed from the overwhelming response from our people who are suffering from food-related diseases and at the same time from those who want to improve their eating habits and lifestyles. We understand it will take generations to change unhealthy eating habits. We recognize the addiction and unhealthy food culture on our nation but this situation does not only exist here—it is worldwide. It may take years to reverse the addiction to unhealthy food, but for now, we are doing what we can as volunteers and as informed tribal citizens to ensure every individual will have access to healthy food, *this is what matters.* We have demonstrated our positivity and optimism in every area of our work and can only continue to be encouraging to those who are discouraged and, of course, to those who do not want to be encouraged. We do not plan to change our approach.

For supporters of our movement, it is essential to continue to advocate for health and wellness to be a top priority on the agendas of our tribal councils throughout Indian Country. For us, it took more than three years to educate the twenty-second Navajo Nation Council about our health, food apartheid and lack of access to healthy food on our nation, epidemics, and unhealthy food inventories. It should not take years to educate and persuade our leaders about the importance of health and well-being when the suffering is evident and overwhelming.

Regarding taxation, this is nothing new to our tribe. Places like Kayenta, Arizona, on our Navajo Nation have utilized their township tax to fund a community park, playground, climbing wall, and skate park. We have been told stories about how historically our elders would tax themselves in times of hardship by limiting their offspring, protecting their families, and making sacrifices necessary to survive. History is repeating itself in the form of taxation to save our tribe—to ultimately allow future generations to live healthier lives. The tax has served as a platform to address suffering, disease, sickness, and unhealthy lifestyles all related to food, and ultimately has been a vehicle to launch health revolutions in our communities.

Moving forward to a healthy food oasis will take many partners and stakeholders. We believe the greatest success from these initiatives is the increased awareness of the expanse of overwhelmingly available processed foods, compared to the little or no availability of fresh, affordable healthy produce. This problem has persisted due to lack of policies or regulations on "junk food" being sold in all areas on the Navajo Nation.

As we move forward in our work, we face more challenges. Although the federal government has funded some tribal health education programs, this approach is only one facet of the solution. These programs also are not uniformly created and pursued with solutions generated by our people from a Diné perspective at the local level. There are clear disconnects between our traditional ways of life, including our traditional food knowledge and practices, and the models pursued by these programs that operate from a non-Diné or non-Indigenous perspective.

We have learned through these legislative processes that health education in our Diné schools is inadequate. Our youth are learning how to live healthy lifestyles, but when they leave school their social and physical environments do not support them in applying these lessons. When they go to the local convenience store or grocery store, they encounter a vast selection of ch'iyáán bizhool. They return home without fresh produce to apply their lessons with their families, and fresh produce is not often available in their homes, as it should be.

At the beginning of our initiatives, we lacked Diné-specific data to support the bills. Fortunately over the years, several Community Health Needs Assessments have been completed, but we still need more data about the health challenges of our people in order to better meet our needs.

Beyond the wellness projects, we now have more opportunities to create dialogues and increase interest in healthy entrepreneurship opportunities—healthy grocery stores, healthy convenience stores, and produce markets that provide more access to fresh and healthful foods. We need continuous support from our tribal leaders and tribal citizens to make a stand to protect our food channels, promote food sovereignty, and pursue other food-related policies.

Through our legislative efforts, we realized that although community advocacy is a relevant focus of public health emphasized throughout higher education, actualizing these goals is a struggle. We need support from public health organizations and schools to engage in Indigenous public health systems, policy development, and program planning, with an emphasis on a model centered around community advocacy, tribal sovereignty, tribal economic development, traditional food sovereignty, sustainable community wellness projects, community empowerment, youth leadership opportunities, and social innovation.

We have learned it takes a community to raise community advocates. We must consider raising young leaders who reflect our matrilineal and matriarchal society. We do not see that tradition reflected in our own Navajo Nation government the way it should be. It is critical for leadership roles to be encouraged at all ages and in all areas.

With the support and guidance of the twenty-second and twenty-third Navajo Nation Councils, former Navajo Nation Presidents Ben Shelly and Russell Begaye, and Vice Presidents Rex Lee Jim and Jonathan Nez, we have created a unique partnership to address the needs of our people while creating opportunities to improve the health of all community members, tribal and nontribal, on the Navajo Nation. Our leaders are listening. They are eating more healthfully and leaders like Danny Simpson are confidently embracing and promoting health leadership. Together, we are setting an example for other tribal leaders and groups around the world that are considering similar policies. The actions of our Diné elected leaders are models to be followed.

Currently, across our Navajo Nation, as our Navajo Chapters initiate and implement Community Wellness Projects, the measures of long-term success will be the sustainability of those projects that flourished from our Unhealthy Foods Tax; the increase in availability of fresh, affordable, healthy foods in all local retail areas; and the healthy, productive lives our Diné will be living. We, the Diné community members, from children to elders, have guided our legislative efforts and successes together from the outset. We have addressed and are continuing to address suffering and we continue to pursue solutions with what our elders taught us, with their stories of survival, and with the visions they gifted us, so that one day, we will be a Healthy Diné Nation.

We all have a civil responsibility to create healthy change. It is time to rebuild #IndigenousPower and regain our identity through healthful food. We need more taste education and culinary experiences to foster a healthy culture with an educated public. The time is ripe to create intergenerational and intertribal synergy and collaboration. We need to continue to connect with our elders and capture their healthy food knowledge to share with the younger generations.

We are the vessels to carry the work forward so that one day we can see our children grow up to be healthy and empowered leaders and change makers. We will continue to be difference makers and grow leaders simultaneously. Why? It is time for Indian Country to practice its sovereignty, heal, and prosper. We are an example for the Indigenous people of the world who are watching us in this movement so that they can create their own laws and policies and reap the health that is overdue to them. We seek to bring more value to our work, to create solutions activated and driven by community members. It is our time to impact the world with visionary change—we cannot wait for others to inspire us, we have already been empowered by our ancestors. We can be dealers of hope and operate from the posture of our hearts with ancestral strength. We can create new environments of change, growth, stability, and sustainability. Our future is in our hands. We need to embrace this stewardship and be a present to the future by being a good ancestor. Together, we can make a history on which a healthy future can be built. *Ahé'hee'* (thank you).

NOTES

1. To read the full text of the Fundamental Law of Diné go to http://www.navajocourts.org/dine.htm.

ROWEN WHITE

7
PLANTING SACRED SEEDS IN A MODERN WORLD
Restoring Indigenous Seed Sovereignty

Across Turtle Island (North America), there is a growing intergenerational movement of Indigenous people proud to carry the message of the grand rematriation of seeds and foods back into our Indigenous communities. Some seeds have been missing from our communities for centuries—carried on long journeys in smoky buckskin pouches or upon the necks of peoples who were forced to relocate from the land of their birth, their ancestral grounds. Generations later, these seeds are now coming back home; from the vaults of public institutions, seed banks, universities, seed keepers' collections, or sometimes dusty pantry shelves of foresighted elders—seeds patiently sleeping and dreaming. Seeds waiting for loving hands once more to patiently plant them into welcoming soil so that they can continue to fulfill their original agreement to help feed the people. I write here to share the hopeful message of the seeds, for these beautiful gems, life capsules of memory, have been my teachers, my guides, my allies, my trellises of hope for nearly two decades.

Traditional seeds have been helping me to find my way home. I descend from the Mohawk people, from the community of Akwesasne. I am Snipe clan, and my traditional name is Kanienten:hawi, which means "she carries the snow," as I was born during a swirling blizzard nearly forty years ago. I am a mother, a farmer, a seed keeper, and a writer. My passion is to keep our traditional seeds alive, along with the vibrant bundle of cultural memory that accompanies them. I call this intergenerational memory the "Seedsongs" that are carried in the blood and bones of our reciprocal relationships with our food and seed plant relatives, who have been helping Indigenous peoples since time immemorial.

For me, it has been a very personal journey to begin to carry the seeds of my ancestors once again. I was raised as a Mohawk woman in an era when my family was very disconnected from the land and the ceremonial cycles, due to the acculturation and assimilation tactics deployed in our communities over the last several centuries. In my family, the last people to farm for a living were my great-grandparents on my father's side. From the time I was a young girl, I always had a curiosity about the natural world, and would find myself playing in the tub of cornmeal in the pantry or digging in the earth outside. As a young woman of seventeen, I made my way to an organic farm in western Massachusetts, where I began to learn about the incredible world of agricultural diversity through the heirloom seeds that my mentor farmer was planting in her greenhouse and fields. Not only did various vegetables come in a prism of colors and a plethora of shapes and flavors, but the seeds themselves had deep connections to specific communities of origin and vibrant bundles of cultural memory and seed stories.

This realization launched me on a powerful and intimate path of inquiry about the foods of my Mohawk ancestors. Where were the seeds my ancestors nurtured? I did not know those seeds' faces, but traces of their sustenance surely made up parts of my own blood and bones. What varieties were and are there? What do their faces look like? Do they still exist? Who stewards them? These questions led me on a two-decade search to reconnect with the elder farmers in our Mohawk and Haudenosaunee communities and find the seeds of my ancestors, to reestablish those vital relationships in my life, home, and hearth. I am happy to say that I am now one of many people in our community who carry a sacred bundle of these seeds, and I have made my life an honoring song to care for these seeds and share their teachings with my children and with future generations.

I received an invitation from the seen and unseen forces in my life to follow this path of food and seed sovereignty. With my children in hand, I have begun to walk the healing path of finding my way home through traditional foods and seeds, a vibrant green and flowering path of remembering my own deeply connected cultural identity as a Mohawk woman. Those seeds have become my teachers, the wise elders in my life, who have taught me so much over many years. They have guided me into homes and hearths and fields within our Haudenosaunee communities where seeds and traditions have been preserved intact. Our elders say that the very food plants we are working with today have been with us since the beginning of creation—corn, beans, and squash emerging from the dying body of the daughter of Original Woman; corn from her breasts; beans from her hands; squash from her umbilicus; tobacco from her mind; strawberry from her heart; sunflowers from her feet. As Akwesasne Mohawk Brenda Lafrance eloquently states, "Corn is the milk from which we sustain ourselves."[1] These stories remind us that the original agreements to steward them run like wild rivers in our blood and bones; when we engage in carrying the seeds from generation to generation, we rehydrate these agreements, and we ensure that the creation story never ends, it keeps unfurling every spring, each new season.

Yes, these seeds are my wise teachers; I often laugh and ask myself, "Am I growing these seeds or are these seeds growing me?" Tending and stewarding them over the seasons have shown me the gaps in my understanding of traditional food culture and seasonal cycles, and they have created my path of inquiry to learn more about what it means to be a Mohawk woman and mother in my full capacity, able to nourish and feed our community and my family. This path has been a rather unconventional rite of passage—as a young woman, I didn't have the opportunity to have a traditional rite of passage. Over the past fifteen years, this journey of seed stewardship has shaped me: corn, beans, and squash have been my aunties and grandmothers who have helped me to find my way into deeper understandings of the sacred life cycles. For this I will be forever indebted to them, and in honor of the grand lineage of seed keepers who have faithfully passed down seeds for our nourishment, my family makes restored commitment to care for these precious seeds for generations yet to come.

For the last eighteen years I have been a passionate activist for Indigenous seed sovereignty. As Indigenous communities begin to revitalize our traditional food systems, there is an urgent need for the restoration of our traditional seeds. In many communities, these traditional seeds had dwindled down to just handfuls or small containers of seeds that had been selected over millennia as regionally adapted and culturally appropriate foods. Foods that restore our health and vitality as Native peoples, foods that help us heal from the legacy of the struggles our people have gone through in the past several centuries. As we heal our relationships to these seeds, we heal ourselves and our relationship to a much larger web of our cosmology. As Mohawk midwife Katsi Cook beautifully states:

> The corn goes into every facet of our lives once you really begin to look at it. I realized in getting involved in the cycle of corn—the blessing of the seed; the preparation of the field; the ceremonies during the corn gestation; and the harvest, storage, and care of the seed—that this is reproduction, this is pregnancy, this is the metaphor. All the ceremonies and the songs, everything we do with our seeds is what we are supposed to do at birth.[2]

Her husband, José Barreiro, reiterates, "Almost everyone who's done these activities (in the cornfield and in our ceremonial longhouse) never forgets it. Corn always stays a part of their identity."[3]

In my own home community of Akwesasne and the surrounding Native communities that sit under the great tree of peace of the Haudenosaunee Confederacy, we have seen a great revitalization of our traditional seed-keeping and agricultural practices. As a result of the historical legacy of traditional agriculture within Haudenosaunee communities and the selection and adaptation brought about by both the land and the hands of farmers, our seed stocks have branched out into a remarkable tree of diversity. Over three dozen distinct bean varieties, more than a dozen distinct corn varieties, and a handful a squash varieties paint a picture of cultural and environmental selection over the last millennium. Haudenosaunee communities are spread out over a large geographic region from southern New York to Ontario and Quebec. Due to the variation of a region that spans across many miles and latitudes, there is a natural heterogeneity that is expressed in varieties of these traditional corns, beans, and squashes between the dozen Haudenosaunee reservations and communities. This further increases the diversity of the gene pool of these cultivars and reinforces the sustainable conservation strategy of redundancy. The solidarity of kinship and trade routes across the Northeast forms a remarkably resilient web of seed security. This seed trade network was intact up until colonial contact, and has endured many waves of revival and decline over the centuries.

Our seeds and the agricultural ways of life that nourish our traditional foods are central aspects of our Haudenosaunee collective identity. These seeds are witnesses to the changes our communities have gone through and are also a testament to the Indigenous resilience of the seeds and the people who survived unspeakable adversities over the last many centuries. On the front lines of colonial pressure since the 1500s, our agricultural lifeways have been deeply disrupted by relocation, displacement, and war. One particular incident stands out in the memories of many families, accounts handed down through many generations of Haudenosaunee people of the time when, under the direction of the new American government Revolutionary soldiers burned a significant number of Haudenosaunee cornfields in retaliation against parts of the Haudenosaunee Confederacy who had opted to side with the Loyalists during the Revolutionary War. In 1779 the Haudenosaunee endured the Sullivan Expedition, a scorched-earth campaign that methodically destroyed at least forty Iroquois villages with their crops and food and seed caches throughout the Finger Lakes region of western New York. This expedition was an attempt to put an end to Haudenosaunee (Iroquois) and Loyalist attacks against American settlements that had occurred in the previous year of 1778. George Washington's orders to General John Sullivan on May 31, 1779, state: "The Expedition you are appointed to command is to be directed against the hostile tribes of the Six Nations of Indians, with their associates and adherents. The immediate objects are the total destruction and devastation of their settlements, and the capture of as many prisoners of every age and sex as possible. It will be essential to ruin their crops now in the ground and prevent their planting more."[4]

The devastation created great hardships for thousands of Haudenosaunee refugees that winter, and many starved or froze to death. This was the biggest assault on our people's capacity to be food sovereign, one that has had lasting effects on today's food systems and land tenure within our Haudenosaunee communities. The growing seed and food sovereignty movement, centuries later, is another testament to the resiliency of our people; we are inextricably linked in identity to our sacred foods, and just as our seeds somehow survived, we have too. As has been told to me by my elders, people who themselves have learned from the seeds as their teachers, the seeds are a reflection of the people. When the seeds are weak and struggling, this means our communities and nations and people are struggling. When our seeds are strong, this means our nations and communities and people are strong and in good health. These sacred and precious seeds carry our story, sprouting alive into new forms to nourish us in many ways. Our beautiful seeds are deeply connected to lineages and specific lands of origin. These foods and seeds are our mirrors, our reflections; their life is our life, we are intimately intertwined with their well-being. We are bound in a reciprocal relationship with seeds that extends back beyond living memory.

Haudenosaunee farmers do not simply maintain one type or variety of corn or bean. Dozens of unique strains were historically selected and grown. Why? Each serves a different function and fills its own environmental, culinary, or cultural niche. This is the biocultural expression of agricultural diversity. This is the palette of our Indigenous resilience, which has allowed us to rise up from the ashes of the burnt cornfields of the Revolutionary War to continue to do what needs to be done to feed our children. The characteristic of any single traditional landrace seed variety encodes the values and environmental pressures of the culture of the people who nurtured and grew that sacred food. Therefore, these seeds are keepers of a record of plant-human relationships predating the written word. The seeds are the witnesses of the past, and also the hope for the future.

In May 2016 we hosted the first annual Haudenosaunee Seedkeepers Society meeting at the central fire of the Confederacy, within the Onondaga community. A number of foresighted individuals and families have been working diligently over the last many decades to ensure that there would be foods and seeds to plant. Within my own home community of Akwesasne, we have seen a collaborative network of organizations and initiatives such as the Haudenosaunee Task Force on the Environment, the Freedom School, Kanenhi:io Ionkwaienthon:hakie, and the Á:se Tsi Tewá:ton (Akwesasne Cultural Restoration Program) come together with the common vision of keeping our seeds alive. In particular, a number of hardworking individuals within the confederacy have kept this cultural vision and spirit fire of seed stewardship alive during the darkest times of our historical trauma. Among them are Steve McComber of Kanawake, Dave and Mary Arquette of Akwesasne, Norton Rickard of Tuscarora, Angela Ferguson and Roger Cook of Onondaga, Terrylynn Brant of Six Nations, Janice Brant of Tyindinaga, Jane Mt. Pleasant of Tuscarora, and many others. Alongside the ancestors who shared their seeds down through the generations, it is in their footsteps that we walk as we carry our sacred seed bundles home.

The diversity of seed crops from Native North America is remarkable; the numerous colors, textures, and flavors of Indigenous resilience are held in purple corns, zebra-striped beans, bright orange squash, and a plethora of other traditional landrace varieties. Yet the diversity that has been cultivated over time in the hands of Native American farmers has been disrupted and displaced over the last century, as tribal communities have witnessed extensive economic, political, ecological, and cultural upheavals. The end result of a large number of intercultural and governmental policy-generated disincentives has been the demise of a large percentage of traditional farming of native crops in

Rowen White holding a braid of white corn at a food summit hosted by the Red Lake Nation in Minnesota, September 2016. *Photo by Elizabeth Hoover.*

tribal communities. In response to this, seed stewards have been working to establish regional seed systems and Indigenous trade routes as a part of the Indigenous SeedKeepers Network (ISKN). As a collective of Native Americans from cultures where agriculture is at the very heart of our cosmologies and lifeways, we understand that seeds are our precious collective inheritance, and it is our responsibility to care for the seeds as part of our responsibility to feed and nourish ourselves and future generations.

The seed movement is growing exponentially within the food sovereignty movement. I have had the great fortune to be a part of the conversation and grassroots action in many Native communities that are working toward seed sovereignty as an inherent part of their food sovereignty initiatives. The foundation of any durable and sustainable food system is held within the seeds. The culmination of the many diverse grassroots seed initiatives and projects is happening; we are working together in a collaborative movement where we assist each other in creating vibrant mentorship networks and seed sovereignty strategies, recognizing that our strength is in our capacity to come together as diverse and unique cultural communities and work together to help rebuild our seed systems once again. This solidarity through a shared common goal is now coalescing into a national network that has many connections with regional and local Indigenous seed projects.

The mission of the ISKN is to nourish and assist the growing seed sovereignty movement within tribal communities across Turtle Island. We are a program of the Native American Food Sovereignty Alliance (NAFSA). As a national network, we leverage resources and cultivate solidarity and communication within the matrix of regional grassroots tribal seed sovereignty projects that are working not only to restore, conserve, and selectively breed culturally appropriate varieties, but to do so within a vibrant cultural context. We accomplish this mission by providing educational resources, mentorship training, outreach, and advocacy support on seed policy issues, and by organizing national and regional events and convenings to connect many communities who are engaging in this vital work. We are publishing a culturally appropriate Seed Sovereignty Assessment toolkit. We aim to create a collaborative framework and declaration for ethical seed stewardship and Indigenous seed guidelines to guide tribal communities as they protect their seeds from patenting and bio-piracy. We support the creation of solutions-oriented programs for adaptive resilient seed systems to enhance our tribal communities' creative capacity to continue to evolve as the face of our Mother Earth changes. ISKN is a shade tree of support to the essential work of regional and tribal seed initiatives, offering a diverse array of resources aimed at nourishing and supporting a vibrant Indigenous seed movement, as a complement to the growing food sovereignty movement within Indian Country.

At the heart of the food sovereignty movement is a shared understanding between many diverse tribal nations that our foods hold a special place in our cultural teachings. There is a dynamic interface between cultural identity and genetic diversity of seeds. At the heart of our mentorship and facilitation work with Indigenous communities actively seeking to restore their traditional seed systems, we examine these questions: What is the importance of cultural memory, tradition, and community in the restoration of regional seed systems that steward agro-biodiversity? How can we leverage culture and community as mechanisms to restore healthy seeds and seed systems? Conversely, how can we strengthen native seeds and traditional agriculture as a mechanism to maintain cultural traditions? How can we work cross-culturally to ensure we have access to our seed resources, while assisting other small farming communities in the deep reconciliation work of sharing seeds with integrity and respect for communities of origin? Can we envision the "seed commons" and coordinate collaborative efforts to care for and protect our seeds in an appropriate relationship to our Indigenous cosmologies? How do we re-create regenerative seed and food economies that treat seeds not as objects or commodities but as living, breathing relatives?

Seed work is slow work, it is intergenerational work that is reliant upon the next generation to continue the stewardship. Seeds also have woven a beautiful tapestry of kinship and Indigenous trade routes across the wide breadth of Turtle Island. Collaborative work with the ISKN and my own work as a traditional seed keeper grow from the vibrant, rich soil made by those who came before us. It is because of these foresighted elders that we still have our traditional seeds. Cultivating heritage and traditional seed varieties is a way we mediate between past, present, and future, and is a powerful mechanism of cultural memory and history. As my elders say, "We do not own the seeds, we borrow them from our children." As Indigenous seed keepers, we see these seeds as our relatives, dynamic and ever-changing, and in carrying them forward to the next generation, we fulfill our responsibility to be good future ancestors and responsible descendants.

Our work is connecting seed projects and communities from Southwest Pueblos to the Paiutes of eastern California; from the Haudenosaunee of the Northeast to the Anishinaabe peoples of the Great Lakes region; from the cornfields of the Pawnees to the desert plains of the Arikara people. Seeds unite people in magnificent kinship routes, and this unification is a beautiful and nourishing process to bear witness to.

Another hopeful trend that is emerging at the cutting edge of the Indigenous food/seed sovereignty and social justice movements is the rematriation of seeds. We are all familiar with the journey to repatriate cultural property within Indigenous communities. Within Native communities we are very familiar with the word "repatriation," which refers to the return of treasures, ancestral remains, and sacred objects of cultural heritage to their communities of origin and their descendants. The displaced cultural property items are physical artifacts that were taken from a given place and people of origin, usually in an act of theft in the context of imperialism, colonialism, or war. The Native American Graves Protection and Repatriation Act (Public Law 101-601; 25 U.S.C. 3001-3013) delineates the rights of Native American lineal descendants, Indian tribes, and Native Hawaiian organizations with respect to the treatment, repatriation, and disposition of Native American human remains, funerary objects, sacred objects, and objects of cultural patrimony, referred to collectively in the statute as "cultural items," with which they can show a relationship of lineal descent or cultural affiliation. In the seed movement, we have begun to use the word "rematriation" in relation to bringing these seeds home again. In many communities, including in my own Mohawk tradition, the responsibility of caring for the seeds over the generations is ultimately within the women's realm. Both men and women farm and plant seeds, but their care and stewardship are part of the women's bundle of responsibility. So the word "rematriation" reflects the restoration of the feminine seeds back into the communities of origin.

Over the last few centuries of disruption to our Indigenous food systems, many of our traditional varieties have left our communities, only to be stewarded by non-Native farmers or seed keepers. Many traditional seeds have been also been stewarded or stored within public or private collections, institutions, and organizations such as public seed banks, universities, museums, and seed companies. As a part of the Indigenous seed sovereignty movement, we recognize the need for these seeds to be returned to a living context. In an era of displacement and acculturation, some of these varieties were completely lost to their communities of origin, and we are now locating derivatives of these seeds in such public and private collections.

Powerful healing work of reconciliation occurs when we work cross-culturally to bring these seeds home to their communities of origin. ISKN is working with communities that are working toward rematriation. We are working with many stakeholders, including Native farmers, gardeners, and representatives from tribal communities; institutions; and organizations that maintain such seed collections; and also other people who can facilitate and lay out the needed framework to assist these seeds in finding their way home. We are working to establish protocols and guidelines for this complex and healing work of seed reconciliation. There are deeply embedded cultural and spiritual aspects of these work, along with legal and political aspects that directly address seed justice.

As Martin Prechtel has described, "It is not enough to save heritage seeds. The culture of those people to whom each seed belongs must be kept alive along with seeds and their cultivation. Not in freezers or museums but in their own soil and our daily lives."[5]

Ultimately, this seed work is one of the most powerful forms of healing for the intergenerational historical trauma that nearly all Indigenous peoples are carrying and trying to metabolize. As our communities begin to rise up and rebuild our traditional foodways, restoring time-honored relationships with our plant relatives, we make powerful strides toward healing the intergenerational wounds that were inflicted when our traditional foodways were violently deconstructed. This work is an antidote to the ancestral memories of General Sullivan burning millions of bushels of seeds in our Haudenosaunee cornfields during the Revolutionary War, to the agony of displacement for southeastern tribes on the Trail of Tears and for the Diné during the Long Walk. An antidote to the separation of people from the land, seeds, and flavors of their birthplaces. This work inevitably unearths deep grief as we acknowledge the pain of having these sacred relationships wounded and buried over generations, but there is no greater healing than the reignition of our spirit fire in the context of our sacred foods.

Mohawk women are taught that when we go into the garden, we have to be of a good mind. We must not speak unkind words to each other or think unkind things. We have to conduct ourselves in a good way around our food relatives. As we begin to carry our traditional seeds again, and begin to eat them as part of our everyday lives, they become the medicine to heal intergenerational grief. A large part of this work is ensuring that we are mentally, spiritually, and emotionally healthy, and are using farming, gardening, seed keeping, and traditional food preparation as a tool for healing. The seeds are our Earth Apothecary.

Only our Mother Earth and her food and seed gifts have the capacity to heal deep ancestral trauma. Through our seed work we return our tears and pain to the earth to be ground and composted into renewed life—rich, earthy loam from which the dreams of our ancestors unfurl into shiny speckled beans and smiles and prayers from the lips of our grandchildren who know no hunger. The seed and food sovereignty movements will play vital and vibrant roles in intergenerational healing; we are already witnessing this. Just as one seed turns into one hundred, and then those hundred seeds turn into thousands, this movement of Indigenous resilience will multiply and grow exponentially. Seeds are going to help us find our way home.

In closing, it is my prayer that all Indigenous peoples will begin to restore the loving and deep-seeded relationships with their ancestral seeds. This is, was, and will be the medicine needed to endure the times ahead. In your own life, water that seed planted deep inside the earth that is your own body, a tiny seed that sings an achingly beautiful song of remembrance, resistance, resilience, redemption, and reconciliation. This powerful seed song is what kept our grandmothers upright, what whispered to them to get up amidst the sorrow to do what needed to be done to tend the earth and feed the children. It was these melodies that guided our grandfathers under the sea of stars as they made their way into new lands to protect the young. This map is written in the seeds and the stars and the waters and the Earth. This seed song is now your heart beating fiercely in promise to uphold the agreements to feed the sacred hungers of time.

NOTES

1. Haudenosaunee Environmental Task Force, *Words That Come before All Else; Environmental Philosophies of the Haudenosaunee* (Cornwall, ON: Native North American Traveling College, 1999), 70.

2. Carol Cornelius, *Iroquois Corn in a Culture-Based Curriculum* (Albany: State University of New York Press, 1999), 255.

3. Cornelius, *Iroquois Corn*, 195.

4. "From George Washington to Major General John Sullivan, 31 May 1779," Founders Online National Archive, https://founders.archives.gov/documents/Washington/03-20-02-0661.

5. Martin, Prechtel, *The Unlikely Peace at Cuchamaquic: The Parallel Lives of People as Plants* (Berkeley, CA: North Atlantic Books, 2012), vii.

PAT GWIN

8
WHAT IF THE SEEDS DO NOT SPROUT?
The Cherokee Nation SeedBank & Native Plant Site

In the spring of 1491, the entirety of the Cherokee people were preparing for the upcoming growing season, as they had done the year before, and in all the previous years. There is technically no end to "gardening season." Soil preparation transitions to planting, which is followed by growing and tending, and finally culminates in harvest—which in turn leads to food storage and seed selection for the following year. Immediately on completion of that process, soil preparation begins to start the cycle anew. Still, the spring planting season must have been a special and magical time for Cherokee people. However, I often wonder if they asked themselves, *What if the seeds do not sprout?*

Fast-forward to 2005: the traditional lifeway cycle had not only been broken, it had been all but forgotten. For the Cherokee Nation's government, agriculture had "evolved" into clear-cutting tribal forest lands and planting nonnative grasses and pine trees under the oversight of the Bureau of Indian Affairs. For the Cherokee people, gardening had become an exercise in purchasing packaged seeds from big box stores and planting those seeds in a semitoxic concoction of commercial potting soil and synthetic chemical plant food. The question of whether the seeds would sprout no longer mattered. One could always either go and buy more seeds or just forget about it and try again next year ... or not.

THE ORIGIN OF A CHEROKEE NATION SEED BANK

In 2006, the world was captivated by the Svalbard Global Seed Vault (SGSV). Conceived and constructed by NordGen, the SGSV arrived on the scene at a time when the news abounded with doom-and-gloom stories that revolved around world famine, widespread crop blights, and agricultural genetic modification run amok. The mission of the SGSV—to preserve seeds of heirloom crops from around the world—was a comforting departure from the pessimism of the day.

During this time, I was a staff-level employee within the Cherokee Nation Natural Resources Department (NRD) and was mostly concerning myself with technical, environmental issues related to the National Environmental Policy Act and various Bureau of Indian Affairs issues. Although I was not a manager, I often had to stand in for my superiors at tribal council meetings to answer questions regarding NRD program activities. In one such meeting, a council member inquired whether I was aware of the SGSV project. Being a lifelong gardener, a biologist, and a budding natural resource manager, I answered that I was vaguely aware of it. The conversation grew and expanded to include most of the tribal council members. The session culminated with direction for NRD to research the process of getting "our" Cherokee Purple Tomato and other "Cherokee heirlooms" placed in the inventory of the SGSV. It seemed the Cherokee people were striving for agricultural immortality and were to reclaim their place as North America's first and foremost agriculturists. *"We were forever going to have our seeds and they would surely sprout."*

Unfortunately, the process of reclaiming our agricultural immortality initially looked to be a short-lived endeavor. We discovered the much-vaunted Cherokee Purple Tomato had a pedigree that barely predated 1900—nowhere near the era of pre-European settlement. The search for other Cherokee heirloom crops grown by either the Cherokee Nation or respected elder Cherokee gardeners yielded a tally of zero. In fact, the research did not uncover a single instance of any heirloom crops being cultivated by the Cherokee people who had traveled the Trail of Tears, and certainly none predating 1492. It appeared that the Cherokee

Nation had bigger worries than whether the seeds did or did not sprout. *"We had no seeds to plant."*

Learning about this lack of seeds would prove to be a watershed moment for me and my coworkers—and the consequences thereof would begin consuming the majority of our professional lives. At this point the entirety of the Cherokee Nation leadership charged us with identifying, locating, procuring, cultivating, and producing seeds from heirloom plants grown by the Cherokees in 1491.

THE QUEST

As classically trained biologists, our initial strategy was to engage the resources of the mighty United States Department of Agriculture (USDA). The USDA operates multiple seed banks throughout North America and beyond, including the National Plant Germplasm System (NPGS). Unfortunately, working with them—as with most federal entities—is not as easy and straightforward as one would like: federal employees are often entrenched in their own bureaucratic systems and are burdened with their own multiple priorities and deadlines. There were to be no seeds forthcoming from the NPGS. However, we were able to extrapolate a few non-federal leads to investigate.

Lead 1: Carl Leon "White Eagle" Barnes, also known as the Corn Man

Just about any research on North American corn turns up the name "Carl Barnes" and describes him as a kind of semi-mythical maize guru. At the time, he was living around four hundred miles away from the Cherokee Nation's capital in a beautiful but desolate area that was once known as "No-Man's Land," but is today part of the Oklahoma panhandle. It was time for a road trip. A seven-hour truck ride later, we arrived at the home of Mr. Carl Barnes, and two things immediately struck us about him: (1) he was exceedingly friendly, warm, and welcoming; and (2) he was indeed the "Corn Man." Mr. Barnes wasted no time inviting us into his home, although it was less of a "home" in the traditional sense and more of a "corn museum." Every nook and cranny of the place was neatly filled with well-documented corn varieties from around the world.

Our research had yielded at least two types of corn that we were in search of—both a "dent" corn and a "flour" corn. As yet, we had no idea how many varieties there might be in each group. Mr. Barnes quickly gave us an education, and we soon learned from him that the dent corn we were looking for was Cherokee White Eagle Corn. He knew of three different varieties of flour corn: Cherokee Colored, Cherokee White, and Cherokee Yellow corn.

Mr. Barnes led us throughout his house, showing us samples of three of the four varieties he had mentioned and generously sharing with us everything that he knew about them. He continued to impress us by generously offering us samples of every Cherokee variety that he had. A quick end to the quest seemed in sight, and things were going well—too well, in fact.

At this point we received our first cruel life lesson in corn. Mr. Barnes explained to me the difference between corn specimen preservation for scientific collection purposes versus specimen preservation for the purposes of maintaining viable seeds that could be grown at a later date. His collection was of the scientific collection variety, and we learned that the germination rate of the sample corn he had given us would likely be quite low. Then, in a fitting send-off, he commanded us Cherokees to go back home and get to the business of growing that corn and to never stop doing so. I truly believe Mr. Barnes was both ecstatic and relieved at our visit, and that he truly wanted and needed his corn to be growing and producing at the hands of the people who had originally perfected the varieties. The Cherokee Nation SeedBank would never have existed were it not for Mr. Barnes.

Lead 2: Kevin Welch, Sara McClellan, and the Eastern Band of Cherokee Indians

Mr. Barnes had provided me with another vital piece of information necessary for the creation of our seed bank: there were a couple of folks back east working with heirloom Cherokee plants in collaboration with the Eastern Band of Cherokee Indians and the USDA, and he thought they could help us with our quest. Upon our return to Tahlequah, we phoned the contacts he had given us, Kevin Welch and Sara McClellan. Once again, fate smiled upon us and a short, informal telephone conversation culminated in Kevin and Sara showing up on our doorstep only a few days later. Despite the eight-hundred-mile "short trip," they arrived happy and full of enthusiasm. More importantly, they came bearing gifts in the form of seeds—lots of seeds.

These two individuals were a true professional-caliber wrestling tag team of heirloom crop specialists. Kevin had a simple rule that he lived by: "Either grow these seeds or get out of my way." Sara was equally passionate, but did infuse a bit more Western science background into the heirlooms they raised. Once again, the Cherokee Nation was blessed with individuals able to provide us with the seeds of numerous heirloom corn, bean, and squash varieties. Although these seeds had been preserved with a priority of high germination rates, Kevin and Sara did warn us that some of the varieties had not been maintained under the strictest "genetic purity" methodology—an issue we would eventually have to address. Yet, again, the Cherokee Nation SeedBank would never have existed if it were not for Kevin Welch and Sara McClellan.

Lead 3: Science Museum of Minnesota Native American Exhibition

Who would have thought there would be a museum in St. Paul, Minnesota, that routinely grew Native American heirloom plants? The Native American Exhibition at the Science Museum of Minnesota maintains a small seed-bank operation. As luck would have it, the exhibition had recently received a very small number of "Cherokee Ceremonial Tobacco" seeds (*Nicotiana rustica*) and was willing to part with "a few." A couple of days later the Cherokee Nation received a package containing nine tiny seeds; only those familiar with the size of tobacco seeds will realize just how small and insignificant nine tobacco seeds would look to the uninitiated. Still, we realized our luck was holding. We now had four varieties of corn, more than twenty varieties of beans, a squash variety, and ceremonial tobacco—all Cherokee heirloom plants from our ancestral homelands. Unbeknownst to us, the quest to secure these seeds would prove to be the perversely simple and easy part of creating what would one day be the Cherokee Nation SeedBank Garden & Native Plant Site.

BUILDING A CHEROKEE NATION SEED BANK

The Cherokee Nation team charged with "building the seed bank" had considerable agricultural, biological, and horticultural experience, but no specific experience in seed banking. As is the Cherokee way, we started by soliciting a wide range of views from employees and citizens—we even held a public meeting. This public meeting was interesting; what was expected to be a small, quiet, and low-key affair actually turned into a loud and enthusiastic crowd of more than one hundred attendees. All seemed quite supportive and eager to assist with the growing/seed-banking project. And not just using the heirlooms we had procured—some citizens also wanted to grow other culturally important native plants. We did not realize at the time that storm clouds were looming.

Hurdle 1: Politics and the Plant World

Since our government had issued a directive to start a seed bank, we were surprised to find agitation and consternation among some people over following that directive. After the aforementioned public meeting, orders came down to scrap the project. Some citizens felt that the government had no business operating a seed bank, and that decisions regarding whether or when to preserve or distribute the seeds of traditional Cherokee plants should be left to individual citizens. After this outcry from a small but vocal minority, the project appeared to be virtually terminated. In an attempt to salvage what work had been completed to date, the tribal leadership decided that having the seeds already collected on hand and properly stored was seed bank enough at this point. Our task, which had not yet begun in earnest, was abruptly finished.

After several months elapsed, a few executive-level officials quietly re-engaged and instructed us to proceed once again with the seed bank. However, the sole purpose of the project was now to keep the heirloom germplasm viable via a plant-grow-harvest-store-repeat process. Our task was not yet finished after all.

Given our new charge, we quickly made the following decisions. First, we would locate the growing site at the main complex. This decision would prove to have both pros and cons, but mostly cons. Next, we decided the facility would consist of a simple patch of dirt fifty feet by sixty-five feet in size. This decision too would prove to have both pros and cons, but again, mostly cons. Finally, we decided that the actual garden site would be fenced, as it was frequented by deer and other animals. Ultimately, this decision would prove vital for the success of the seed bank.

Hurdle 2: Swamps and Trash

The Cherokee Nation's tribal complex was constructed in the late 1960s atop the best swampland and rubble that money could buy. Soil is the foundation upon which gardens are formed. Unfortunately, our selected and approved site had approximately two to four inches of infertile clay atop piles of waste asphalt and concrete rubble. Not ideal, but with copious infusions of organic matter and soil amendments we rectified this issue. Within weeks, we had twelve inches of rich, loamy topsoil. Unfortunately, due to the production requirements placed on the garden site, this work of soil amendment and conditioning is a continual affair (remember, plants "eat" dirt).

At this time, the sole purpose of the garden site was to grow the heirloom crop seed stock on a frequent enough basis to keep a fresh, viable supply of seeds for subsequent replanting and refreshing of the genetics. Thus, the entire seed-bank site was a small, fenced garden spot located within an idle, grassy field. Its only "neighbor" was a trash- and brush-filled drainage ditch. The site's poor aesthetics and limited expansion potential were of no consideration at the time, but would prove to be enormous challenges as the project evolved.

2007: THE FIRST PLANTING

With seed and soil ready, it was now time for the fun part—getting our hands dirty. Thus, in May 2007 a small contingent of Cherokee Nation employees planted four rows of Cherokee White Eagle Corn and a few rows of heirloom beans and squash, albeit with a bit of argument and disagreement as the work progressed. At this point, there was one universal question in all of our minds: *"What if the seeds do not sprout?"*

In my humble opinion, gardening is as much art as it is science. Despite our attempts to control the process and either claim glory or accept responsibility for defeat, all of the truly wondrous work of actually creating a seed, initiating germination, and growing a plant is in the hands of a higher power. The human element in gardening is basically reduced to the need for a modicum of intelligence regarding moisture, nutrients, temperature, and solar radiation factors. As previously stated, the seed-bank staff had experience in gardening and horticulture, so we had access to the necessary water, sufficiently prepared soil, 65°F+ temperatures at four-inch soil depths, and plenty of sun . . . so now we waited. Luckily for the Cherokee Nation, Carl, Sara, and Kevin had provided us good product—better than they knew or had predicted. We had a nearly 100 percent germination rate. *"The seeds did sprout."* Future questions such as growing, flowering, pollination, and so on, remained, but *"sprouting"* we had achieved.

This brings me to the main requirement for successful gardening—sweat. Just about every aspect of gardening requires a significant amount of this commodity. If one is willing to sweat every day, one can and will be a successful gardener. Corn is particularly easy, especially these heirloom varieties. A mere handful of seeds yielded more than two gallons of harvested kernels. The squash was equally successful, and we were able to break even with the beans—a miracle we did not fully appreciate at the time. *"The seeds did sprout, grow, and yield."* The Cherokee Nation now had a functioning seed bank.

EXCESS SEED

Following our first successful crop, we harvested, prepped, sorted/selected, and stored our heirloom seeds. Once those tasks were completed, we were left with a quantity of excess germplasm. Resisting all of the "plant politics" from the year earlier, the seed-bank staff embraced an "ask for forgiveness rather than permission" strategy and opted to offer the seeds to Cherokee citizens for free if they would only promise to plant them. Whether or not this was the best decision is a matter for others to ponder, but I do know that this decision would ultimately create an immense amount of labor for staff that would consume immense amounts of their time.

With the help of the Cherokee Nation Communications Department and word of mouth, we made the offer of Cherokee Heirloom Seeds to Our Tribal Citizens. The response was overwhelming and demand outstripped supply in short order; the rest is history. Now, the Cherokee Nation SeedBank & Native Plant Site provides approximately five thousand packages of seed each year to Cherokees and educational institutions that cater to Cherokees. As the name suggests, the SeedBank would eventually expand into the husbandry of wild plants. The native garden site that supplies the SeedBank maintains more than one hundred varieties of plants. Tribal citizens place orders to the SeedBank via a dedicated website and online ordering system that automatically verifies Cherokee tribal citizenship. Cherokee Nation SeedBank staff provide tours and presentations to hundreds of people each year. The Cherokee Nation administration touts the SeedBank's success, and the Communications Department writes numerous stories about it. Ours is a "feel-good" program that does much to preserve and propagate both plants and culture.

Each year, the SeedBank staff battles Mother Nature, politics, and numerous other foes in an effort to fulfill our mission of providing culturally relevant plant materials to Cherokees practicing their culture. Because our citizens have now come to expect this service, failure is not an option—although it is an ever-present possibility.

WHAT IF THE SEEDS DO NOT SPROUT?

Even after ten years of successful growing seasons and numerous programmatic accolades, each spring we ask ourselves the same question, *"What if the seeds do not sprout?"* Mother Nature's continual reminder of exactly who is in charge presents a constant challenge for the SeedBank & Native Plant Site. In the past decade of growing we have experienced

- Record climate extremes of all varieties . . . sometimes within the same season
- Arachnid and bacterial pests too numerous to list, some of them imported from other continents
- Theft and vandalism of the human variety
- Theft and vandalism of the nonhuman variety

- Incidental chemical warfare from well-meaning neighbors
- Botanical failures that seem to be devoid of any rational explanation

And the list goes on and on.

Based on the tenets of statistics, one could expect that each year of success or partial success would only increase the odds of failure in successive seasons. Were the Cherokees of 1491 such astute agriculturalists that "crop failure" was not even in their lexicon? Or, as occurs at the SeedBank & Native Plant site of today, was failure of one variety offset by increased production of another variety?

Growing hundreds of species of plants in an area to which they are not native is no easy task. Aside from the plethora of expected issues listed earlier, the Cherokee Nation SeedBank & Native Plant Site Staff have also had to quickly develop skill sets in unexpected fields:

Proper genetic expression. The Cherokee Nation is the only entity growing many of these crops. As we select for the best or most-desired traits, the morphology of these species is bound to "drift"—a phenomenon we have already witnessed. Are the seed selection activities we conduct for "improvement" purposes properly part of our mission?

Hybridization. Corn varieties readily cross with one another. We call it "hybridization" even though use of that term is inaccurate within an intra-species context. Regardless, the propensity for crossing explains why there are countless varieties of corn. Cherokee White Eagle corn was one of the initial varieties we obtained; it is a bluish purple "dent" corn—meaning each kernel has a dent or dimple on its superior surface—and often it is marked by the silhouette of a flying white eagle on each side. When we received our initial stock of this corn seed we had few "eagles" and many kernels were dent-less (a flour corn trait). Our flour corn varieties likewise were contaminated with dented kernels. We have "corrected" these issues in our fields, but other gardens in close proximity to our facility and overachieving pollinators are major threats to our genetics. Deer can be fenced out. Bumblebees and red-wing blackbirds... not so much.

Climate change. Oklahoma can have very cold winters (–30°F is possible) and usually has very hot summers (+115°F is possible). Additionally, the weather can turn from very wet to very dry in the matter of just a few days. Thus, the Cherokee plants we obtained from the mountainous area of North Carolina have had to endure an instantaneous change in climate conditions.

Our corn varieties—and likewise squash—thrive in the Oklahoma heat, as long as proper irrigation procedures are followed and the temperature does not spike to 115°F. Beans are a different story; the North Carolina varieties thrive in a growing season characterized by temperatures between 70°F and 80°F and moisture conditions that are never overly dry or saturated. In Oklahoma, expecting such conditions is just not realistic. Our transition from winter to summer can often be measured in hours (not days, let alone weeks) and moisture is usually a "feast or famine" situation. Planting our beans for a spring crop often yields either mold-riddled plants that never pass the cotyledonous (that is, embryonic) seed stage or vigorous vine growth with no flower pollination due to excessive heat. Planting a fall bean crop is a sure bet for an early frost (Murphy's law). Thus, we have to double our efforts and labor, planting half a crop in spring and again in fall. Even then, we often are only able to "hold serve"—bumper crops are seldom realized (exasperating our citizens trying to order seeds from the website). We are hoping that eventually our efforts will lead to a bit of Oklahoma climate tolerance . . . but then one has to question whether that is that proper genetic expression of the variety. It is quite the botanical and cultural conundrum.

Many of the wild plants of cultural importance to the Cherokees are shade-loving species requiring or preferring wetlands and adapted to cool mountainous locales. The Cherokee Nation SeedBank Garden & Native Plant Site is a sun-scorched, lowland locale. Irrigation systems and planted shade assist in keeping certain species alive, but making them thrive is a challenge to say the least.

And then there is subject of anthropogenic climate change. Since this topic often leads to vitriolic arguments, discord, and strife, I will limit my comments here. I will say, however, that the climate does seem to be changing, and in a manner that is not beneficial to the operations of the Cherokee Nation SeedBank Garden & Native Plant Site.

Cherokee Nation SeedBank scale of operations. The Cherokee Nation Seed-Bank began as an operation to grow and maintain a few heirloom crop varieties. Shortly thereafter, a simple, seemingly innocuous suggestion to plant a few stalks of Rivercane (*Arundinaria gigantea*) has snowballed into a multi-acre Cherokee Nation park with hundreds of plants needing tedious daily care. Each year, our elders request that other plants be added to the site. Simply displaying culturally important plants is then no longer enough; culturally appropriate propagation is required to allow for annual harvesting and use to perpetuate the traditions. Human and financial resources are, of course, always limited; balancing this equation is becoming more and more problematic.

Politics. Politics pervades everything. Plants are neither Democrats nor Republicans, but the people who grow them are, which becomes problematic.

Political issues abound within the plant world:

- Access—should Cherokee heirlooms be made available to non-Cherokees (other tribal affiliations, nontribal individuals or entities)?
- Operational costs—is it culturally acceptable to profit, or at least recover costs, via the sale or commercialization of SeedBank products?
- Genetically modified organisms, theft/vandalism, anthropomorphism, supply utilization priorities... the list goes on and on.

ONCE THE SEEDS HAVE SPROUTED, NOW WHAT?

Today, the Cherokee Nation is operating the SeedBank Garden & Native Plant Site, and thus far "the seeds have been sprouting." Tomorrow, one of many potential calamities, ranging from budgetary woes to extreme climate events, could change everything. In the spring of 1491, the entirety of the Cherokee People were involved in preparations for the upcoming growing season... just as they had done the year before and in each previous year before that.

Were the agricultural challenges and problems faced by the Cherokees then any more or less precipitous than those we face today? Today, a failed crop would be an embarrassment; in 1491 a crop failure could result in far more serious and deleterious outcomes, all having grave and dire consequences. Thus, I am certain that the Cherokee gardeners in 1491 sweated as if their lives depended on it—because their lives *did* depend on it. Today, we still sweat, but I am not sure we *really* sweat. Still, the Cherokee Nation SeedBank Garden & Native Plant Site staff sweat to keep these seeds and the traditions they represent alive.

An old Cherokee saying goes, *No self-respecting Cherokee would ever be without a corn patch.* I think the Cherokees of 1491 did indeed ask themselves, *What if the seeds do not sprout*? I believe they asked that question because it was the plants that kept them alive and, in fact, what made them Cherokee. One could not exist without the other. Today when we ask the same question, I hope we arrive at the same conclusion.

DENNIS WALL
VIRGIL MASAYESVA

PEOPLE OF THE CORN

Teachings in Hopi Traditional Agriculture,
Spirituality, and Sustainability

This chapter describes aspects of a unique relationship between an ancient agricultural practice and the culture that it sustains. An agricultural technique known as "dry-farming" because it relies strictly on precipitation and runoff water (along with hard work and prayer), has kept the Hopi culture intact for nearly a thousand years. In addition, aside from the sustenance it provides the people of the high desert of northern Arizona, corn enters into nearly every aspect of traditional Hopi life, contributing to the development of values, the sharing and passing on of tradition, and the celebration and connection with the Great Mystery.

The authors of this article are members of the staff of the Institute for Tribal Environmental Professionals (ITEP), a tribal training and support organization based at Northern Arizona University in Flagstaff. ITEP's mission is to help tribes to build capacity in their environmental management programs. Its work centers on air quality management training but also addresses other media, including drinking water, wastewater, and solid waste, as well as challenges that tribes face with environmental toxins such as nuclear waste and heavy-metal deposition.

The late Virgil Masayesva, founding director of ITEP, was a member of the Hopi Tribe raised in the village of Hotevilla on Third Mesa on his family's farm, located in a valley his family called Hopaq. Virgil was a katsina dancer and a student of and active participant in Hopi spiritual practices. Much of the material in this essay is based on the teachings of elders under whom Virgil studied. Indented extracts of the Hopi origin story are Virgil's retelling of the stories he was taught. Dennis Wall, an Arizona native, is an author, longtime freelance writer-photographer, and former ITEP editor.

Virgil Masayesva relates the story his elders taught him about the emergence of the Hopi people and their subsequent settlement on the Hopi mesas of northern Arizona:

> After their Emergence into the Fourth World, the clans that would one day comprise the Hopi people approached the Guardian Spirit, Masaw, in the region that is now northwest Arizona and asked his permission to settle there. Masaw recognized that the clan people's former life, which they knew was not bringing them happiness, had been given over to ambition, greed, and social competition. He looked into their hearts and saw that these qualities remained, and so he had his doubts that the people could follow his way. "Whether you can stay here is up to you," he told them.
>
> Masaw warned the clan people that the life he had to offer them was very different from what they had before. To show them that life, Masaw gave the people a planting stick, a bag of seeds, and a gourd of water. He handed them a small ear of blue corn and told them, "Here is my life and my spirit. This is what I have to give you."

There is a distinction between the one true Hopi, Masaw, and the people who follow his way. Masaw is the true embodiment of a Hopi; the people who follow his way are merely Hopi Senom, or People of the Hopi. Following common tradition, however, we refer to members of the Hopi Tribe as Hopi.

To be Hopi is to embrace peace and cooperation, to care for the Earth and all of its inhabitants, to live within the sacred balance. It is a life of reverence shared by all the good people of Earth, all those in tune with their world. This manner of living lies beneath the complexities of *wimi*, or specialized knowledge, which can provide stability and wisdom but when misused can also foster division and strife.

Deeper still in the lives of traditional Hopi people lies the way of Masaw, a way of humility and simplicity, of forging a sacred bond between themselves and the land that sustains them. Masaw's way is embodied in corn. At the time of the Emergence, Masaw offered the clan people a manner of living that would not be easy. Dry-farming in the high desert of northern Arizona, relying only on precipitation and runoff water, requires an almost miraculous level of faith and is sustained by hard work, prayer, and an attitude of deep humility. Following the way of Masaw, the Hopi people have tended to their corn for nearly a millennium, and the corn has kept them whole.

For traditional Hopis corn is the central bond. Its essence, physically, spiritually, and symbolically, pervades their existence. For the people of the mesas, corn is sustenance, ceremonial object, prayer offering, symbol, and sentient being unto itself. Corn is Mother in the truest sense—the people take in the corn and the corn becomes their flesh, as mother's milk becomes the flesh of the child. Corn is also regarded as a child, as when a farmer's wife tends to the seeds and newly received harvest, blessing and ritually washing the corn, talking and singing to the seeds and ears. The connection between people and corn is pervasive and deeply sacred. In a remarkable symbiosis between the physical and the spiritual, the Hopi people sustain the corn and the corn sustains Hopi culture.

Victor Masayesva Sr. remembers as a young man in the 1920s and 1930s working just after dawn in his family's cornfields north of the village of Kykotsmovi. He could hear other farmers up and down the valley, a place his family calls Hopaq, as they sang to their corn plants. "That's how you take care of the plants," he says. "You sing to them, because they're just like humans, they have their own lives, and they like to hear you singing to them."

The Hopis' intimate relationship with corn is a bond that reaches back many centuries (terraced fields near the village of Paaqavi have been farmed since at least AD 1200). That bond reflects their profound reliance on the plant to sustain them in both good and difficult times. Even in this century, says Masayesva, there have been winters when corn, dried and carefully stored, was essentially the only food available to the Hopi people.

In the early 1940s, when his two brothers were drafted into World War II, Masayesva considered moving to Phoenix to continue his work in highway construction. He spoke with his father, who told him, "I'm getting old, and soon I won't be able to take care of our fields any longer. I want you to take over the farming. This is your decision. If you choose to be a farmer, you won't get rich, but

The mealing trough—Hopi, circa 1906. *Photo by Edward S. Curtis. Edward S. Curtis Collection, Library of Congress, LC-DIG-ppmsca-05085.*

you can sustain not only yourself but your family, and there are other benefits. It's going to take a lot of hard work. You have to be able to accept that responsibility."

Masayesva spoke with some of the elders at Hotevilla, his village on Third Mesa. They told him that to be a farmer would be a good thing—the fundamental Hopi way. After careful consideration he made his decision, and he has been tending his family's fields, probably the largest fields remaining at Hopi, ever since.

"One of the things that Masaw wanted the people to do was to plant, to be farmers," Masayesva says. "A long time ago I spoke with one of the last major chiefs in this village. He told me he wished he were in my place. He had certain religious responsibilities, obligations, and he and the other priests were concerned that they might not fulfill those obligations. He told me, 'You're like a child, you don't have these things weighing down on you.'" The priest told Masayesva that he was living the simple life that Masaw had offered the people, and in doing so, he was blessed. Masayesva says that farming is a crucial element in a way of life that binds the people, the corn, and the Great Mystery. Hopi farmers believe that singing to the plants is much like photosynthesis, that the songs energize and rejuvenate the plants. "It's all tied together. When you first plant your seeds, you take very good care of them, and when the plants come up, you go and sing to the plants, and the plants dance in rhythm to the song. That's how we were taught, and it is a practice we continue."

To test the strength and character of the clans, Masaw instructed them to travel in the four directions, to make their way in a difficult world and face the hardships that would determine whether they might come back and follow the life he offered to them. He told the clan people that at some point he would signal them that it was time to return.

Thus began the Migration Period, marked now throughout the Southwest and beyond by stone ruins and other structures, by petroglyphs and pictographs, and on a more subtle level by the spirits of the people who lived and died along the way.

During their journeys the clan people relied on corn as a primary means of survival. The varieties of corn they carried and cultivated were uniquely suited to the harsh, unforgiving environment in which they would eventually settle. During their travels they learned how to plant, cultivate, and protect the corn; how to use carefully developed techniques to sustain the plants, channel water, and discourage pests. Along with these things they learned precise methods of prayer and ceremony to ensure a harvest that might mean the difference between survival and starvation. No one can fathom the hardships they faced or which clans were left behind; those who were unable to embrace Masaw's way probably did not survive.

Sometime later Masaw sent out his signal, and slowly the surviving clans began moving back toward the Hopi Mesas. First to return were the people of the Bear Clan, who were told that their land included all that lay between the Colorado and Rio Grande Rivers—with three mesas forming the spiritual center point. Soon they had founded and settled the village of Oraibi on Third Mesa. More clans followed, and in time a new community, a new tribe called Hopi, was formed.

One by one, other clans came to the mesas and demonstrated their special skills and talents. Over time, through careful negotiation among the different groups, a covenant was established that they would reject the old practice of clan selfishness and instead contribute to the whole of the newly forming tribe.

The clans settled in their separate villages on the mesas. In doing so they became one, the People of the Hopi. In valleys and canyons they planted their fields with the small, resilient corn that Masaw had given them, corn as hardy and sturdy as the people; and the corn and the people were able to survive. The harshness of the land was indeed the reason that Masaw had provided this place for the clan people, for in such a place only a life of humility, balance, and hard work would ensure their survival. Their shared hardship was the prime bond that held them together.

The Hopi developed ceremonial and spiritual practices common to all the clans, though they also kept their unique clan ways. The villages grew in number, the people kept to Masaw's way, and the corn kept the people whole, sustaining for a thousand years a culture unique in its richness, diversity, and pervasive spirituality.

Hopi corn farming is an endless cycle; the very seeds used today to plant blue, red, white, and yellow Hopi corn arise from a lineage that reaches back many centuries. The tough, smallish plants have been bred to provide sustenance in an unforgiving environment. Cultivation methods developed by the Hopi people—such as planting the seeds deep in the soil and tending them carefully by hand throughout the growing season—have resulted in an agricultural efficiency known in few other places on Earth. Hopi farming endures strictly through the bounty of the universe—primarily on whatever precipitation falls in a given season.

Prayer and supplication, embodied most publicly in the dances of the katsinas, are religious-cultural practices woven deeply into the daily lives of traditional Hopis. Through ritual and ceremony the people entreat the spirits of the earth, sky, mountains, and clouds to bring the rain, to tame the wind, to provide a bounty in the fields year after year. This all-embracing focus on sacred ceremony is a powerful cultural binder, guiding the people in common purpose as it sustains a rich cultural tapestry of spirituality, work, and tradition.

WINTER AND EARLY SPRING

To choose an arbitrary "starting point" for a year's agricultural cycle at Hopi, one can look first to late winter when the katsinas descend from the San Francisco Peaks north of Flagstaff to enter the villages and dance for rain and regeneration. The katsinas dance for the vitality not only of the Hopi people and their crops but of the bird world, the insect world, the reptile and amphibian worlds—the

Cornfield, Indian Farm near Tuba City, Arizona, in Rain, 1941. *Photo by Ansel Adams. US National Archives and Records Administration, Still Picture Records Section, Special Media Archives Services Division (NWCS-S), NAID #519988. Courtesy of Wikimedia Commons.*

worlds of plants and animals and humans everywhere on Earth. They dance so that the living world will continue.

In the kivas in late winter, special ceremonies are performed, including the planting of bean sprouts. The kiva chief monitors the growth of the plants with a close eye, as the sprouts are harbingers of the level of success the people can expect in the fields during the coming year. The relationship between the bean sprouts and the fall harvest exists on many levels, some of which cannot be shared with those outside the societies. Perhaps the most understandable is the practical relationship between plants and tenders: If the sprouts grow strong and hardy, those responsible for cultivating them will probably exercise equal care in the cornfields. A duty of the kiva chief is to admonish those whose sprouts are allowed to dry out or come up weak and spindly.

In their homes farmers ask their wives what will be needed in the coming year's harvest. The wives are responsible for drying and storing seed stock from the previous year, for securing the seeds and dried corn from rodents and deterioration, and for keeping track of each year's planting needs—for both food and ceremony. These are skills and knowledge they pass on to their daughters, nieces, and granddaughters. The contribution of Hopi women to the longevity of these hardy varieties of corn cannot be overstated; through their understanding, keen eyes, and careful genetic selectivity, Hopi women have kept the corn extant for centuries.

Zetta Masayesva, who has resided in the Hopi village of Hotevilla for many decades, describes her intuitive approach to selecting seed stock: "When I choose the seed corn I don't care if the ear is long or short, as long as the kernels look hard. Those are the ones that will come up. You can tell which ones are weak. We pick the ones that are strong, that will germinate. We know how to pick the ones that are not so good."

A traditional Hopi farmer married to a Hopi woman does not plant for himself but for his wife's family. Each year before planting begins, his wife advises him on the quantities and types of corn needed to provide for the food and ceremonial needs of her family and perhaps others as well. The man, in turn, tells his wife how many gunnysacks of each type of corn seed he will need to plant his fields, and she prepares them.

The corn planted each year will be used for a variety of purposes: for food, for ceremonial use, to contribute to weddings and other social events, for use during prayer, and as material for rituals performed by Hopi secret societies that cannot be shared with outsiders or even with other Hopis who do not belong to those societies. A farmer's wife must have a clear sense of these varied needs and how best to satisfy them in the coming harvest. Her understanding of the different needs for corn requires intimate knowledge of Hopi culture and religious and ceremonial practices. A woman who can determine the quantities and types of corn needed for the coming year holds a bounty of general knowledge of the Hopi way. Over time she will pass that knowledge on to her sons and daughters, nieces, nephews, and grandchildren.

Before the seeds leave the home, the woman blesses them with prayer and a symbolic washing, a sprinkling of water. She talks to the corn seeds, wishing them good fortune as they grow into new plants. She tells them that she looks forward to seeing them again when they return at harvest time. Zetta Masayesva describes her relationship to the seed stock: "It's kind of like a mother taking care of a child. You take special care, you wash their hair, talk to them, prepare them—in this case, for planting." Victor Masayesva visits his fields in March to study them and prepare for cultivation. How much moisture has been retained in the soil? Has runoff caused erosion damage that must be repaired? Are there worms present? Some farmers plant in March, gambling on a frostless early spring. Sometimes they are lucky; other times the seedlings are frost-burned and killed off, though most of the hardy plants will regenerate new seedlings within a few weeks. Masayesva generally plants in mid-May, pursuing a conservative, reliable method that has never failed him.

Spring is a time when the family comes to Hopi from all over to assist in the planting. This period is for renewing family bonds, for sharing stories and experiences, for working together toward a common, important purpose. In earlier times only the men would plant. These days female family members assist in the fields as well. The women are also responsible for providing food for everyone during the laborious planting process. Leigh Kuwanwisiwma, the Hopi Tribe's former cultural preservation officer and resident of Paaqavi village on Third Mesa, recalls when he was a boy in the 1950s, around the end of the horse-drawn plow era of Hopi farming. He says farmwork creates good children and responsible adults:

> Back when agriculture was widespread—and unfortunately that is declining these days—part of a boy's role was to get out there farming, learning the hardships, dealing with the environment, listening to his grandfather and father and uncles. I remember watching my grandfather saddling up the burro early in the morning. If my grandmother was packing a noon snack, we knew we would be out all day. At that moment my heart sank. It was hard work, and there were times when I hated it. But if grandmother was just packing water, I would be so happy because we would be home by noon and I could play!

The hardships of his youthful farming days, Kuwanwisiwma says, may have seemed like heavy burdens then, but they have come to shape his adulthood, instilling in him an appreciation for hard work, for patience and faith, and for being able to put off the rewards of success in favor of duty and responsibility.

In the fields of his youth, as he prepared the soil and repaired drainage channels, as he planted the corn seeds and offered them their first small taste of water, Kuwanwisiwma was developing a fundamental connection with the earth. His labor involved hoeing the soil, checking for and removing worms, thinning the plants, channeling runoff, helping to erect windbreaks, and building stick shelters around the plants to keep crows and coyotes away. As he performed these tasks, he learned about farming, nature, the animal world, weather, and wind, the rhythms of life. Now he teaches the same knowledge and skills and, he hopes, the deeper lessons they hold, to his young son.

LATE SPRING AND SUMMER

After planting, a traditional Hopi farmer spends much of his time in the fields, tending each plant with loving care. It is a labor-intensive way of farming. Often family members work alongside him; in these days of outside jobs and distant residences, their times together on the farm are perhaps more important for family cohesion than ever before.

There are numerous concerns in spring and summer: Caterpillars are a constant threat and must be removed by hand from individual plants; ravens and other animals must be discouraged through various means; and if the field lies in a runoff path, the shaping of channels and dikes is ongoing, especially in a wet year (the current multiyear drought has presented its own challenges). In the fields the farmer relies on knowledge, faith, and prayer. In the villages the katsinas dance to bring rain that will allow the plants to germinate and grow to fruition.

During late summer and early fall sweet corn is sometimes harvested and roasted in stone-lined, underground steaming ovens. Victor Masayesva's oven is located on the edge of one of his fields alongside a shallow arroyo. The oven is primed with a wood fire then the corn is heaped inside and the oven sealed. The steaming process takes all night. Masayesva arises from bed several times during the night to check that the oven remains completely sealed, ensuring a well-steamed batch of corn which, when ready, will be shared generously by family members or dried for later use.

AUTUMN AND THE HARVEST

When the corn plants have grown to four or five feet tall, the ears are filling out, and their husks have begun to crack with dryness, harvesting begins. This is another time when the family gathers. The plants are knocked down and the ears harvested and tossed into truck beds to be carried home to the women of the family.

In Hopi tradition it is never proper for a man simply to bring the corn into the home, set it down somewhere, and tell the woman, "There you are." Instead, the man presents the corn directly to his wife, and she steps forward to receive it. Her act of receiving is a way of honoring him, as his personal, respectful presentation honors her.

When the corn harvest arrives, Zetta Masayesva welcomes the ears into her home, thanking them for growing well and providing food for her family. Her long-held tradition, common among traditional Hopi women, is to handle each ear separately, greeting and talking to it as she examines it for quality and firmness.

There is another reason why the woman spends so much time touching and examining the harvest: She is searching for a small number of perfect ears, which she will set aside for ceremonial and ritual use. The ears she seeks are generally small and always elegant in form: large at the bottom, tapering smoothly to narrow tops, the end kernels arranged in perfect symmetry. These "Corn Mothers" will serve a variety of ceremonial needs—and not merely as symbols, for to the Hopi people, corn is Mother in a very real sense.

Drying and storage of corn are the domain of Hopi women. Ears are sorted into stacks. The Hopi tradition is to stack the ears neatly in the home in overlapping form. These days, some women use boxes to store their corn; Zetta Masayesva frowns on that practice, but modern ways have crept into Hopi agricultural practices, and many have shifted to more labor-efficient methods.

WINTER AND THE VARIETIES OF CORN-BASED FOOD

Soon after the harvest is separated, dried, and stored, winter sets in. During the winter season, which can be bitterly cold on the mesas, and throughout the year, corn is a basic dietary component for the Hopi people.

A staple for Hopi meals is *piki*, a paper-thin, layered bread made most often from ground blue corn, water, and ashes, cooked by hand on a special flat stone (other varieties of corn are also used for piki, including a delicious red-corn piki mixed with chili peppers). Zetta Masayesva says she, like most Hopi females, has been making piki since childhood. "I pat it down with my palms onto the hot stone. I've done it all my life, like most Hopi women, so we don't feel the heat of the stone on our hands."

Corn is either steamed and dried or simply allowed to dry. It is often eaten directly off the cob or used to make hominy; for pudding (red corn is generally the variety used), which is heaped into a bowl and taken in pinches by the diners; for *somiviki*, small balls of cooked cornmeal wrapped and tied inside husks; and eaten in various other forms, including steamed and roasted sweet corn.

For centuries corn grinding has been a formative social experience for Hopi girls. The work, performed by hand using grinding stones sometimes referred to by the Spanish terms "mano" and "metate," is grueling, akin to the difficult labor that boys perform in the fields. Grinding is a social bonding event for girls; as they work side by side they talk, joke, tell stories, and share cultural knowledge. Most important in terms of Hopi culture, grinding corn is a time when young girls learn to take their traditional place in the family, to accept their gender-based role as provider and nurturer in the home, and to learn the value and necessity of hard work.

CEREMONIAL USES OF CORN

The sacred nature of corn is reflected in its pervasive use in Hopi society, not only for food but as ritual and ceremonial material. Secret Hopi societies use corn in a variety of ways, but outsiders are not privy to those uses. Across the many clans that make up the Hopi people, however, corn has universal uses related to celebrating, praying, and maintaining the people's connection to the Infinite.

In ground form white corn is the variety most commonly used for ceremony and ritual. White cornmeal and seeds are used in kiva rituals throughout the villages, and cornmeal is offered to each clan's guardian deity, represented by an icon in the home. Traditional Hopis carry white corn powder, or *homa*, in a pouch for use in a variety of prayer offerings.

During Powamu (winter solstice) *pahos*, or prayer feathers, are given to people throughout the villages. Pahos serve multiple purposes, but most typically they are a way in which the people offer prayers for good health, a long life, and goodwill and happiness for all living creatures. In the hours before sunrise, pahos are deposited in special places and individual prayers are made, followed by the scattering of homa on the paho. This is followed by offerings to the rising sun.

Homa also plays a role during the katsina ceremonies. It is used to "feed" and bless the katsinas as prayers are made that the katsinas will reward the people with an abundance of rain and a strong harvest of crops for the benefit of all people. As the katsinas begin their song, the katsina chief sprinkles homa on the dancing spirits with deliberate and passionate instructions that the dance be performed in harmony with the Earth and with vibrancy and a good heart. Some katsinas wear garb that is adorned with parts or symbols of the corn plant.

Childbirth and the Naming Ceremony

Corn is used ceremonially to mark significant milestones in the lives of the Hopi people. The ceremonial significance of corn is demonstrated from the moment a child is born, when a Corn Mother is placed beside the child, to remain with him or her for the first twenty days after birth. White corn is used in infant naming ceremonies twenty-one days after a child is born. The ceremony is a combination of festivity, prayer, and family unity. Most importantly it is a time to give the child a name that will stay with him or her for the remainder of life.

Mothers, grandmothers, and aunts gather on this special occasion, each prepared to offer a name that somehow reflects their clan lineage. Before the rising of the sun the newborn is first given a bath, then the hair is washed, usually by the maternal grandmother. She is the first in line to begin the naming ceremony, followed by other grandmothers and a succession of aunts.

The white corn is gently brushed over the baby's naked chest, with words spoken from the heart, eternal words that are offered to the child: "Your name shall be (name). You shall carry this name through the rest of your life, in sickness and in health. You shall carry this name through your adulthood until the day that you shall sleep in peace." After the naming ceremony, at the breaking of dawn, the newborn is taken outdoors to face the rising sun, and the identification of a new child has begun.

Initiation

As in many cultures worldwide, an initiation ceremony is held to mark the transition point when a child begins moving into adulthood. At about the age of twelve, Hopi boys and girls take their place in one of two Hopi societies. Before their initiation begins, each child is given a Corn Mother. The ear of white corn is never large; clutching on to the largest ears would be contrary to the Hopi way of humility. The initiate holds the ear of white corn close throughout the long initiation ceremony. Afterwards, parents sometimes plant the kernels, bringing forth new plants that hold special meaning for those involved.

End of Life

On the third day after death—the day before the spirit of the departed is released from the physical body—relatives take food to the burial place. At that time cornmeal is laid down along a ceremonial path to help guide the departing spirit on its way to the Grand Canyon, which the Hopi people consider their spiritual home. And so, a life that has been linked to corn on every level from the very moment of birth now follows a trail of cornmeal to the final spiritual resting place.

> When the clans accepted Masaw's way, they asked him to lead them. Masaw told them that would not be possible because, he said, he recognized that they carried a lot of knowledge and that they would eventually

be controlled again by their own ambition—perhaps to some finality. "At that time," Masaw told them, "I will return to you." To that, the people of the clans assured Masaw, "We will remember our past and try not to repeat it, and we will continue to learn from experience."

In this era of heightened mobility and pervasive mass communication, the remoteness of the Hopi villages, which cushioned them for centuries from the impact of Spanish colonialism and Euro-American incursions, no longer represents the barrier it once was. Times are changing, and Hopi culture is stressed today as never before.

Kuwanwisiwma says that Masaw's spirit still abides at Hopi. The question now, he says, is whether the people can continue to hold to the old ways, to remain people of the corn. Says Kuwanwisiwma:

> The generation of Hopis today, lives in the real world. No longer can we say this is a "white man's world" and we're up here separate from that world on these mesas. We're part of the dominant culture. We too have become influenced materialistically through the cash economy, with different kinds of value systems that have become our way of life. That is, I suppose, good to some extent: You work hard for something and you gain materially. But at the same time it's impacted our culture; we now rely on other forms of survivability.
>
> The ceremonial cycle may be ongoing—though much has been lost already—but the strength of the culture is under strain because the corn, and the way of the farm, have slowly been impacted. When I got married in the early 1970s, my father gave me a piece of his cornfield, and he said, "You take care of it, grow your corn for your family and in-laws." That's how I assumed responsibility as a husband and father. And I think that kind of social responsibility to family can be strengthened if younger people can appreciate what it means to be part of the Hopi way through farming.
>
> Participation in the ceremonies, as we see now with younger kids being initiated and participating, is important. They need to be told in the kivas, in the homes, that the corn is the way the Hopis have chosen; it goes back to our Emergence. As Hopi people, we are fortunate to have survived this long. It is a privilege to be a part of this complex Hopi community of clans living under this one philosophy of corn, of humility. I think if we can continue to teach that, we'll strengthen the culture as it stands.

DEVON A. MIHESUAH

10

COMANCHE TRADITIONAL FOODWAYS AND THE DECLINE OF HEALTH

Some of the Comanches were short, but slim. Nowadays Comanches are real fat. Probably they were slimmer then because they had to work all the time on their farms like we did. People were thin 'cause they worked and we ate fruits and vegetables and meat and we didn't have chips and pies and junk from the grocery store like today. —Comanche full-blood Henry Mihesuah, 2001

Comanches traditionally hunted wild game (most notably bison), traded and raided for agricultural foods, gathered wild plants and, by necessity, were physically active. Their varied diet and physical lifestyle kept them free from obesity, diabetes, high blood pressure, and related maladies. They contended with wounds, insect stings, animal bites, and European-brought diseases, but other than periodic malnutrition from lack of resources, they did not face diet-related illnesses until after they were forced onto reservations in the 1870s. At Fort Sill, their enforced sedentary lifestyle and inability to hunt and roam combined with poor government rations resulted in physical and emotional problems. Comanches are no longer confined today, of course, but because of socioeconomic conditions, lack of access to nourishing food, inadequate education about nutrition and exercise, apathy, and no agricultural tradition to revitalize, many Comanches have practiced poor lifestyle choices that have resulted in unprecedented health problems.[1]

EARLY DAYS

Comanche sociocultural history is multifaceted.[2] Comanches, who call themselves Nʉmʉnʉʉ (The People), may once have been part of the Shoshone people who lived along the Gila River in Arizona. Those who migrated north to Utah around 500 BC are known as the Sevier Complex. They moved farther north to the Great Basin north of the Great Salt Lake during a thirteenth-century drought and were known as Shoshones. A group that became the Comanches splintered off around 1700 and moved to the Rocky Mountain areas of Colorado and Wyoming. They split into more than a dozen bands and many more extended family groups that occupied twenty-four thousand square miles of varied terrain.[3] After the 1680 Pueblo Revolt, Comanches appropriated horses that the Spanish left behind, and they continued raiding horses after the Spanish reconquest of New Mexico a decade later. Horses changed Comanche culture; the animals expanded the Comanches' hunting and raiding range, carried more food and goods, and allowed the people to move with the seasons. The tribe's horsemanship and raiding prowess allowed them to dominate the Spanish Southwest and later their "Comanchería," the southern plains. Their success in warfare, raiding, kidnapping, and horse raising earned them the labels "Lords of the Southern Plains," "Lords of the Plains," "Spartans of the Plains," "Terror of the Southern Plains," and "Cossacks of the Plains."

The Quahada (also Kwahadi or Kwaharu) Band (antelope eaters) roved the Llano Estacado (sometimes called Staked Plains) in northwest Texas and northeast New Mexico, south of the Canadian River. The Kotsoteka (or Kutsutuuka) Band (buffalo eaters) roamed the region around the Red, Canadian, and Arkansas Rivers in Oklahoma and Kansas; while the Peneteka (or Penatuka) Band ("honey eaters," or "quick striking") ranged farther south into the headwaters of the Sabine and Trinity Rivers in central Texas. The "Middle Comanches": the Nokoni (or Nokoninuu) Band ("those who travel around," or "those who turn back"), Taninuu Band (liver eaters), and Tenewa (those who stay downstream) ranged in the Cross Timbers area of northern Texas to the New Mexico mountains. The Yamparika or Yaparuka Band (root eaters) ranged north to the Arkansas River.[4]

Because of the many splintered groups, one cannot properly say that all Comanches behaved in the same manner or even looked the same. In addition, what Comanches ate and how they prepared their foods depended on their environments, the seasons, and the groups with whom they traded and raided.

It was not until they were united at Fort Sill in the late 1870s that one could make more generalized statements about the Comanche diet.

While living in the Rocky Mountain area, the tribespeople hunted game, most notably bison. Also on the menu were other large game animals: antelope, black bear, elk, mule, and white-tailed deer; wild fruits, including juniper berries, grapes, mulberries, plums, persimmons, and prickly pear fruit; as well as seeds, roots, and wild onions. They were not discriminating when resources were scarce, consuming armadillos, frogs, insects, lizards, rats, snakes, skunks, and tree bark.

As the people dispersed, they found a variety of foods, but their mainstay was bison. Antelope could be found in eastern New Mexico, where Comanches hunted by using the "Magical antelope surround," in which a medicine man or woman organized an antelope dance to last several days, then transferred some power to a few men. Comanches made two lines facing east. The medicine person stood at the center with the two men with power on horseback on either side. Using medicine sticks mounted with antelope hoofs, the medicine man or woman crossed his or her arms, a move that disallowed the antelope from escaping. One rider took a stick and rode northeast while the other rode southeast, forming a circle where they passed each other, thus creating a barrier the antelope could not escape. The lines of people then traced the riders' paths to form a large circle and slowly moved inward, trapping the scared animal until it dropped dead.[5] Other sources describe antelope hunts in which the animals were simply chased down over a period of hours.[6] Another source describes a basic "antelope surround" in which hunters surrounded the animals and rode around them in progressively smaller circles, shooting any antelope that attempted to flee.[7]

Like other tribespeople prior to colonization, Comanches stayed perpetually active. In the 1830s in north Texas, painter and traveler George Catlin observed the "Camanchees" engaging in "ball-play" and reported that their ability to ride horses was not "equalled by any other Indians on the continent." He stated that horseracing was "a constant and almost incessant exercise." Among their talents they could hold onto a horse with their legs as their body dropped down to one side, effectively using the horse as a shield; all this while also carrying a shield, bow, and lance. As a horse and rider rode past him, Catlin saw only the rider's foot atop the horse's back. Granted, Catlin discovered that the riders used a woven hair halter that helped support their bodies when they slipped to the side of their horse, but the strength and agility needed to accomplish this feat is still considerable.[8] Army Captain Randolph B. Marcy observed the same skills in the 1850s, recounting that a Comanche "throwing himself entirely upon one side of his horse, and discharging his arrows with great rapidity towards the opposite side from beneath the animal's neck while he is at full speed—is truly astonishing." Marcy also claims to have watched women on horseback at full speed chase and capture antelope by lassoing them.[9]

Comanches supplemented their diets by stealing from the sedentary Pueblo tribes that grew corn and squashes and raised sheep, as well taking goods from the formidable Apaches, Navajos, Pawnees, Utes, and Osages. Comanches also raided Texan and Mexican settlements, taking cattle and other foodstuffs that they either traded with New Mexicans for tools and household goods, or quickly consumed while on the run.[10] Catlin explained that about ninety miles from the "Camanchee" village, on the Red River, sat a Pawnee Pict settlement with cultivated areas of maize, pumpkins, beans, melons, and squashes. He suggested that they had an alliance with the "Camanchees, hunting and feasting together," which presumably included trading garden produce. He also noticed that "Camanchees" decorated their deerskin and elk skin dresses with elk teeth (probably buglers); either these were trade items or perhaps they also had access to elk meat.[11] Marcy observed that those Comanches who lived around the Red River in the 1850s consumed very few plants but "enormous quantities" of fresh meat.[12] In the Texas Cross Timbers, the Arkansas, Canadian, Washita, Red, Pecos, and other rivers cut east to west, fed by streams and springs of drinkable water. Comanches in this area hunted and gathered wild fruit such as grapes and plums. They sometimes ate bear, although they did not favor the fatty meat and primarily used the oil to prepare animal hides.[13]

Comanches sometimes killed bison with bows made of bois d'arc, also known as Osage orange (*Maclura aurantiaca*), wrapped with deer sinew and measuring around three feet in length; the arrows were twenty inches long.[14] They spitted meats with a stick and roasted them over a fire of elm, mesquite, oak, willow, bois d'arc, or cottonwood. Sometimes they boiled and stewed meats in the paunch of an animal. After acquiring metal pots from traders, they mixed a variety of animal cuts together, stirred them with a stick, and removed the cooked bits by hand. Some cooked meat in a pit covered with leaves, straw, and dirt.[15]

The 59,020-acre Wichita Mountains National Wildlife Refuge in southwestern Oklahoma is today home to hundreds of species of mammals, birds, reptiles, amphibians, and fish. Thomas C. Battey, a Quaker who visited among the Kiowas and Comanches in the late 1870s, traveled this area and reported seeing bison, elk, antelope, deer, and wolves. He observed Kiowas utilizing the "bean-like seeds" of mesquite trees. They pounded the beans into a course meal, added sugar and water, allowed it to ferment, then dried it into "cakes" that Battey described as having a "pleasant vinous taste." These cakes also could be broken down and boiled with meat to create a mush.[16] The use of mesquite is confirmed by Comanche informants living around Fort Sill in 1933. Mesquite pods are slightly sweet, so they reported removing the seeds from the ground pods and mixing them with cornmeal to sweeten the meal, or mixing the corn and mesquite meals with water to create a drink.[17]

Comanches always shared food. Battey observed, "No one is deficient in food if another has it." People sat or squatted on the same matting they slept on. A board or thick hide was placed in front of each person, and one woman removed pieces of meat from the kettle and distributed it along with "bread" and cups of coffee. The diners ate with their hands and sometimes a knife. Battey asserts that they did not wash their hands and did not object to meat "that had been carried thirty or fifty miles, swinging and flopping upon the sides of a mule, until covered with dust, sweat and hair; it needs no washing, or at least gets none before being put in the camp-kettle. If the hair, boiled into strings and served up with the beef, is unpalatable, it is quietly taken out of the mouth and thrown away." Battey muses on what some might view as unsanitary eating conditions; "If dirt is, as has been defined, matter out of place, there is none in an Indian camp; for what can be out of place where nothing has a place?"[18]

Several outsiders claimed to know what Comanches ate and refused to eat, but sometimes those observations were contradictory. General R. A. Sneed, a trader who operated at Fort Sill and Anadarko between 1885 and 1890, stated that Comanches would not eat anything with feathers, scales, or fins: "Of fish, they knew nothing and consequently they paid no attention thereto."[19] Ernest Wallace and E. Adamson Hoebel wrote (with no attributed source) that Comanches did not eat turkeys because they associated turkeys' running from pursuers as cowardly.[20] Battey also wrote in 1876 that Comanches and Kiowas did not eat birds or fish, "neither does the Kiowa eat the flesh of the bear." He asserted that these animals "are forbidden," "unclean," "tabooed, in short, bad medicine." Animals that "were made for the injury of men," such as panthers or "venomous serpents and reptiles" reportedly were not eaten either.[21] Captain R. G. Carter, in contrast, observed that Comanches did eat turkeys. After being cleaned, the birds were thrown onto a fire whole, then after adequate scorching the skin and feathers were peeled off.[22]

Wallace and Hoebel and Battey assert that Comanches did not eat fish, but Carter claims to have seen Comanches in the 1870s coat fish with mud then cook them under hot coals. The hard mud was broken off and the fish salted before being eaten.[23] There is also a Comanche Band known as Fish Eaters (*Pekwi Tuka* or *Pekwi Tuhka*). The Mihesuah family also refutes the no-fish-eating assertion. Henry Mihesuah (c. 1922–2010) of the Quahada Band recounted that his grandfather Mihesuah (who, along with Quanah Parker and others, resisted relocation to Fort Sill until 1875) had no qualms about eating fish and birds, and neither did his full-blood parents, Joshaway and Carrie. Wallace and Hoebel cited Herman Lehmann, a man who had been kidnapped as a child by Apaches and lived for a few years with Comanches, as their source for asserting that Comanches did not consume frogs (except in times of scarcity, such as winter) and swine because they were associated with mud. In fact, however, Lehmann only mentioned that Apaches did not eat frogs and that he, personally, did not eat swine.[24] Numerous outsiders did observe that Comanches ate turtles in the same manner as the Five Tribes: by burning them alive in a fire, then cracking their shells to remove the meat.[25]

Wallace and Hoebel also write that Comanches drank the blood of large game animals and ate the raw brains mixed with bone marrow, which they learned to extract without splintering the bone. The authors assert that Comanches especially liked raw liver doused with bile, and ate cooked liver covered with marrow. Tripe was often eaten raw, but the Indians did clean it first and sometimes cooked it over a fire. But the authors contradict themselves a few pages later by writing that Comanches did not wash entrails, kidneys, and kidney suet before eating the uncooked offal.[26] Displaying their flair for drama, the two authors claim that Comanches took "the greatest of pleasure" in sucking out milk through cuts made in udders. Neither of their cited sources, however, make mention of this behavior. They also take from Lehmann that udder milk from a dead animal was a great dining "pleasure." Again, Lehmann was living with Apaches, not Comanches, when he had these dining experiences. Noah Smithwick mentions that Comanches served him a "dish" of curdled milk from fawns and suckling calves, but Wallace and Hoebel added to this story without any substantiating source that the Comanches considered this dish a "delicacy." Lehmann stated that one time he and the Comanches whom he rode with killed a cow, roasted the meat, and covered it with a sauce made of suet, honey, and water. This is the only time this particular combination is mentioned in literature or oral testimonies.[27]

In 1933, anthropologists Gustav Carlson and Volney Jones interviewed elder Comanches in Indiahoma, west of Cache and south of the Wichita Mountains in southwestern Oklahoma. The unnamed informants identified many food and medicinal plants but the anthropologists do not say when these foods were used. The man who assisted Carlson and Jones in finding Comanches who possessed knowledge about plant use was between fifty and sixty years old at the time of the project. Surmising that these informants were at least sixty in 1930, it might be assumed these plants were used during their parents' and grandparents' lifetimes, from around 1870 (and possibly earlier) up to the time of the interviews. At the same time that Comanches were forced to rely on government rations, they probably still collected plants when they could.[28] It appears the people had a wide variety to choose from. Most of these plants are seasonal, however, so they would not have been accessible year-round unless they were dried and stored.

Comanche informants reported using a plant like a wild onion, but sweet, with blue flowers. This may be camas lily (*Camassia esculenta*), a plant that is currently present in Comanche County, Oklahoma.[29] Camas was a crucial food of the Plateau and northern Plains tribes. The root-like bulbs were a highly sought-after trade item, as they are nutritious, and when roasted or boiled, taste similar to pumpkin or sweet potato but much sweeter. The sun-dried bulbs can be mashed, shaped into "loaves," then baked again for storage. Camas was almost eradicated in the Northwest around the 1870s because settlers used the camas prairies for their hog herds, but it is now the focus of restoration programs in Plateau region of Washington State and Idaho.[30]

Various wild grapes (*Vitis* spp.) that Comanches either ate fresh or dried for later use are trickier to identify. Grapes known to Oklahoma tribes as the fast-growing, tart, and seedy "possum" or winter grapes (*V. cinerea*) are not particularly tasty. Muscadines (*V. rotundifolia*) also grow in the area and are nutritionally rich. Today, wild grape foragers use them to make wine, jam, and preserves and utilize the grape leaves as wraps. Informants mention an unidentified fruit that is described as black and round like a grape and might be serviceberry (which turns dark purple).[31]

Comanches ate small hackberries (*Celtis laevigata*), either raw or crushed and rolled with fat into a ball, then roasted on a spit over a fire. Alternatively, they dried the fruit and ground it into flour.[32] Hackberries are rich in minerals (mainly calcium carbonate), protein, carbohydrates, and antioxidants. They have also been used by numerous tribes to treat diarrhea and heavy menstrual bleeding, and the bark was used for sore throats.[33] Comanches were admonished to "shout like a wolf" before eating the ripe fruit of hawthorns or haws (*Crataegus*) to avoid constipation and stomach pain, and they used the inner bark of a hawthorn as a chewing gum.[34]

Comanches, along with Hopis, Navajos, Bella Coolas, and Cahuillas, and perhaps other peoples consumed juniper berries (*Juniperus virginiana*) and used the seeds, bark, leaves of Pinchot's juniper (*J. pinchotii*) for medicine and dyes.[35] Comanches also competed with animals and birds for mulberries (*Morus rubra*), which they ate fresh. Mulberries grew throughout the plains states and eastward in valleys and floodplains.[36] The fruit ripens in summer and falls to the ground with a shake of the tree. Mulberries are rich in antioxidants; vitamins A, B, C, and E; iron; potassium; and manganese.

Comanches used two kinds of onions. Sweet onion (*Allium*) they ate raw, mixed with other foods for flavor, or braided then roasted over a fire. They also utilized a smaller onion with red flowers, although that was less popular. Tribes across the Southeast and in Indian Territory ate wild onions on a regular basis. Despite the strong smell and indigestibility (for some people), wild onions contain considerable amounts of vitamins C and A, and many people today value them for their taste and the aboveground greens that appear in spring.[37] For example, Choctaws hold springtime "wild onion dinners" where the onions are most commonly added to scrambled eggs and pork dishes. (Caution: take care not to gather wild onions in cow pastures or in fields treated with pesticides due to the danger of contamination.)

Many tribes consumed either fresh or dried persimmons (*Diospyros virginiana*), which are found throughout the southern plains, along with the Mexican persimmon (*Brayodendron texanum*). When the fruits ripened in fall, Comanches pounded them, removed the seeds, then dried the paste for future use.[38] People who harvest persimmons today typically eat them raw or slice and dry them. As with other fruits Comanches use, persimmons are rich in vitamins and minerals. The Mihesuahs still use persimmons that grow on the family allotment outside of Duncan.

Comanches ate fruits they called plums that grew throughout Comanchería. Some were probably chokecherries. The sweet Chickasaw plum (*Prunus angustifolia*), sometimes called sand plum, was eaten fresh or dried for winter then reconstituted in boiling water. Another plum referred to as *su:kui* was sundried for winter. An alternative to gathering plums was to raid pack rat nests.[39] Comanches occasionally used what they referred to as "fall plums," probably the black or yellow *Prunus angustifolia*, which ripens in late summer and into fall.

Like other tribes of the Southwest, Comanches consumed the tart prickly pear cactus fruits (*Opuntia engelmannii*). After removing the small spines, Comanches either ate the fruit fresh or removed the seeds and dried the fruits in the sun. Informants also described another (unnamed) cactus fruit with a red juice that they rubbed on their foreheads before eating to ward off diarrhea.[40] Cactus fruits are rich in fiber, B-complex vitamins, vitamin C, copper, and magnesium.

Comanches made use of a variety of nuts, including pecans and black walnuts, both of which they stored for winter. Many tribes utilized acorns, but they require much preparation before consumption due to toxicity. Comanches used acorns from the blackjack oak, and instead of painstakingly shelling, drying, skinning, grinding, leaching, and washing, they shelled and boiled them before eating them whole.[41] Another red and sweet root, which informants said "grows like a radish and has a root like a radish" was peeled and eaten raw. This probably is the nutritious, perennial prairie turnip (*Psoralea esculenta*) also known as "Indian bread root," which numerous tribes have used extensively as a staple. The root was often boiled and mashed, then mixed with meat or fat, or was dried to store for times of famine. The purple- or blue-flowered prairie turnip is found in southeastern Oklahoma.[42]

Children reportedly liked sour sumac fruit (*Rhus glabra*) in summer.[43] Sometimes known as the "lemonade tree," the staghorn sumac produces bright red cones at the ends of its branches that resemble fuzz on a stag horn. Each cone is a single-seeded fruit (drupe). The seed inside can be ground and used a spice, or a tea can be made by steeping one or two tablespoons of drupes in hot water.[44]

Comanches refer to a plant that resembles a sunflower and has edible roots that they used to boil or roast. This could be the Jerusalem artichoke—which is neither an artichoke nor from Jerusalem; rather, it is a sunflower.[45] The tubers, known as "sunchokes," are fibrous, taste nutty, and are rich in minerals. Comanches chewed the roots of several other plants for flavor, including the button snakeroot, also known as rattlesnake master (*Eryngium yuccifolium*), and the purple prairie clover (*Dalea purpurea*); the former has a fatty taste and the latter is sweet. Another food was the young, green shoots and roasted seeds of the American lotus (*Nelumbo lutea*).[46]

Despite the nutritional value of these plants—even if they were continually available—after their confinement to a reservation Comanches could not have subsisted on fruit, onions, and nuts alone. The informants did not identify a single plant that could be considered a "leafy green." Protein was in short supply during the reservation years, and the people's later reliance on processed foods took a dramatic toll on their health.

PHYSICALITY

Officer Marcy described Comanches in 1854 as being of "medium stature, with bright copper-colored complexions and intelligent countenances, in many instances with aquiline nose, thin lips, black eyes and hair, with but little beard ... the women are short with crooked legs."[47]

In the 1830s, George Catlin deemed the Camanchees as "rather low" in stature and "approaching corpulency." He claimed to have observed an obese man named Tawahquenah (Mountain of Rocks) whom he estimated to weigh at least three hundred pounds. After denouncing the Camanchees as "one of the most unattractive and slovenly-looking races of Indians that I have ever seen ... almost as awkward as a monkey on the ground," he contradicts himself by writing that a man becomes "handsome" after he mounts a horse and "gracefully flies away like a different being." In another, equally contradictory passage, he wrote, "The Pawnee Picts, as well as the Camanchees are generally a very clumsy and ordinary-looking set of men, when on their feet; but being fine horsemen, are equally improved in appearance as soon as they mount upon their horses."[48]

It could be that Catlin exaggerated his descriptions of corpulence in order to drive home his point about Comanche horsemanship. Moreover, Catlin weighed only around 135 pounds so he may have perceived anyone bigger than he was as unusually large. John Holland Jenkins, in contrast, recalled that the Comanche men who rode with Buffalo Hump around 1830 "were almost without exception large, fine-looking men, displaying to the very best advantage the erect, graceful, well-knit frames and finely proportioned figures."[49]

In *The Comanche Empire*, Pekka Hämäläinen briefly discusses what he suggests were overfat Comanches in the early 1800s. He mentions a man named "A Big Fat Fall by Tripping" who was so obese that he had to be dragged by a travois. (The Comanche's name actually was "A Big Fall by Tripping"). Hämäläinen also discusses Tawahquenah, the man described by Catlin; as well as a "large and portly" Comanche in Texas named Pahayuko and yet another "corpulent" Comanche described by Captain Marcy. Yet another man, Tutsayatuhovit, was said to have been of "gigantic stature but also of great breadth."[50]

Hämäläinen asserts these men were all "overweight" and physically unable to fight, so they found other ways to gain prestige and wealth, such as by amassing large numbers of horses. This may very well be true; but the pertinent question is, how could these men have become so obese on a diet of game meat? At that time, Comanches did not have much if any bread or flour. Nor were sugary foods available. Any corn they may have raided or traded for was not the sweet yellow corn we know and crave today. If the corn had not already been ground into meal, someone had to exert considerable physical energy to prepare the kernels for eating. And, even if the corn were ground, if the meal had not been prepared with wood ash, anyone who consumed quantities of it might suffer from pellagra, a disease associated with niacin deficiency.

One possibility is that these men may not have been as large as described; rather, they were simply larger than their fellow Comanches and therefore stood out to observers. Hämäläinen took the vignette about "A Big Fat Fall by Tripping" from the often unreliable Wallace and Hoebel, who tell readers only that they got the story from unidentified "modern Comanches."[51] The Tawahquenah vignette comes from Catlin, who as we have seen, may have been quite biased. The story about Tutsayatuhovit comes from Kavanagh, who got it from Joe Attocknie, who actually said, "He could not come into the tipi . . . he had to stoop to get his head inside . . . they just took the front part of the tipi apart . . . he was a very, very large man." Attocknie never said he was "fat"; he may simply have been tall.[52] Regardless of whether these men were indeed extremely large or were just very tall, they were the exceptions, not the rule. If they really were obese, there could have been physical problems (thyroid, for example) that accounted for their size.

The foods we eat affect every aspect of our physical and mental health. Numerous studies have explored the connection between nutrition and the heights of peoples around the world, including of specific tribes in the United States. The theory is that taller people consume a more nutritious and varied diet than those who are shorter. How does this idea connect to the Comanches?

One hundred years after Catlin and others made their observations about Comanche physiques, Marcus Goldstein measured 1,956 Comanches and found the average height was 5 feet, 6 inches for men and 5 feet even for women.[53] His study, however, did not measure all Comanche bands, only those who lived around the Lawton, Oklahoma, area. Even then, not every Comanche in that area was, or is, short. For example, my full-blood Comanche father-in-law, Henry Mihesuah, stood 6 feet, 2 inches. My husband and his male cousins also stand over 6 feet. Peta Nocona, the father of Quanah Parker, has been described in racist terms as a "great, greasy, lazy buck."[54] Quanah was purported to be six feet tall, but his mother, Cynthia Ann Parker, was a white woman captured by Comanches in 1836, and obviously her genetics factored into his.[55]

Many reports describe Comanches as short in comparison to northern Plains tribes. Catlin argued that the Osages who lived around 1844 were the "tallest race of men in North America, either of red or white skins; there being very few indeed of the men, at their full growth, who are less than six feet in stature, and very many of them six and a half, and others seven feet."[56] Again, Catlin likely exaggerated, but clearly the Osages were taller than other tribespeople he observed. Using data collected in the 1890s by Franz Boas on Natives born between 1830 and 1872, anthropologist Richard Steckel argued in 2001 that Cheyennes and Arapahos who lived in the mid-plains area in the 1800s were possibly the tallest people on earth, in part because of their varied diet of game meats and native plants. Cheyennes traded with other tribes for a variety of foodstuffs and other goods. Steckel says that tribes that lived farther to the north or south—which would include the Comanches—were shorter because they had less available biomass to consume.[57]

Height and nutrition are complex topics, and there are arguments about the accuracy of the data and analysis. For example, John Komlos and Leonard Carlson argue that Steckel's estimate is about 1.2 inches too tall.[58] Moreover, other writers have asserted that Plains tribes were dependent on bison and rarely consumed fruits, vegetables, or roots.[59] The data used in Steckel's study were also collected generations after the Comanche confinement to Fort Sill, where they were forced to consume less-than-adequate rations. The lack of bison—which their ancestors had consumed almost every day—as well as other game meats and plants they had previously gathered could account for their shorter stature.

Under the theory that some tribespeople were tall due to their diet, environment, and lack of exposure to stressors such as disease, hard labor, and malnutrition, then the "short Comanches" merit further consideration. The Comanches were not a unified group; numerous bands roamed tens of thousands of acres within varying environments and ecosystems. Some Comanche bands' pre-reservation diets were possibly more varied than the diets of the reportedly taller Osages, Arapahos, and Cheyennes. All these tribes consumed game meats and gathered native plants, but Comanches also accessed various plant foods by raiding agricultural tribes that cultivated maize, squash, and beans, and perhaps other plants. Comanches occasionally entered the Cross Timbers region in Texas, and they either traded with or raided the agriculturalist Caddos and Wichitas.[60] Comanches also had access to Confederate cattle, which they either consumed or sold to *comancheros* (traders), who in turn sold the animals to Union agents.[61]

Prince and Steckel argue that one factor accounting for the shorter heights of Comanches and Kiowas was the stress they faced from contending with intrusive whites in the 1840s,[62] but similar stressors also affected the taller Cheyennes, who warred with Crows, Lakotas, Comanches, Kiowas, Apache bands, Blackfeet, and Shoshones, in addition to dealing with US encroachment. Stress, anguish, and depression over population, land, and cultural loss came at different times to tribes depending on when they encountered colonizers, but this situation was common to all Plains Natives by around the 1870s.

A factor that is barely touched on in all of these height studies is that Comanches captured American, Mexican, and Spanish people, as well as members of various tribes. Individuals from Mexico, especially, tended to be shorter than North American tribes.[63] Through the decades, numerous outsiders have commented that Comanches appeared to be part Mexican. General Philip Sheridan, for example, remarked in the mid-1880s that Cheyennes and Arapahos look "more purely Indian and the finest models" and he assessed Kiowas and Lipan Apaches as the next most "Indian" in appearance. He believed Comanches were "adulterated" with Mexican blood.[64] James Mooney commented in 1896 that 25 percent of the Comanche population reflected race mixing with their ancestors' captives.[65] In 1927, Clark Wissler estimated that almost 63 percent of Comanches were full-bloods, but based his assessment only on "hair dress, costume, and posture."[66] Only four years later, however, Marcus Goldstein estimated that a mere 10 percent were full-bloods.[67] Comanches today are generally aware of the racial mixture in many of their family trees.

TREATY RESTRICTIONS

In October 1867, Comanches signed the Treaty of Medicine Lodge Creek, which gave Comanches, Kiowa-Apaches, and Kiowas a reservation between the Red and Washita Rivers, an area General Sheridan deemed as "the finest lands in the Indian Territory." The treaty stipulated that they could hunt south of the Arkansas River and would receive clothing for thirty years. The treaty also required that tribes send their children to school so that they would learn to farm. These were not popular provisions, but the tribes agreed because they could hunt buffalo, albeit with military escort. The treaty said nothing about supplying food to the tribes other than to provide seeds, farming tools, and monies for those who "grow the most valuable crops."[68]

Around four thousand Natives settled at Fort Sill. Not every Comanche agreed to the treaty and some had no intention of living on the reservation. Those who did settle there contended with a lack of medical services and poor rations. Some members of the Quahada Band, such as Quanah Parker and Mihesuah, refused to settle and preferred to raid and hunt as they had traditionally. This proved difficult because white hunters continued to slaughter bison for their hides and tongues, then leave the meat to rot. Prior to 1800, there may have been 60 million bison. By 1840 that number had dropped to almost 37 million. By 1870 there were 5.5 million, and in 1889 there were only 541 animals left.[69] And, white settlers continued to intrude on to tribal lands guaranteed by treaties.

On June 26, 1874, buffalo hunters at the Battle of Adobe Walls defeated seven hundred Indians, including Comanches and allied Arapahos, Kiowas, Kiowa-Apaches, and Southern Cheyennes. Three months later Colonel Ranald Slidell Mackenzie of the Fourth Cavalry attacked the Comanches in their Palo Duro Canyon. Comanches escaped, but Mackenzie slaughtered at least 1,500 of the tribe's horses, burned their tipis, and took their supplies. Without their horses and bison, Comanches faced starvation and finally surrendered at Fort Sill.[70]

At the fort there were not enough buildings and never enough resources. The government attempted to turn Comanches into cattle ranchers, but conditions were so bad that the starter herd of 340 cattle was butchered for food. The agents persisted with the cattle, and the people still resisted. Dogs chased the herds and boys used them for target practice. In an effort to regain their past hunting glory, men engaged in mock bison hunts.[71] Officer Hazen hired white men to plow and fence 1,200 acres, then plant small gardens, corn, and fruit trees. Initially, a few Peneteka men and women worked the fields.

The government issued corn to the Comanches, but they had no way to grind it. Therefore, Agent Tatum allowed them to sell it to whites so that Comanches could buy their preferred but nutritionally void sugar and coffee, or they fed the unwanted corn to their horses. Tatum also expressed concern at the "low grade" of flour sent. Bacon often contained rocks and barrels of pork contained coal in order to make the foods sold by weight heavier, and therefore more profitable to those who sold to the contractors. And it was not just the Natives who had to go without. The officers and workers also depended on the rations.[72] In August 1869, rations were supposed to consist of one-and-a-half pounds of beef, three-quarters of a pound of corn or meal, one-quarter pound of flour, four pounds of sugar, and two pounds of coffee. Often the supplies did not arrive. Sometimes bacon was substituted for beef and occasionally the reservation population received tobacco and soda, the later suggesting they were supposed to make bread. Unfortunately, beef consisted of Texas longhorns that had been healthy in fall but became so starved and malnourished by winter that they could barely walk to Fort Sill. Luckily for Comanches, they could still hunt bison, at least for a few more years.[73] Often rations were delayed because of poor roads and swollen waterways with no bridges.

In spring of 1876, Comanches and Kiowas were presented with sheep and goats that had been driven from New Mexico. The tribes did not like mutton, so the animals were used instead for archery practice, and wild animals enjoyed the remains.[74] The repercussions of living sedentary lifestyles and not being able to outrun diseases as they had done previously continued through the early 1880s, when many died from whooping cough and malaria.[75] The advent of epidemic diseases was not surprising. Marcy wrote in 1885 about the Northern Comanches' ability to escape disease because they roamed the plains, which he described as a "vast district" where "perhaps no part of the habitable globe is more favorable to human existence, so far as the atmosphere is concerned... free from marshes, stagnant water, great bodies of timber, and all other sources of poisonous malaria, and open to every wind that blows, this immense grassy expanse is purged from impurities of every kind, and the air imparts a force and vigor to the body and mind which repays the occupant in a great measure for his deprivations."[76]

The bison continued to dwindle and without meat to supplement their meager rations, Natives at Fort Sill more often than not went hungry. By 1881, bison were almost gone, yet the government decreased the amount of food sent to tribes "one-fourth by insufficient appropriations."[77] When rations did arrive, hungry Natives tended to consume them immediately. In addition, in 1881, a severe drought caused complete crop failure.[78] Not surprisingly, an Indian agent noticed that many Comanches had become addicted to the hallucinogenic "mescal beans."[79]

In 1882, at the Kiowa, Comanche, and Wichita Agency the 1,400 Comanches, as well as smaller numbers of Kiowas, Caddos, and Seminoles, went without food several days a week and resorted to butchering their breeding cattle. The Indian agent suggested the disturbing theory that some in the government were advocates of that "starving process" as a motivator for the Indians to learn to raise cattle and to adopt "the ways of civilized life."[80]

Inspector General Tinker commented that at Anadarko in 1891 the Comanches often chased cows like they had hunted bison, then they skinned the hapless cows before they were dead. This was a sight that bothered many whites, especially when they watched mothers feed their children pieces of raw liver dipped in blood.[81] Wallace and Hoebel exaggerate this reality, writing that

"Children crowded around the butcher and begged" for the "finest delicacy" of liver covered with bile.[82] The authors assume that this behavior spurred the Commissioner of Indian Affairs to order slaughterhouses be built and that women and children be disallowed from witnessing the slaughters and that no one be allowed to consume blood and intestines.[83] Despite the order, Comanches at Anadarko continued to kill cattle in the same manner.[84]

In June 1889, the *St. Louis Post-Dispatch* newspaper wrote an editorial complaining about the rich farmland set aside around Fort Sill for the Comanches. The land sat unused, claimed the paper, in "the most flagrant case on record of fine farming lands wasted—the garden spot of the Territory withdrawn from agriculture." The writer referred to the Comanches as "a few thousand of the most wretched Indians to be found in the territory" and "vagabond relics" of a once powerful Comanche Nation. He complained that the Comanches did little or no hunting, did not farm at all, and were "entirely dependent on government rations."[85] The "garden spot" descriptor was a bit of an exaggeration, because Indian Agent James F. Randlett wrote one year later that although the bottomlands were excellent for farming, crops often failed because of hot winds and irregular rainfall. Kaffir corn grew well, but it was mainly fodder for cattle.[86]

Regardless of the quality of the farmland, most Comanches still did not care to try their hand at farming. They preferred eating meat and foraging, when allowed, but usually had to make do with government rations, or if they happened to have money, food they bought. Sneed, the trader at Fort Sill, stated that he purchased goods for his store in St. Louis, Chicago, and Fort Worth, and that the Indians particularly liked dried fruits such as dates, figs, raisins, and prunes, so he was sure to order "large quantities." He did not, however, comment about how much they purchased.[87]

Comanches continued to receive flour, but did not know what to do with it. Francis E. Leupp, commissioner of Indian Affairs from 1904 to 1909 recounted in 1911 what Quanah told him about flour and salt:

> That was the first time any of our band had seen either flour or salt in these forms. We knew the taste of the alkali of the desert, but salt in grains like this we could not understand. One Indian after another opened his bag at the top. Ran his hand in, and put a fistful of the contents into his mouth. Then it was funny to see them go spitting and sputtering about, trying to get rid of the taste and the burning on the back of the tongues. To flour we were equally strangers. We knew something of grain foods, but our

"Butchering Beef and Loading the Meat." *Photo by Hutchins & Lanney, 1891. Courtesy of Amon Carter Museum of American Art, Fort Worth, Texas, P1967.167.*

flour or meal was made by the rude process of smashing and rubbing between stones, and was consequently coarse and gritty; but here was a substance so soft that it seemed to disappear as we put it by little pinches into our mouths; it had no flavor, and when, in our effort to get enough to extract some tastes from it, we filled our mouths, we nearly choked, and then found our teeth and tongues gummed up with a thick paste which was even harder to get rid of than the salt. But Indians are resourceful; and as those settlers, crossing the next rise of ground looked back at us, they saw every Indian engaged in slitting his bags at the bottom, emptying the salt and flour on the ground, and drawing the bags, now open at both ends, over his calves for leggings.[88]

There is no evidence that Comanche fried the flour they received at Fort Sill. The "fry bread myth" is most commonly associated with Navajos, many of whom assert that fry bread is what allowed their ancestors to survive their confinement at Bosque Redondo from 1864 to 1868. Research reveals that no one made fry bread there; rather, fry bread emerged as a common food among tribes after 1960.[89] Still, Comanches could have mixed flour with water and dropped the balls onto coals to cook, or perhaps made unleavened bread.

Comanches had to have eaten foods besides rations, such as the plant foods they sometimes gathered. In addition, Fort Sill is situated near stands of timber and Cache and Medicine Bluff Creeks. The nearby Wichita Mountains, waterways, and grasslands supported antelope, deer, and smaller animals such as rabbits and squirrels, in addition to a variety of fish and the previously mentioned nuts, fruits, and bulbs.

After the reservation lands were opened to whites eager to take any resources they could, Comanches attempted to settle on their allotments, but they no longer received government rations, and many leased their lands to whites in order to purchase food. Many eventually sold their allotments for pittances. The 1928 "Meriam Report," a lengthy study of conditions on Indian lands, revealed serious problems such as poverty, poor health care, inadequate housing and education, and lack of self-determination.[90] Indeed, many people were poor—including whites and blacks—and not everyone had access to nutritious foods. Some however, managed because they knew how to farm and cultivate gardens.

Around 1871, the Fort Sill Indian School opened four miles south of Fort Sill and a mile from Cache Creek. One young man who was taken from his parents and delivered to Fort Sill was a son of Mihesuah, who went by that name until he was given the name Joshaway. He and his brother Topachi ran away, but were caught and returned by soldiers. Teachers cut his long braids and banned him from speaking Comanche, and missionaries at the nearby Deyo Mission Christianized him.[91] Like many of the other boys at the school, Joshaway found out that one of the only professions open to him was farming. In the 1920s, Joshaway hired black sharecroppers to help work the fields on his allotted land, and white neighbors called on him regularly for assistance in feeding their families. Joshaway cultivated eighty acres of cotton and maize with a team of horses and a single plow on the Mihesuah home allotment and the allotments that belonged to his sisters. Joshaway and his son, Henry, took turns following the plow, holding onto its two handles and with a strap around their necks, and they harvested and shucked the corn by hand. The Mihesuahs also grew a large family garden in the front yard and picked from fruit trees surrounding the property. Joshaway rotated his crops and depended on rain, their well, and the nearby creeks for water. They raised chickens, hogs, and turkeys—the turkeys were also driven between the rows of plants to eat the grasshoppers. Rabbits often ate the young cotton plants, and the Mihesuahs replanted more than once per season. They hunted squirrels and rabbits with bows Joshaway made from bois d'arc wood and deer sinew, and arrows made from dogwood and turkey feathers. The local creek supplied them with fish. Wild turkeys and deer had become scarce by the 1940s, but there were plenty of cottontails, jackrabbits, coons, opossums, bobcats, squirrels, quail, snakes, and coyotes, which people referred to as wolves. When Joshaway killed a cow or steer, all his neighbors came to visit. They consumed almost all of the animal, including the cleaned and boiled entrails they called *quetucks*. During the Great Depression, the Mihesuahs consumed and canned their garden produce, ate beef when they could get a cow, as well as having pork cured with brown sugar. They ate chitterlings, saved the grease, and saved hog lard in iron pots.[92]

By 1910, Joshaway was among the 15 percent of Comanches who were successful farmers. In 1925, there were only 500 farmers out of 5,023 Comanches, Kiowas, and Kiowa-Apaches, and six years later, there were 599 farmers out of 5,500.[93] Many Comanches had sold their allotments. Clearly, not everyone was eating garden produce and game meats.

DIABETES

Kelly M. West asserts that diabetes was unknown among Indian Territory/Oklahoma tribes until 1940. He cites nurses and doctors who, in the 1930s, could not recall any cases of diabetes among Natives, nor any references to blood glucose tests performed in the early twentieth century. This suggests that there were no cases of diabetes prior to that time. However, West only used sources from the Lawton Indian Hospital, Shawnee Indian Clinic, North Carolina Cherokees, and Mississippi Choctaws.[94]

West does not take into consideration that Indian Territory newspapers contained thousands of advertisements peddling cures for diabetes from 1880 through Oklahoma statehood in 1907, and thousands more up to 1940. Among other signs of widespread ailments, there were at least three thousand advertisements for dentists between the start of the Civil War and Oklahoma statehood in 1907, in addition to numerous comments about health issues by boarding school doctors, in reports of the commissioners of Indian Affairs, and in oral testimonies of Native and non-Native residents of the Territory.[95] There were plenty of ailments that suggest many residents—most notably members of the Five Tribes (Cherokees, Choctaws, Chickasaws, Mvskokes, and to a lesser extent, Seminoles)—were on their way to becoming diabetic after the Civil War: piles (hemorrhoids), "bowel problems," "liver complaint," appendicitis, sour stomach, indigestion, "wind on the stomach," bloating, colic, stomachaches, diarrhea, and biliousness of the blood.[96] And, the foods that white Americans ate, Indians also consumed.

Unlike western and plains tribes, many of which were able to continue hunting and gathering, members of the Cherokee, Chickasaw, and Choctaw tribes had been consuming white flour, sugar, and lard prior to their removal to Indian Territory in the 1830s. Many tribal members, especially relatively affluent mixed-bloods, ate the same foods that white residents consumed, including large quantities of white flour in the form of biscuits, gravy, and pancakes, in addition to cakes, pies, cookies, candy, lard, and butter. After the Civil War, at a time when Comanches still hunted bison and other game animals, and were not yet eating white flour and sugar, the physician of the Cherokee Male and Female Seminaries commented that some of the Cherokee students were growing obese from their fatty, sugary, and salty diet.[97] Illustrating how often the school served dishes made of flour, in 1893, an order was placed for sixteen thousand pounds of white wheat flour for fewer than three hundred students.[98]

In his article West also questionably states, "Oklahoma tribes had, for the most part, considerable medical attention in the 19th and early 20th centuries, often by the same physicians who were finding diabetes common in their white patients."[99] The reference to "physicians" is a misnomer. White physicians treated some Natives, but by no means all of them. And, many of those "physicians" had only a modicum of training. For example, my great-grandfather Thomas Abbott Sr. lived on the Choctaw Nation and worked as a physician for the mining towns around what is today McAlester, including modern-day White Chimney, Scipio, Indianola, Canadian, and Krebs. He attended medical school at Paducah, Kentucky, for one year, then learned more medical care as an "apprentice" to two German doctors. He passed the modest tribal medical test then served as physician for the Bolen-Dornell, Dan Edwards, McEvers, and Galveston coal companies.[100] Semi-trained physicians like Dr. Abbott would have had little knowledge of diabetes, much less of blood sugar measurements. Therefore, I consider it very possible that many Natives suffered from diabetes or prediabetes before the Civil War, and some probably did after the 1830s removals.

The health of Oklahoma Indians continued to decline through the twentieth century. In the 1970s, a team of researchers discovered that 75 percent of 2,095 Oklahoma Indians from fifteen tribes weighed more than 160 percent "of standard" weight. Seventy-five percent of the Comanches interviewed were obese.[101] Kelly West argued that such a prevalence of obesity was not new, claiming Comanches historically viewed corpulence as desirable based on the example of "Chief Big Fall by Tripping" (he added "Chief" to the name to prove his point). He further wrote of studying a group of women who were "exceedingly fat." These women stated that they believed they should weigh fifty pounds "more than what would be considered ideal by women from the upper-class white cultures of Oklahoma and the United States." He then asserted that the Oklahoma Indian women he spoke with "worry a little when my diets cause them to fall below 170 pounds. And, many say they don't feel as well when they are so peculiarly 'thin.' They tell me of concerns of friends and family to whom they may appear dangerously frail at weights under 200." West's claim has influenced other researchers to conclude that modern Comanches strive for obesity as a standard of normalcy.[102]

At the end of West's paper about obesity being a desirable trait among Plains tribes, he writes cavalierly, "The purpose of this essay has been more to entertain than to instruct or enlighten."[103] Clearly, he did not write this piece for an Indigenous audience. In addition, West's claim is disconcerting because if obesity is normative among Comanches, then attempts to rectify the obesity and diabetes epidemics might be futile. West's assertion has also influenced health researchers who have not checked his trail of investigation.

In the early 1990s, Anne Walendy Davis interviewed eleven full-blood Comanches about their attitudes on health and food. All the people she met took medications for high blood pressure, diabetes, arthritis, or emphysema, and interviews revealed that what they consumed was directly related to their health problems.[104] Shirley C. Laffrey argued in 1985 that in order to formulate realistic health goals for a culture, one must first understand what that culture believes constitutes good health.[105] To do that, however, one must first define the "culture." Not all members of a tribe think the same way. Value differences based on cultural beliefs permeate all tribal societies. Some tribes have been factionalized since the eighteenth century, with the Cherokees being a prominent example. Political and economic arguments have swirled for centuries based on views about intermarriage with non-Indians, conversion to Christianity, adherence to traditional ceremonies, honoring male and female cultural roles, and other factors. Many members of tribes across the country, even full-bloods, know little if anything about their traditional tribal culture, while some mixed-bloods speak their tribal language and participate in tribal religious ceremonies. Although all members of a tribe may have a common history and proudly declare their tribal citizenship, they no longer adhere to common culture.

Tribal traditions have been altered, and in the case of the Comanches, the nomadic hunter lifestyle changed after the demise of the bison and confinement to Fort Sill. There are at least 17,000 enrolled citizens of the Comanche Nation of

Oklahoma. Almost eight thousand live around the tribal headquarters, located in Lawton, in southern Oklahoma, but other members of the nation live throughout the country. As is true of other tribes, some Comanches speak their language and are knowledgeable about their history, whereas others have thoroughly adopted the mores and worldviews of the mainstream society. These differences in worldviews are reflected in several studies.

In 1986, Lee and colleagues relied on West's assertions in writing that "fat or abundant women were considered beautiful"; that middle-aged Plains Indian women believed they should weigh about fifty pounds more than what upper-class white women considered ideal; and that Comanche women expect to gain "considerable weight" between the ages of twenty and forty. They are suggesting, as did West, that Comanche women strive toward obesity or, at least, to look that way.[106] In another study, one Comanche woman commented, "I think Indians are not so concerned about being skinny, you know, they enjoy eating and, you know, they won't sacrifice what it takes to . . . you know, what other people thinks makes 'em look nice. You know, makes 'em look slim and trim and pretty and all that to Indians, that's insignificant. . . . I wanta be healthy, but I'm not gonna put so much emphasis on my bein' tall and thin . . . that's just not important to me . . . non-Indians probably do a lot more in that area than Indian people do, I feel like."[107] Lee and colleagues come to the unsubstantiated conclusion that so many Comanches are overfat because they believe that "fat or abundant women were considered beautiful."[108] These are not expressions of traditional culture; rather, they are individual beliefs. I know numerous Indigenous women who would not consider these statements to be standard. Numerous Indigenous women are physically active, maintain gardens, and monitor their nutritional intake.

In another instance, Davis interviewed a woman who stated: "I tell you what, when I eat beef, I have to have tallow with it—the fat. Even if I had a piece of steak . . . they trim the meat too much and to me, that's Comanche. See. That's something I don't worry about."[109] The Comanches were traditionally nomadic hunters. Unlike hundreds of other tribes who cultivated corn, squash, beans, peppers, and other crops, Comanches have no agricultural tradition, so the stance that eating beef fat is a Comanche tradition might make sense to this woman. The problem is that beef is not a traditional food for Comanches. They did not hunt cattle. The bison they consumed contained less saturated fat than beef, as well as fewer calories. In addition, Comanches did not traditionally eat the floury, sugary, salty, and fatty foods that have greatly contributed to the diabetes explosion.

It is true that some Comanches joke about being called "Lards of the Plains." That attitude, however, is not the same as believing a body weight as described by West is desirable. The Comanche Nation is attempting to solve serious diabetes and obesity problems among its people. The Comanche Nation Diabetes Awareness Program has a "mission to increase awareness & prevent diabetes in American Indian and Alaska Native communities through exercise and nutrition"; there are also programs to combat drug use, domestic violence, child abuse, and suicide.[110] The nation has a fitness center that offers a "Workout Warriors" exercise program, along with Zumba, aerobics, and jiu-jitsu courses. Diabetes awareness initiatives include health fairs, a Nike shoe program, low-impact and chair exercises, and walking programs. Yet, like other tribes, the Comanche Nation often sends out messages that contradict those health-promotion efforts. The monthly *Comanche Nation News* often contains articles about exercise and healthy diet, while simultaneously publishing recipes for nontraditional and unhealthy foods containing highly processed ingredients.

Lack of an agricultural tradition to revitalize, limited resources, cultural disconnect, and no plans for food initiatives are among the reasons why Comanches suffer from high rates of diabetes, obesity, high blood pressure, and substance abuse and the problem shows no sign of abating among full-blood, mixed-blood, traditionalist, and thoroughly acculturated Comanches alike. Many tribes with agricultural legacies, such as Choctaws, Cherokees, Anishinaabes, Pueblos, and Iroquois, are now fighting their health problems by "recovering their ancestors gardens"; that is, by researching their traditional foodways, growing gardens, and incorporating traditional dishes into their diets.[111] Those Natives whose ancestors were primarily hunters cannot hunt game animals every day, even if they have the wherewithal to procure hunting licenses. Instead, they can choose to incorporate into their diets foods their tribe historically gathered or traded for, in addition to foods that they perhaps did not consume historically but that are indigenous to this hemisphere. Backyard and community gardens—even small ones—are becoming popular ways to produce food, to stay active, and to learn to cook healthy dishes.

Tribal councils are supposed to help safeguard the health and well-being of their constituents. Tribal members have the power to vote for and hold accountable tribal leaders who promote access to nutritional education, health care, nutritious and sustainable foods, garden tools and seeds, and support for elders who can impart traditional knowledge about food and medicinal plants to the younger generations. Savvy and determined people comprise the Comanche Nation. I hope that someday they will find solutions to their health crisis.

NOTES

Epigraph: Henry Mihesuah, *First to Fight*, edited by Devon Abbott Mihesuah (Lincoln: University of Nebraska Press, 2002), 8.

1. Oklahoma State Department of Health, "Behavioral Risk Factors among American Indians in Oklahoma," https://www.ok.gov/health2/documents/HCI_BRFSSNative American2004.pdf.

2. For general works on the Comanches, see John Francis Bannon, *The Spanish Borderlands Frontier, 1513–1821* (New York: Holt, Rinehart and Winston, 1970); T. R. Fehrenbach, *The Comanches: The Destruction of a People* (New York: Alfred A. Knopf, 1974); Albert S. Gilles Sr., *Comanche Days* (Dallas, TX: Southern Methodist University Press, 1974); William T. Hagan, *United States–Comanche Relations: The Reservation Years* (New Haven, CT: Yale University Press, 1976); Pekka Hämäläinen, *The Comanche Empire* (New Haven, CT: Yale University Press, 2009); Ernest Wallace and E. Adamson Hoebel, *The Comanches: Lords of the South Plains* (Norman: University of Oklahoma Press, 1952); Elizabeth A. H. John, *Storms Brewed in Other Men's Worlds: The Confrontation of Indians, Spanish, and French in the Southwest, 1540–1795* (College Station: Texas A&M University Press, 1975); Thomas A. Kavanagh, ed., *Comanche Ethnography: Field Notes of E. Adamson Hoebel, Waldo R. Wedel, Gustav G. Carlson, and Robert H. Lowie* (Lincoln: University of Nebraska Press, 2008); Kavanagh, *The Comanches: A History, 1706–1875* (Lincoln: University of Nebraska Press, 1999); Stanley Noyes, *Los Comanches: The Horse People, 1751–1845* (Albuquerque: University of New Mexico Press, 1993); S. C. Gwynne, *Empire of the Summer Moon: Quanah Parker and the Rise and Fall of the Comanches, the Most Powerful Indian Tribe in American History* (New York: Scribner's, 2010); W. W. Newcomb Jr. "Comanches: Terror of the Southern Plains," in *The Indians of Texas: From Prehistoric to Modern Times*, edited by W. W. Newcomb (Austin: University of Texas Press, 1961), 155–91.

3. E. A. Hoebel, *The Political Organization and Law-Ways of the Comanche Indians* (Millwood, NY: Kraus International, 1940), 12–13; Thomas W. Kavanagh, "Comanche Population Organization and Reorganization, 1869–1901: A Test of the Continuity Hypothesis," Pt. 2, Memoir 23: *Plains Indian Historical Demography and Health. Plains Anthropologist* 34, no. 124 (May 1989): 99–111.

4. See Randolph B. Marcy, *Exploration of the Red River in Louisiana, 1812–1887* (Washington, DC: A. O. P. Nicholson, Public Printer, 1854), 93–96, for details about the ranges of Comanche groups.

5. Wallace and Hoebel describe this in *Comanches*, 67–68, but give no citation.

6. Robert Harry Lowie, "The Northern Shoshone," *Anthropological Papers of the American Museum of Natural History*, vol. 2, pt. 2 (New York: Trustees of the American Museum of Natural History, 1909), 185.

7. See F. M. Buckelew, *The Life of F. M. Buckelew, the Indian Captive, as Related by Himself* (Bandera, TX: Hunter's Printing House, 1925), 99–100.

8. George Catlin, "Letter No. 42: Great Camanchee Village," in *Letters and Notes on the Manners, Customs, and Conditions of the North American Indians: Written During Eight Years' Travel (1832–1839) Amongst the Wildest Tribes of Indians in North America* (New York: Dover, 1973), 2:65.

9. Marcy, *Exploration of the Red River in Louisiana*, 103–4.

10. "Governor Anza Dictates Comanche Peace 1786," in *Forgotten Frontiers: A Study of the Spanish Indian Policy of Don Juan Bautista de Anza, Governor of New Mexico, 1777–1787*, edited and translated by Alfred Barnaby Thomas (Norman: University of Oklahoma Press, 1932), 71–83; Wallace and Hoebel, *Comanches*, 69.

11. Catlin, *Letters and Notes*, 2:79, 83.

12. Marcy, *Exploration of the Red River in Louisiana*, 111.

13. Nelson Lee, *Three Years among the Comanches: The Narrative of Nelson Lee, the Texas Ranger* (Albany, NY: Baker Taylor, 1859), 114–15.

14. Marcy, *Exploration of the Red River in Louisiana*, 106.

15. Wallace and Hoebel, *Comanches*, 71; Fannie Hudson interview, April 27, 1937, Indian and Pioneer Papers (IPP), 45: 41–42; Nathan J. McElroy interview, May 11, 1937, IPP 58: 110–11, Western History Collections, University of Oklahoma, Norman.

16. Thomas C. Battey, *The Life and Adventures of a Quaker among the Indians* (Boston: Lee and Shepard, 1876), 283.

17. Gustav G. Carlson and Volney H. Jones, "Some Notes on Uses of Plants by the Comanche Indians," *Papers of the Michigan Academy of Science* 25 (1940): 530. For uses for mesquite, see Ken E. Rogers, *The Magnificent Mesquite* (Austin: University of Texas Press, 2010).

18. Battey, *Life and Adventures of a Quaker*, 322–33.

19. Gen. R. A. Sneed, "The Reminiscences of an Indian Trader," *Chronicles of Oklahoma* 14, no. 2 (1936): 145.

20. Wallace and Hoebel, *Comanches*, 70.

21. Battey, *Life and Adventures of a Quaker*, 323, 330; Nearly sixty years later Sneed wrote virtually the same thing, and it appears he copied from Battey. See "Reminiscences of an Indian Trader," 154–55.

22. Robert Goldthwaite Carter, *On the Border with Mackenzie; or, Winning West Texas from the Comanches* (n.p.: Antiquarian Press, 1935), 279; Melburn Thurman has already pointed out inaccurate assertions in Wallace and Hoebel's work in "Nelson Lee and the Green Corn Dance: Data Selection Problems with Wallace and Hoebel's Study of the Comanches," *Plains Anthropologist* 27, no. 97 (1982): 239–43. Noting the authors' use of "a fraudulent Comanche source" regarding the Green Corn Dance, among other things, Thurman stated: "Quite clearly, as Ethnohistory, Wallace and Hoebel's book is methodologically unsound, as seen in their use of generalized assertions based on data agreement to the exclusion of further analyses of conflicting cases" (242).

23. Carter, *On the Border with Mackenzie*, 279.

24. Herman Lehmann, *Nine Years among the Indians, 1870–1879: The Story of the Captivity and Life of a Texan among the Indians*, edited by J. Marvin Hunter (Albuquerque: University of New Mexico Press, 1993), 206.

25. Battey, *Life and Adventures of a Quaker*, 323; Carl Coke Rister, *Border Captives: The Traffic in Prisoners by Southern Plains Indians, 1835–1875* (Norman: University of Oklahoma Press, 1940), 18–19.

26. Wallace and Hoebel, *Comanches*, 71.

27. Wallace and Hoebel, *Comanches*, 72; Lehmann, *Nine Years among the Indians*, 115, 116, 182; Noah Smithwick, *The Evolution of a State* (Gammel Book Co., 1935), 178–79.

28. Carlson and Jones, "Uses of Plants by the Comanche Indians," 517–42; Herman Asenap interview, October 30, 1937, IPP 3:185–90. The age of Asenap, the man who assisted Carlson and Jones, is unclear. The IPP says he was fifty-five in 1933, whereas the birthdate on his headstone would have made him fifty at that time; see *Find a Grave*, http://www.findagrave.com/cgi-bin/fg.cgi?page=gr&GRid=50066129. For a list of plants Comanches used as food and medicine, see the Native American Ethnobotany online database at http://naeb.brit.org/ (type "Comanche" in the search box).

29. Carlson and Jones, "Uses of Plants by the Comanche Indians," 528–29. See also *Oklahoma Vascular Plants Database*, http://www.oklahomaplantdatabase.org/.

30. Molly L. Sultany, Susan R. Kephart, and H. Peter Eilers, "Blue Flower of Tribal Legend: 'Skye blue petals resemble lakes of fine clear water,'" *Kalmiopsis* 14 (2007): 28–35; Briony Penn, "Restoring Camas and Culture to Lekwungen and Victoria: An Interview with Lekwungen Cheryl Bryce," *Focus Magazine* (June 2006): http://www.firstnations.de/media/06-1-1-camas.pdf.

31. Carlson and Jones, "Uses of Plants by the Comanche Indians," 526–27.

32. Carlson and Jones, "Uses of Plants by the Comanche Indians," 527.

33. Frank G. Speck, "A List of Plant Curatives Obtained from the Houma Indians of Louisiana," *Primitive Man* 14 (1941): 57; James William Herrick, "Iroquois Medical Botany" (PhD diss., State University of New York, Albany, 1977), 306; Leland C. Wyman and Stuart K. Harris, *The Ethnobotany of the Kayenta Navaho* (Albuquerque: University of New Mexico Press, 1951), 18.

34. Carlson and Jones, "Uses of Plants by the Comanche Indians," 526, 531.

35. Carlson and Jones, "Uses of Plants by the Comanche Indians," 527; David E. Jones, "Comanche Plant Medicine," *Papers in Anthropology* 9 (1968): 3.

36. Carlson and Jones, "Uses of Plants by the Comanche Indians," 526.

37. Carlson and Jones, "Uses of Plants by the Comanche Indians," 520. Kelly Kindscher, *Edible Wild Plants of the Prairie* (Lawrence: University Press of Kansas, 1987), 12–17.

38. Carlson and Jones, "Uses of Plants by the Comanche Indians," 526.

39. Carlson and Jones, "Uses of Plants by the Comanche Indians," 526; Kindscher, *Edible Wild Plants*, 169–75.

40. Carlson and Jones, "Uses of Plants by the Comanche Indians," 523, 527.

41. Carlson and Jones, "Uses of Plants by the Comanche Indians," 531. See the recipe for Luiseño "Weewish" in D. Mihesuah, *Recovering Our Ancestors' Gardens* (Lincoln: University of Nebraska Press, 2005), 142–44.

42. Carlson and Jones, "Uses of Plants by the Comanche Indians," 529; Kindscher, *Edible Wild Plants*, 183–89; Ernest Small, *North American Cornucopia* (Boca Raton, FL: CRC Press, 2013), 545–49.

43. Carlson and Jones, "Uses of Plants by the Comanche Indians," 527.

44. John K. Crellan and Jane Philpott, *Reference Guide to Medicinal Plants: Herbal Medicine Past and Present* (Durham, NC: Duke University Press, 1990), 417–19.

45. Carlson and Jones, "Uses of Plants by the Comanche Indians," 529–30; Kindscher, *Edible Wild Plants*, 129–33.

46. Carlson and Jones, "Uses of Plants by the Comanche Indians," 531–32.

47. Marcy, *Exploration of the Red River in Louisiana*, 98.

48. Catlin, *Letters and Notes*, "Letter No. 42," 65, 67; and "Letter No. 43," 73–75.

49. John Holland Jenkins, *Recollections of Early Texas: Memoirs of John Holland Jenkins* (Austin: University of Texas Press, 1958), 8.

50. Hämäläinen, *Comanche Empire*, 259, 424 n. 38.

51. Wallace and Hoebel, *Comanches*, 39–40; Kavanagh, *Comanche Ethnography*, 309.

52. Joe Attocknie interview, June 17, 1969, Doris Duke Collection, T-448, p. 12, Western History Collections, University of Oklahoma. Kavanagh, *Comanche Political History: An Ethnohistorical Perspective* (Lincoln: University of Nebraska Press, 1996), 147.

53. Marcus S. Goldstein, "Anthropometry of the Comanches," *American Journal of Physical Anthropology* 19 (July–September 1934): 289–319.

54. James T. Shields, *Cynthia Ann Parker* (St. Louis, MO: Chas. B. Woodward, 1886): 32.

55. There are many real and imagined stories about Cynthia Ann Parker's life, but the best researched is Margaret Schmidt Hacker's *Cynthia Ann Parker: The Life and the Legend* (El Paso: Texas Western Press, 1990).

56. Catlin, "Letter No. 38" in *Letters and Notes*, 2:40, also available at http://www.xmission.com/~drudy/mtman/html/catlin/letter38.html.

57. Richard H. Steckel, "Inequality Amidst Nutritional Abundance: Native Americans on the Great Plains," *Journal of Economic History* 70, no. 2 (2010): 265–86; Joseph M. Prince and Richard Steckel, "Nutritional Success on the Great Plains: Nineteenth-Century Equestrian Nomads," *Journal of Interdisciplinary History* 33, no. 3 (2003): 353–84; Steckel and Prince, "Tallest in the World: Native Americans of the Great Plains in the Nineteenth Century," *American Economic Review* 91, no. 1 (2001): 287–94.

58. John Komlos and Leonard A. Carlson, "Toward the Anthropometric History of Native Americans, c. 1820–1890," *Social Science Research Network Working Paper Series* (August 2010): 5.

59. See, for example, R. B. Marcy and G. B. McClellan, "Exploration of the Red River of Louisiana in the Year 1852," US 32d Congress, Senate Executive Document 54 (Washington, DC: Government Printing Office, 1853), 102; Clark Wissler, *The American Indian: An Introduction to the Anthropology of the New World* (New York: Oxford University Press, 1917), 19.

60. Elizabeth A. H. John, "Portrait of a Wichita Village, 1808," *Chronicles of Oklahoma* 60, no. 4 (1982): 416, 417, 418, 421–22; Pekka Hämäläinen, "The Western Comanche Trade Center: Rethinking the Plains Indian Trade System," *Western Historical Quarterly* 29 (1998): 485–513.

61. Frances Levine, "Economic Perspectives on the Comanchero Trade," in *Farmers, Hunters, and Colonists: Interaction between the Southwest and the Southern Plains*, edited by Katherine A. Spielmann (Tucson: University of Arizona Press, 1991): 155–70.

62. Prince and Steckel, "Nutritional Success on the Great Plains," 379.

63. See John H. Moore and Janis E. Campbell, "Blood Quantum and Ethnic Intermarriage in the Boas Data Set," *Human Biology* 67, no. 3 (1995): 499–516; S. V. Subramanian, Emre Özaltin, and Jocelyn E. Finlay, "Height of Nations: A Socioeconomic Analysis of Cohort Differences and Patterns among Women in 54 Low- to Middle-Income Countries," *Public Library of Science* 6, no. 4 (2011): e18962; R. L. Jantz, D. R. Hunt, A. B. Falsetti, and P. J. Key, "Variation among North Amerindians: Analysis of Boas's Anthropometric Data," *Human Biology* 64 (1992): 435–61.

64. De B. Randolph Keim, *Sheridan's Troopers on the Borders: A Winter Campaign on the Plains* (Philadelphia: David McKay, 1885), 195.

65. James Mooney, "The Aboriginal Population of America North of Mexico," *Smithsonian Miscellaneous Collection* 80, no. 7 (1928): 13.

66. Clark N. Wissler, *North American Indian of the Plains* (New York, 1927): 144, 148.

67. Goldstein, "Anthropometry of the Comanches," 290.

68. Keim, *Sheridan's Troopers on the Borders*, 295; "Treaty with the Kiowa and Comanche, 1867," October 21, 1867, 1 Stats, 581, ratified July 25, 1868, proclaimed August 25, 1868; and "Treaty with the Kiowa, Comanche and Apache, 1867," October 21, 1867, 1 Stats, 589, ratified July 25, 1868, proclaimed August 25, 1868, in Charles J. Kappler, *Indian Affairs: Laws and Treaties* (Washington, DC: Government Printing Office, 1904), 2:977–82; 2:982–84.

69. On the demise of the bison, see Andrew C. Isenburg, *The Destruction of the Bison* (New York: Cambridge University Press, 2001).

70. Colonel W. S. Nye, *Carbine and Lance: The Story of Old Fort Sill* (Norman: University of Oklahoma Press, 1937).

71. Eugene E. White, *Service on the Indian Reservations. Being the Experiences of a Special Indian Agent while Inspecting Agencies and Serving as Agent for Various Tribes* (Little Rock, AR: Diploma Press, 1893), 246–47.

72. Tatum to CIA, July 24, 1869, Record Group 62: Letters Received by the Office of Indian Affairs, 1824–81, M234, roll 376: 291, National Archives.

73. Hagan, *United States–Comanche Relations*, 64–65.

74. James Mooney, *Calendar History of the Kiowa Indians*, Seventeenth Annual Report of the Bureau of American Ethnology (Washington, DC: BAE: 1895–96), 339–40.

75. *Annual Report of the Commissioner of Indian Affairs* (hereafter *ARCIA*) for 1878, 46; *ARCIA* for 1883, 70. All reports available at http://digicoll.library.wisc.edu/cgi-bin/History/History-idx?type=browse&scope=HISTORY.COMMREP.

76. Marcy, *Exploration of the Red River*, 102–3.

77. *ARCIA* for 1881, 138.

78. *ARCIA* for 1882, 64–65.
79. *ARCIA* for 1889, 191.
80. *ARCIA* for 1882, 64–65, 66–67.
81. Hagan, *United States–Comanche Relations*, 185.
82. Wallace and Hoebel, *Comanches*, 72.
83. *ARCIA* for 1890, cxlvi. In *United States–Comanche Relations*, 185, Hagan also includes the phrase "filth quarter" as part of the CIA's order, but the CIA report contains no such phrase.
84. Hagan, *United States–Comanche Relations*, 185.
85. *St. Louis Post-Dispatch*, June 15, 1889.
86. *ARCIA* for 1890, 331–32.
87. Sneed, "Reminiscences of an Indian Trader," 139–40.
88. *Boston Evening Transcript*, February 25, 1911.
89. See D. Mihesuah, "Indigenous Health Initiatives, Frybread, and the Marketing of Non-Traditional 'Traditional' American Indian Foods," *Native American and Indigenous Studies* 3, no. 2 (Fall 2016): 45–69.
90. Lewis Meriam, *The Problem of Indian Administration: Report of a Survey made at the request of Honorable Hubert Work, Secretary of the Interior, and submitted to him, February 21, 1928* (Baltimore, MD: Johns Hopkins University Press, 1928).
91. According to a twenty-eight-page booklet in the collection of Henry Mihesuah, Fort Sill Indian School opened in 1871 (Centennial Booklet Committee, *The Fort Sill Indian School Centennial, 1871–1971*, n.p., n.d.). The latest date listed in this booklet is 1980. Hagan, writes in *United States–Comanche Relations*, 199–200, that the school opened in 1892. Henry has always been told by family members that his father, who was born in 1874, was taken as a small child to attend Fort Sill Indian School. This makes the 1871 founding date more likely. Numerous other family members also attended the school. Harry Wahhahdooah, ed., *Deyo Mission* (Lawton, OK: C & J Printing, n.d.). The Baptist Deyo Mission was established in the fall of 1893, eight miles west of present-day Lawton, Oklahoma.
92. H. Mihesuah, *First to Fight*, 21–22.
93. Annual Statistical Reports for 1925 and 1931, Kiowa Agency Records, 1893–1950, Annual Statistical Reports File, Record Group 75: BIA Records, National Archives, Southwest Region, Fort Worth, Texas.
94. Kelly M. West, "Diabetes in American Indians and Other Native Populations of the New World," *Diabetes* 23, no. 10 (October, 1974): 841–55.
95. For discussion of West's theory, see D. Mihesuah, "Historical Research and Diabetes in Indian Territory: Revisiting Kelly M. West's Theory of 1940," *American Indian Culture and Research Journal* 40, no. 4 (2016), 1–21.
96. For information on the Five Tribes specifically, see D. Mihesuah, "Sustenance and Health among the Five Tribes in Indian Territory, Post-Removal to Statehood," *Ethnohistory* 62, no. 3 (2015): 263–84.
97. D. Mihesuah "Sustenance and Health."
98. *Cherokee Advocate*, August 26, 1893.

99. West, "Diabetes," 841.

100. *Pittsburg County, Oklahoma: People and Places* (McAlester, OK: Pittsburg County Genealogical and Historical Society, n.d.), 1–2; Clyde Wooldridge, *The Capital of Little Dixie: A History of McAlester, Krebs, and South McAlester* (Krebs, OK: Bell Books), 35, 48, 63, 65, 204; "Dr. William Elliott Abbott," typed pages on file at the Pittsburg County Genealogical and Historical Society.

101. Elisa Lee, Paul S. Anderson, John Bryan, Carman Bahr, Thomas Coniglione, and Mario Cleves, "Diabetes, Parental Diabetes, and Obesity in Oklahoma Indians," *Diabetes Care* 8, no. 2 (1985): 109.

102. Kelly M. West, "Culture, History, and Adiposity, or Should Santa Claus Reduce?" *Obesity/Bariatric Medicine* 3, no. 2 (1974): 49, 51–52. See also Lee et al., "Diabetes, Parental Diabetes, and Obesity in Oklahoma Indians," 112; Dennis W. Weidman, "Diabetes Mellitus and Oklahoma Native Americans: A Case Study of Culture Change in Oklahoma Cherokee" (PhD diss., Department of Anthropology, University of Oklahoma, 1979), 214; Felecia G. Wood, "Ethnic Differences in Exercise among Adults with Diabetes," *Western Journal of Nursing Research* 24, no. 5 (August 2002): 505.

103. West, "Culture, History, and Adiposity," 52.

104. Anne Walendy Davis, "Discovering Comanche Health Beliefs Using Ethnographic Techniques" (PhD diss., Texas Women's University, 1992), 56–58.

105. Shirley C. Laffrey, "Health Behavior Choice As Related to Self-Actualization and Health Conception," *Western Journal of Nursing Research* 7, no. 3 (1985): 291.

106. Lee et al., "Diabetes, Parental Diabetes, and Obesity in Oklahoma Indians," 112.

107. Davis, "Discovering Comanche Health Beliefs," 75.

108. Lee et al, "Diabetes, Parental Diabetes, and Obesity in Oklahoma Indians," 112.

109. Davis, "Discovering Comanche Health Beliefs," 72.

110. The Comanche Nation is hardly alone in having a diabetes program; so do the Choctaw Nation, Cherokee Nation, Klamath Tribes, Pawnee Nation, Seminole Nation, Hualapai Tribe, Confederated Tribes of Siletz Indians, and others.

111. *American Indian Health and Diet Project*, http://www.aihd.ku.edu/; *Indigenous Eating*, https://www.facebook.com/pages/Indigenous-Eating/478119175562387?ref=hl; *From Garden Warriors to Good Seeds: Indigenizing the Local Food Movement* (an exploration of an array of farming, gardening, and food sovereignty initiatives across the country), http://gardenwarriorsgoodseeds.com/; Decolonizing Diet Project (DDP) blog, http://decolonizingdietproject.blogspot.com/.

GERALD CLARKE

BRINGING THE PAST TO THE PRESENT
Traditional Indigenous Farming in Southern California

For thousands of years, the Indigenous people of the lands known today as California developed a relationship with the land, plants, and animal life that enabled Native communities not only to survive but thrive. This symbiotic relationship was undoubtedly developed through keen observation, generations of experimentation, and the passing down of empirical knowledge.

Most historians and anthropologists mark the development of agriculture as the trigger that ended the hunter-gatherer way of life for some cultures and ushered in the dawn of civilization. Yet hunter-gatherer societies continue to exist in the world today. One aspect of such societies that has been generally ignored until recently is the modification of "nature" to encourage beneficial outcomes for human survival. These cultivating practices served to ensure the survival of these societies, but also benefited the environments in which they lived. One only needs to look up the definition of "farm" to see why anthropologists and historians did not recognize these practices as agriculture when they encountered the Indigenous peoples of California.

> Farm: an area of land and its buildings used for growing crops and rearing animals, typically under the control of one owner or manager.

This definition reflects a couple of key differences in the views of agriculture between Native Californians and their European counterparts. First, "growing crops and rearing animals" implies a great deal of control over the "product," whether plant or animal. The farmer actively directs what, how, when, and where these products are produced, from sprout or birth until harvest or death. Second, "under the control of one owner or manager" implies the idea of land or resource ownership. For the early European explorers, landownership was a centuries-old concept that they used to justify claiming land even when Indigenous people were living on it.

This perspective kept the Europeans from understanding that California's Native people did create cultivated areas that one could recognize as "farms." In fact, if we were to adopt a more liberal understanding of the definition of a farm, we could easily conclude that California's Native people not only developed agriculture, but possibly developed a more efficient and sustainable system than that of the Europe model. If we can understand that "growing crops and rearing animals" could include the pruning of oak trees to benefit acorn production, the burning of meadowlands to produce grassy pastures that support animal populations, and the selective harvesting of plants to encourage plant production, then we can conclude that California was indeed *one big farm*. And while the tribes of California had no concept of legal landownership, tribal territories, clan areas, and familial gathering sites *did* exist and continue to exist today. With this understanding, we can assert that the traditional clan or familial gathering areas cultivated by Indigenous people could be considered farms.

California's Indigenous peoples are continuing these traditional practices, or in some cases are reclaiming these practices, as part of a larger effort to reclaim their cultural sovereignty, improve the health of their people, and assert an alternative approach to modern life. In this chapter, I focus on the traditional foods, plant use, and environmental stewardship I see occurring in southern California, relying on a combination of academic articles and other publications, interviews with knowledgeable tribal people, and my personal experience. While I mention aspects of Indigenous plant use throughout California, a survey of the entirety of California's tribes, their environs, plant use, and traditions is beyond the scope of this chapter.

CALIFORNIA AT CONTACT

While accounts vary as to who might have been the first European explorer to reach California, it can be asserted with good confidence that substantial contact between Europeans and the Indigenous people of California occurred during the eighteenth century. With the founding of San Diego de Alcalá in 1769—the first of twenty-one missions established by Franciscan priests in California—a consistent presence of European culture existed side by side with the Indigenous communities nearby. These Indigenous communities, which anthropologist Alfred Kroeber labeled "tribelets," consisted of several families grouped together for their common welfare. They shared language, customs, and the daily tasks necessary for survival.

In writing about the Native people of northern California commonly known as the Pomo, Kroeber notes:

> In any strict usage, the word "tribe" denotes a group of people that act together, feel themselves to be a unit, and are sovereign in a defined territory. Now, in California, these traits attached to the Masut Pomo, again to the Elem Pomo, to the Yokaia Pomo, and to the 30 other Pomo tribelets. They did not attach to the Pomo as a whole, because the Pomo as a whole did not act or govern themselves, or hold land as a unit. In other words, there was strictly no such tribal entity as "the Pomo"; there were 34 Pomo miniature tribes.[1]

This description applies equally to various tribal groups in southern California, including the Cahuilla, Luiseño (Payómkawichum), Kumeyaay, and others. It is estimated that at the time of contact, California was home to more than five hundred distinct tribelets who spoke more than three hundred dialects of approximately one hundred distinct languages.

We must consider the geography of California to fully understand the necessity of living within a tribelet system. From open, arid desert regions to narrow valleys cut by ocean-bound rivers to coastlines that alternated between sandy beaches and steep cliffs, California's geography is not conducive to the large, densely populated settlements founded by the Aztecs in central Mexico or the Mayas in the Yucatan Peninsula. Instead, California's Native peoples organized themselves in small communities and created a system of sustainable living based on a relationship with the natural world that resulted in mutual benefits for the people and the environment.

Although there were not large Indigenous "cities" in California, numerous journal entries by early Spanish explorers report encountering Indigenous communities along California's coast that sometimes numbered as many as eight hundred individuals or more. A community of this size would require a coordinated and specialized approach to resource management to support its population.

While the geography and the available flora and fauna of the area were important, so too was what anthropologist Lowell Bean characterized as a "reciprocal obligation" belief system he found among the Cahuilla.[2] Reciprocal obligation stemmed from a belief that humans were only a part of a larger interactive system composed of cooperating entities that shared responsibility in the workings of the universe. These beliefs fostered an ecological ethic of gathering only what one needs and always leaving some for the other beings. These actions, combined with the performance of appropriate ceremonies and rituals ensured that the universe would remain in balance and that humans along with other entities could continue to sustain life.

For many tribes in California, a belief in the interconnectedness of various life forms and the shared responsibility of these beings formed the basis for their resource management. The concepts of competition and subjugation of nature were mostly absent from traditional beliefs about the world and a human's place within it. For generations, a system of best practices was passed down, tested, and added to for the benefit of all interconnected entities.

Today, these beliefs are commonly referred to as "traditional ecological knowledge" (TEK), referring to ecological management practices formed over generations and based on observation, experimentation, and long-term relationships with plants, animals, climate, and environment. The results of these management practices were extensive. For example, the promotion of grasslands not only provided seeds for human consumption, but also provided key species of wildlife (quail, rabbit, deer, antelope, elk) with a food supply that sustained healthy and available animal populations. From those healthy animal populations, in turn, Native people gained meat for nourishment, furs for clothing, and bone and sinew for tools, regalia, and other ceremonial and utilitarian objects. In her book *Tending the Wild*, M. Kat Anderson describes the relationship between Indigenous people and the environment as "a knowledge built on a history, gained through many generations of learning passed down by elders about practical as well as spiritual practices."[3]

European explorers themselves commented positively in their journals on the condition of the land they encountered. In the sixteenth century, the English explorer Sir Francis Drake sailed up the Pacific coast of North America, searching for the Northwest Passage, a water route thought to cross the North American continent. While historians do not agree on the extent of Drake's exploration of California, while sailing along the California coast he wrote in his journal "infinite was the company of very large and fat deere, which there we sawe by thousands, as we supposed, in a heard."[4]

During 1769, Gaspar de Portolá led an expedition of soldiers up from Mexico through southern California on their way to San Francisco Bay. Also on the journey was Franciscan missionary Fray Juan Crespi. On Tuesday, July 18, 1769, near the San Luis Rey River just north of present-day San Diego, Crespi wrote,

> A little after three in the afternoon we set out to the north. We climbed a hill of good soil, all covered with grass, and then went on over hills of the same kind of land and pasture. We must have traveled about two short leagues, when we descended to a large and beautiful valley, so green that it seemed to us that it had been planted.... There are, indeed, pools of good water, with tules [bulrushes] on the banks. The valley is all green with good grass, and has many wild grapes, and one sees some spots that resemble vineyards.[5]

The observations of Fray Crespi hint at one of the most prominent misconceptions that Europeans had about the lands in California. To them, this fertile land was their reward from God for the risks they had taken, for spreading the word of Christianity, for the hunger and hardships they experienced. They were being rewarded with a pristine, Garden of Eden-like landscape to claim as their own.

What these early explorers were actually witnessing was centuries of careful ecological management that involved the coppicing, pruning, harrowing, sowing, weeding, burning, digging, thinning, and selective harvesting of plants to maximize each species' productivity, usefulness, and contribution to the interconnected interests that made up the ecological system: plants, animals, climate, weather patterns, soil, humans.

Through careful management, the Natives of California were able to harness the full potential of their available resources. The full extent of plants utilized by these communities may never be known. I include here descriptions of just a few that are in use today as part of the effort to reclaim Indigenous health and sovereignty.

ACORNS AND OTHER FOOD STAPLES

Much has been written regarding California Indians' utilization of the acorn as a major food staple. It is easy to see why. California is home to eighteen species of oak that occur throughout most of the state. After gathering acorns in the fall, tribes had various ways to dry and store them until needed. The fact that they could be stored for later use made acorns a reliable staple that could be utilized year-round. By some estimates, unshelled acorns could be stored for up to ten or twelve years.

Nutritionally, acorns provide much of what was needed for humans not only to survive, but to flourish. One researcher on the subject has suggested that oaks could have supported population densities fifty to sixty-five times higher than those estimated at the time of contact. Additionally, modern nutritional analysis of acorns show that some species contain up to 18 percent fat, 6 percent protein, and 68 percent carbohydrate, with the remainder being water, minerals, and fiber. By comparison, modern varieties of corn and wheat have about 2 percent fat, 10 percent protein, and 75 percent carbohydrate. Acorns are also a good source of vitamins A and C and many essential amino acids.[6]

Acorns were typically gathered in late September and October. Groups would camp among the oak groves, and it has been said that various clans would claim specific hereditary gathering areas. Today, bedrock mortars or grinding holes can be found near these sites. Many of the tribes in southern California preferred the acorn of the black oak (*Quercus kelloggii*), but other varieties were also utilized. Black oak trees, sometimes as tall as one hundred feet, drop their ripened acorns with help from the Santa Ana winds that occur at that time of year. Acorns that did not fall naturally were knocked down with long sticks and gathered from the ground.

The gathering and grinding of the acorns was only the beginning of the process. Unprocessed acorn meal is bitter and unpalatable to humans due to a high tannic acid content, and substantial leaching or washing is required to remove it. Some tribes leached the tannic acid from the pulverized acorn meal in sandy basins pressed into shallow streambeds. Water from the stream would cover and lightly flow over the acorn meal, washing away its bitterness. Over time, some tribes in California developed basket-weaving traditions that produced some of the finest examples of basketry in the world. They developed a technique of filling finely and tightly woven baskets with acorn meal and placing them in lightly running water for leaching. Some tribes in northern California adopted a subterranean leaching process. Whole, shelled acorns were buried in sand, and water was periodically poured over the area until the acorns were judged palatable and ready to be dug up. Sometimes acorns were boiled or roasted prior to burial.[7]

The leached acorn meal might be cooked until it thickened to a gelled consistency, incorporated into a soup, or formed into small cakes. At tribal gatherings today, attendees judge acorn mush (*wii'wish* in the Cahuilla language, *shaw'ii* in the Kumeyaay language) on its smoothness of texture, nutty flavor, and absence of any bitter taste. At a tribal gathering I attended on the Morongo Reservation recently, I was served roasted turkey with acorn gravy.

Anthropologist Alfred Kroeber, who studied California tribes extensively, estimated that more than 75 percent of Native Californians relied on acorns for food on a daily basis. However, Willie Pink, a Cupeño/Payómkawichum descendant and Indigenous plant expert, thinks anthropologists have placed too great an emphasis on acorns as a food source. He stresses the importance of the holly leaf cherry as equally important to tribal communities in central and southern California. Pink points out that ancient village sites in southern California tended to be situated near stands of wild cherry trees as opposed to oak groves. The cherry tree he refers to is *Prunus ilicifolia*, commonly referred to as the holly leaf cherry or simply wild cherry.

HOLLY LEAF CHERRY

Prunus ilicifolia is a drought-resistant, evergreen plant that produces fruit which, unlike its domesticated cousins, has very thin flesh but a large, prominent pit that contains a large seed. The flesh of the cherry is sweet when ripe and was certainly eaten. Today, it is not uncommon for tribal people to make jelly from the wild cherry, although its thin flesh provides little juice. A small jar of jelly requires many cherries and quite a bit of effort.

For California's early tribal communities, however, it was the seed within the large pit that was of greater importance. The cherries were gathered in late summer or early fall when they ripened. Once dried and shelled, the seeds could be stored for long periods to provide for year-round use. Wild cherry is not a large tree but a modest-sized bush, which made harvesting cherries easier than harvesting acorns.

The seed, like acorn, can be ground into a meal, but also requires leaching in order to eliminate naturally occurring hydrocyanic acid (cyanide). While some people simply leach or wash the seeds, others roast them prior to leaching in order to rid them of the cyanide. Willie Pink roasts cherry pits in a modern oven, but the same could be done in a clay pot near a campfire. Once leaching and grinding are completed, the meal is cooked until it thickens to a pudding-like consistency. In my experience, the prepared cherry-seed mush looks similar to cooked acorn mush and has a subtle, nutty taste with a slightly fruity flavor.

In addition to using holly leaf cherry as a food source, Kumeyaays and Cahuillas were among the tribes that treated colds and coughs with infusions made from the roots or bark of the tree. This infusion was typically ingested as a type of cough medicine. During the spring and summer, while the sap of the tree was running, the infusion was made from the bark. During the winter, while the tree was dormant, the roots were used to make the infusion.[8]

It should be noted that both acorns and wild cherry seeds were widely traded throughout California and beyond. Both could be dried and preserved, allowing communities with excess supplies to transport and trade them with tribes that had little access to the growing areas of these foods. Jan Timbrook, an anthropologist and ethnobiologist who specializes in the Chumash culture, states that cherry kernels were an important trade item for Chumashes and "were measured with women's basketry hats; one hat full of cherry seeds worth two of acorn."[9]

PINE NUTS

Many of the varieties of pines that grow in California produce edible pine nuts. Some examples include the Jeffrey pine (*Pinus jeffreyi*), gray pine (*P. sabiniana*), the four-leaf pinyon pine (*P. quadrifolia*) and the single-leaf pinyon pine (*P. monophylla*). While pine nuts from these species were important to the tribal communities, so too was the pine pitch or sap. Pitch was used as an adhesive and waterproofing agent on various utilitarian objects, such as to waterproof baskets or adhere gourds for rattles.

For the tribes of eastern and southern California, the four-leaf or single-leaf pinyon pine were accessible varieties that could be found growing in the Mojave Desert, eastern Sierra Nevada, and Tehachapi Mountain Range, as well as in southern California's coastal mountain ranges. These varieties typically grow at elevations between 4,000 and 7,500 feet above sea level. The single-leaf pinyon pine will generally begin bearing cones at about 35 years of age, begins producing good seed crops at about 75 to 100 years, and reaches maximum production at about 160 to 200 years.[10]

Like the wild cherry seeds and acorns, pinyon pine nuts are a good source of protein, unsaturated fats, and carbohydrates. They can be dried and stored for long periods, allowing use throughout the year. However, pinyon pine nut production varies from year to year due to climatic conditions and is not as consistent or predictable as the holly leaf cherry. Yet despite its inconsistent availability, the pinyon pine nut has been important to many Native communities of California. From ancient times, archaeological evidence points to its utilization as a food source, and it is mentioned in some tribes' traditional songs and stories.

Newcomers to California observed the importance of pine nuts to the local tribes. In the mid-1800s, John Charles Frémont—a lieutenant in the Army Topographical Corps—was commissioned by the US government to explore and map the Pacific Northwest. Returning from this expedition through the eastern Sierras, Frémont recorded the Indigenous people's utilization of pine nuts in his journal: "These [pine nuts] seem now to be the staple of the country; and whenever we meet an indian, his friendly salutation consisted in offering a few nuts to eat and trade."[11]

YUCCA

Various types of yucca grow in southern California as well as Baja California, Mexico, at elevations from sea level up to eight thousand feet. Yuccas provided the tribal communities of these areas with both food to eat and fibers for utilitarian use. The late Cahuilla elder Alvino Siva stated that the Cahuilla preferred the fibers of the Mojave yucca (*Yucca schidigera*) for making rope and cordage, but found the chaparral yucca (*Hesperoyucca whipplei*), also called our Lord's candle or *panu'ul* in the Cahuilla language, preferable for food. This spiny green yucca typically grows in rocky soils and produces a stemless cluster of long, rigid leaves that end in sharp points. The leaf edges are finely saw-toothed and, along with the sharp points, can easily draw blood from a person untrained in its proper harvesting.

When it matures after five to seven years of growth, the chaparral yucca shoots up a large stalk from the center of the plant that can attain twelve feet in height. Toward the top of this large stalk grow clusters of flowers that may be white, off-white, or white edged with violet. After the flowers are pollinated, the petals drop and the plant develops seedpods that eventually dry and open up, spreading seeds around the area for future plant propagation.

The chaparral yucca offers a few different opportunities for food gathering. First, the young stalk can be harvested before it flowers. Eaten raw it tastes like a cross between an apple and celery. If cooked, it can be eaten by itself or puréed and used in breads or cakes. Today, a number of tribal families also gather the flower petals (*se'ish* in the Cahuilla language). After being cleaned, then parboiled two or three times, with the water poured off and replaced each time, the petals can be eaten, usually with a little added salt. To me, the cooked petals taste like a "floral cabbage." Nowadays, people also often sauté the cooked petals in olive oil with onion, minced garlic, and even bacon! While not entirely traditional, this dish eaten with wild cherry or acorn mush and roasted wild rabbit serves as a healthy alternative to your local fast-food supersized meal.

Finally, once the yucca petals have naturally dried up and fallen off, the plant produces edible seedpods. The seedpods resemble mini-cucumbers and have a similar texture and consistency. They can be bitter if allowed to grow too large, so the smaller ones are preferable for eating raw. Some people have even begun to pickle the pods and are canning jars of "yucca pickles."

However, the yucca is vulnerable to overharvesting. Harvesting the stalk or petals means little or no seed would be produced and could lead to the plant's eradication from an area. I have come across yucca stalks that had been cut down in order to gain access to their blossoms. This approach to gathering is unsustainable. I have taught my own children to limit the number of yucca stalks they harvest per year and not to take more than one-quarter of the blossoms from any one yucca plant.

ELDERBERRY

Elderberry (*Sambucus mexicana*) is a fast-growing large shrub or small tree that grows throughout California. It is drought tolerant, grows to heights between eight and twenty-five feet, and features bright green foliage and yellowish flowers that mature into clusters of small, dark blue, almost black, berries.

Elderberry is important to wildlife and numerous bird species. Its dense, multistemmed structure provides cover for numerous small mammals, birds, and reptiles. The flowers and berries also serve as an important food source for wildlife, especially in the dry months of late summer.

For California's Indigenous people, elderberry has multiple uses for food, medicine, and utilitarian or ceremonial objects. In excavations of ancient village sites, archaeologists have found evidence that California's earliest inhabitants used elderberry. The archaeologist Paul Schumacher reported that he often found elderberries growing near ancient Native American settlements and gravesites.[12]

Among elderberry's utilitarian uses, the young shoots were used for arrow shafts and larger branches for bows. Fibers from the inner bark were also used in making women's skirts or aprons, and the leaves rendered a black dye for basket-making fibers, which was the source of the bold designs often found in California Native basketry. The woody portions of the elderberry were used to make pipes for smoking Indigenous tobacco, as well as to make flutes and clapper sticks. Elderberry branches feature a soft and pithy center that made hollowing out the branch for these uses easier.

Medicinally, Indigenous people brewed a tea from elderberry blossoms that is said to reduce fever and help fight off colds. Contemporary herbalists proclaim that elderberry tea boosts the immune system, alleviates allergies, lowers blood sugar levels and blood pressure, moderates the digestive process, and even slows down the spread of cancer. While these claims have not been proven by medical science, California Native people's centuries-old use of elderberry as a medicine points to its effectiveness as a remedy for some illnesses.

Elderberry fruit is high in antioxidants and numerous vitamins, including 60 percent of the recommended daily allowance of vitamin C. The berries are also a good source of essential nutrients like calcium, potassium, manganese, iron, and phosphorus. Historically, the berries were eaten raw, cooked, or made into a cider.[13] Drying the berries allowed them to be stored, transported, and used at a later time, allowing tribes to benefit from their nutritional content throughout the year.

MESQUITE AND SCREWBEAN

I've grouped mesquite (*Prosopis glandulosa*) and screwbean (*P. pubescens*) together because they are related plants that typically grow in arid desert environs below three thousand feet in elevation. Both grow throughout the eastern deserts of southern California, but the mesquite has a far greater range and can be found down through Baja California and along the US-Mexican border to the Gulf of Mexico.

Tribes used these trees for a multitude of purposes. The Desert Cahuilla used hollowed-out mesquite stumps for grinding mortars, and larger limbs were used for building material. The wood was also used as fuel for cooking and for firing pottery. Mesquite branches were used for bows and the thorns were used as tattooing needles. Farther south, the Paipai, Southern Kumeyaay, and Kwatsáan used mesquite and screwbean wood for similar purposes.

Both plants offered desert tribal communities a dependable supply of edible flowers and seedpods, typically gathered from June through August. The long, beanlike pods of the honey mesquite were once the most important food source for these desert communities, and mesquite groves were often burned to increase production. From a nutritional standpoint "the honey mesquite compares favorably with barley in which approximately 8 percent is crude protein, and 54 percent carbohydrate and a little over 2 percent fat."[14] The blossoms of the honey mesquite could be dried and used in a tea or cooked and formed into cakes. The seedpods could be eaten fresh or pulverized into a meal, which could be added to water for a nourishing drink or be baked into cakes or mush. Screwbean was treated similarly, but required additional processing to rid it of bitterness. A subterranean leaching process was used for this purpose. Both mesquite and screwbean could be dried and stored for long periods, allowing for their use throughout the year.

Today, it is not uncommon for tribal cooks to include these traditional foods as ingredients in modern dishes. Either by necessity or by choice, current tribal communities use a diverse number of indigenous and nonindigenous plants as ingredients in their kitchens. These contemporary practices are evidence of California's Native people's ability to adapt new approaches to food as part of their efforts to regain food sovereignty.

EFFORTS TODAY

In the spring of 2017, I was contacted by a local man wanting to know if I was interested in taking a Slovenian documentary crew out into the local hills to gather plants that provided some of the traditional foods of the Cahuilla. Their documentary series, *Cestou Necestou*, featured international tourist sites and cultural interactions with various peoples around the world with an emphasis on food. This episode, to be shown on Slovenian television, was dedicated to the cultures of southern California.

We drove into the local hills of the San Bernardino National Forest, where I have a permit for traditional plant gathering and use. My four-wheel-drive truck bounced up what was once a fire access road and we stopped a good way from the highway and began our hike. The site we visited—and which I have visited often— hosts a number of different edible plants. We gathered yucca blossoms, white sage, and Indian potato (*Dichelostemma capitatum*). I also pointed out the chia plants that were not yet ready for harvest.

It was a warm day, but in spite of the heat, I knew the importance of wearing pants to protect my legs from the thick underbrush and dead branches. Then I noticed the cameraman was wearing shorts, and his legs were scratched and bleeding. I paused and explained how my ancestors used fire and controlled burns to "clean" the hillsides. The thick chaparral would be transformed into nutrients to nourish next year's seedlings, and the people would be able to traverse the hillsides with ease. Our modern perspective is skewed to think of ancient people as living a hard life of subsistence and survival. Our schools, books, and movies all tend to feature Indigenous cultures as "roughing it" in an unforgiving wilderness. But these were not the conditions that met the eighteenth-century Spanish explorers. In their diaries they mention the ease with which they were able to make their way through the meadows, forests, and chaparral of the California landscape. It was the ecological practices of the Native population that fostered the European idea of California as a "pristine" wilderness!

Yet, now, fences, permits, private property, overgrowth, habitat loss, and loss of the ancestral land base have made the gathering of plants highly challenging for California's Native people. Daniel McCarthy understands firsthand the difficulties imposed on Native people by the federal government. As a former archaeologist and tribal liaison to the San Bernardino National Forest, he worked for several years on obtaining "administrative passes" for tribal people who wanted access to forest lands for traditional gathering purposes. Just as non-Native people can purchase recreational permits for hiking or camping, Daniel advocated for Native access to forest resources and for a new category of acceptable activity: cultural gathering. In 2006, he began working to educate Forest Service officials about the need for such a permit and in 2007, the regional forester and the state director of the Bureau of Land Management (BLM) approved the policy.

Why were National Forest administrators resistant? Clearly, there exists a fundamental misunderstanding of what traditional gathering practice is. No doubt fears of overharvesting, selling of resources, and negative environmental consequences were major concerns. However, these factors are not evident in historic accounts of California's Indigenous plant-gathering practices, pointing to the need to educate government agency employees.

In their journal article "Native American Land-Use Practices and Ecological Impacts," M. Kat Anderson and Michael J. Moratto assert that the wild plant management practices of California's Indigenous populations involved "the human manipulation of native plants, plant populations, and habitats, in accordance with ecological principles and concepts, that effects a change (either beneficial or negative) in plant abundance, diversity, growth, longevity, yield, and quality to meet cultural needs."[15] How was this knowledge derived? Anderson and Moratto contend that these practices were and are "based upon traditional knowledge of natural processes gained over the millennia, [and] were applied to increase the quantity and improve select qualities of focal plant species.... This traditional knowledge, which permitted the adaptive success of large human populations and the maintenance of Sierran [Sierra Nevada Range] environments for more than a hundred centuries, must not be dismissed."[16] This perspective is mirrored in the beliefs and practices of those California Indigenous people today who carry traditional plant knowledge and continue to gather plants and promote environmental stewardship within their respective communities. While tribal people within California's borders encounter the same life challenges as many other Americans—careers and employment, getting their kids to school, sitting in traffic, paying bills, and so on—today there are tribal people making efforts to preserve traditional plant knowledge and practice ecological stewardship.

Richard Bugbee, a Payómkawichum descendant, actively began pursuing his interest in Indigenous plant use in his early twenties. He worked at a health food store and eventually attended herbology classes at Palomar College. "I've always had an interest in the plants and animals" says Bugbee. On hunting trips with his grandfather, Bugbee learned of the connections among plants, animals, and the environment. This is what biologists call a trophic structure: the feeding relationships between organisms of a community, similar to a food chain. "My grandfather looked at coyote poop before we went rabbit hunting. If it was full of manzanita seeds, then he knew there were no rabbits in the area. If the coyotes needed manzanita berries to fill their stomachs, there were no rabbits around to be hunted."

What Bugbee was learning was ethnoecology in its natural and cultural form—along with the role of humans in this symbiotic relationship. "Take the chia plants" he explains, "when we were actively harvesting these plants, we would use our seed beaters and knock the seed from the tallest plants and into our baskets. Some of the seeds would miss the basket and fall to the ground to germinate the next year. Now that we haven't been harvesting in the old way, the birds have been eating the seeds from the tall plants, and only the shorter chia gets reseeded. Chia is getting shorter and shorter now without human interaction."

In addition to his herbology coursework, Bugbee also completed courses in anthropology, archaeology, and botany. He learned all the accepted scientific names for various plants, but found that these names would change over time as various plants were subjected to DNA and other tests. This confusing system of classification and naming wasn't conducive to Native communities that were actively teaching and passing down their plant-use traditions.

Although Bugbee learned from many different individuals, he considers Jane Dumas as his mentor. She was a Kumeyaay woman from the Jamul community who took notice of his interest in plants. She asked him to accompany her on her gathering trips to a site where her mother had gathered plants. "At first, I was just her lookout," Richard says with a smile. "Back then, you had to gather plants wherever they grew, and sometimes that was on private property!" Over time, Dumas began to instruct Richard on gathering practices, plant uses, and environmental factors that had been taught to her decades before. Richard even began attending Kumeyaay-language classes in order to understand the Kumeyaay names for plants and places. He prefers the Indigenous names because they have been used consistently for thousands of years.

Lorene "Lori" Sisquoc is another person seeking to reclaim her community's Indigenous cultural sovereignty. She is an enrolled member of the Fort Sill Apache Tribe of Oklahoma, but was born and raised in California and is also descended from the Mountain Cahuilla of southern California. Sisquoc attended Sherman Indian School, a Bureau of Indian Affairs (BIA) boarding school located in Riverside, California. She now works at Sherman as a dorm mom and serves as culture traditions leader for the school. Understanding the damage that the US government boarding school system had on tribal people historically, she says, "I work hard to make up for the damage that the boarding schools have caused to Native people and traditions being passed down." One way she is doing this is to share her knowledge of traditional plant use with her students.

She began her journey into traditional plants, foods, and medicines in her teens. In 1973, after she got into some minor trouble, her mother sent her to the Red Wind Foundation, an intertribal organization situated in the Los Padres National Forest near San Luis Obispo, California. There, she lived in a tipi, gathered foods, learned songs, and was introduced to an Indigenous perspective of the environment. It was to be her awakening to the value of a traditional lifestyle. Although most of what she learned was introductory, it induced a passion for learning about traditional plant use, songs, stories, ceremony, and healing. From there, Sisquoc has learned from whomever was willing to teach her. While she lists Barbara Drake (Tongva), Donna Largo (Mountain Cahuilla), and Ester Quintano (Desert Cahuilla) as her main teachers, she has learned from many others, as well as from tribal gatherings, language classes, and workshops. She recalls a campout sponsored by the Indian Health Services where the late

Annie Hamilton, a Cahuilla tribal member, instructed participants in language and plants.

As for gathering plants for food or medicine, she believes the environment is "set up to be used." She says she feels bad about allowing plants to go unpicked because the plants will feel uncared for. She remembers hearing of a tribe whose word for an oak grove translated into "they wait for us." Sisquoc stresses the importance of gathering to us as human beings. When gathering, it's important to pray, be thankful, and be respectful. "We need this. Healing the earth is healing ourselves." "Because everything is connected" Sisquoc says. "It's not divided. Our language, spirituality, survival, our stories; it's all connected."

When she is out gathering, she is seldom alone. Her children and grandchildren typically accompany her on gathering trips. This generational learning environment is a contemporary reflection of how information has been passed down for centuries. Young and old work together for the benefit of their family, clan, tribe, and nation, ensuring that each new generation will be able to continue traditional gathering and stewardship practices. The elder with the knowledge of what plants to gather, where those plants grow, and when to gather them directs the younger people whose youthful agility enables them to climb trees, dig roots, or scale hillsides.

Concerning the difficulties that tribal people face in reclaiming cultural knowledge, she points out that many of the knowledgeable tribal elders have passed away. Some elders underestimate the value of the knowledge they hold. "Maybe an elder only knows one thing," says Sisquoc, "they don't think they have knowledge, but they do! Their memories are so important and valuable." Another problem is access. Many of the traditional gathering sites are now on private property or on state or federally controlled lands. While there are now agreements between federally recognized tribes and the National Forest Service and the BLM, a number of non-federally recognized tribes like the Tongva and Acjachemen do not have access to those lands.

Access was one of the issues that led to the formation of the Chia Café Collective, a grassroots organization of southern California tribal members and collaborators committed to the revitalization of Native culture with an emphasis on plants, foods, and medicines. The members of the collective come from a variety of tribal and nontribal communities. They conduct educational workshops about edible and medicinal plant use, advocate for the protection and restoration of California native plant environments, and provide opportunities for mentoring and apprenticeships for interested tribal youth. The Chia Café

Collective has presented to high schools, universities, and nontribal community organizations, as well as tribal gatherings throughout southern California.

Each spring, the Malki Museum, located on the Morongo Reservation near Banning, California, holds its annual Agave Harvest and Roast. This event offers an opportunity to harvest, prepare, and taste agave hearts as they have been harvested and cooked for thousands of years by southern California's Indigenous peoples. In addition to roasted agave hearts, cholla cactus buds, stinging nettles, acorn mush, and elderberry tea are typically on the menu. Members of the Chia Café Collective participate in the Malki agave roast and similar events that promote awareness of a traditional Indigenous plant diet, emphasize the health benefits of eating traditional foods, and advocate for a return to such a diet.

In working with tribes directly, the collective has cooperated with tribally run Temporary Assistance to Needy Families (TANF) programs to promote cultural revitalization and in support of healthy nutrition initiatives to fight rising obesity and diabetes rates among Indigenous people. Craig Torres, a cofounder of the collective, speaks of "decolonizing the diet." "Our palates have become so desensitized because of a lot of the processing that [modern] foods contain with the added white sugar and flour" he says. The late Cahuilla elder Alvino Siva echoed Torres's belief about modern foods. He believed that processed foods have ruined our sense of taste. In a discussion we had a few years ago, Siva stated that salt and sugar is so prevalent in modern food that we can no longer taste the subtle flavors of our traditional foods.

One approach Indigenous foods advocates have taken to the reintroduction of traditional foods is to create contemporary dishes that incorporate traditional Indigenous ingredients. These Indigenous "fusions" are a way to get people accustomed to the tastes and textures of Indigenous ingredients. The hope is that over time people will gradually incorporate more Indigenous ingredients into their diets.

In addition to returning to traditional foods, Barbara Drake, Tongva descendant and cofounder of the Chia Café Collective, advocates for a return to using native plants for teas and medicines. "Native plant gardens are popular now because of the droughts," she asserts, "and people are learning that they can have [grow] medicines right in front of them. We encourage people to grow their own," says Drake. However, members of the collective are clear on their message to novice plant enthusiasts. They do not encourage wild plant gathering and caution against their use as food or medicines except under the guidance of a knowledgeable person.

In *The Ethnobotany Project: Contemporary Uses of Native Plants; Southern California and Northern Baja California*, members of the collective warn of the federal, state, county, and city laws that govern the gathering of wild plants, but also caution readers about the plants themselves: "Some [plants] are toxic or have toxic parts and require knowledge that is not provided in this book—knowledge that was handed down through countless generations of Native people."[17]

California Native plant expert Nicholas Hummingbird (Cahuilla/Apache) is founder and manager of Hahamongna Nursery in Pasadena, California. The nursery has under cultivation four dozen species of plants that are native to the Los Angeles Basin and surrounding hills. The nursery provides low-cost native plants for restoration, conservation, and park uses. Hummingbird warns that people should not take native plants for granted. "In our culture, we earn our knowledge, we earn our relationships . . . because then it's valued more" he stresses. "When you see our Indigenous people gathering, they're doing so with a knowledge and a relationship of a thousand years. What we do enhances the very environment that we're interacting with." Hummingbird's views are shared among many Native communities today. This contemporary belief is a reflection of the traditional concept of reciprocal obligation that was cited earlier, the ecological ethic that one gathers only what one needs but always leaves some for other beings. "We give. We give a song. We give tobacco. We give water in honor of what we're taking. [It's] an honor that we are responsible for what we take. We don't just take because it's there. That life form has every right to exist as much as we do," says Hummingbird.

ENVIRONMENTAL REHABILITATION

For more than a century, California's Native people have been denied access to much of their traditional homelands. Thereby, the landscape has been denied the restorative management practices that Native people have performed for centuries. According to Daniel McCarthy, tribes in California actively managed pinyon pine groves by pruning, weeding, and conducting controlled burns in and around the groves to control undergrowth that competed with the pines for water. Thick undergrowth also fueled more intense wildfires, allowing the flames to burn up into the tree's canopy, typically killing the tree. In a condition of reduced undergrowth, the low flames from wildfires would simply burn underneath the trees, allowing the trees to survive.

The Tribal Forest Protection Act of 2004 allowed the Forest Service and the BLM to place tribally proposed stewardship projects on their list of priorities.

McCarthy reached out to tribal elders in the region and learned about the importance of tribal reliance on the pinyon pine. This led to the creation of the Parry Pinyon Pine Protection Project, which sought to reduce fuels in and around pinyon pine groves to protect them from wildfire and reduce stress from drought. In 2006, McCarthy and numerous Native and non-Native volunteers gathered at Pinyon Flat Campground just East of Ribbonwood, California, to manually clear brush from areas around the trees. While clearing overgrowth was the primary goal, the direct interaction with the trees and their immediate surroundings was educational for participants. "I really knew nothing before we started," says McCarthy. "As dense as the vegetation was . . . those trees thrived. I saw that the seedlings need a nurse plant and protection to grow." This realization has guided McCarthy's replanting efforts ever since.

Other species of trees are also suffering from the lack of traditional management. By 2015 it was estimated that the bark beetle had killed more than twenty-nine million trees in California. In Southern California alone, the invasive goldenspotted oak borer beetle has killed more than one hundred thousand old-growth oak trees. Although recent drought conditions in the western United States have certainly affected the overall health of the trees, experts agree that another contributing factor in the unusually large die-offs is an unnaturally dense forest. The dense stands of trees in Californian forests are a direct result of the National Forest Service wildfire suppression efforts over the last century. These suppression efforts, combined with the absence of burning, thinning, and pruning activities that were part of Indigenous management practices, have created a forest full of dead plant matter and debris that fuel today's extremely devastating wildfires.

Gone are the days when tribal groups had the freedom to perform selective tree thinning and controlled burns to prevent these types of infestations and die-offs. Instead, government agencies are dealing with "super fires" that burn vast acreages with so much heat that no plant survives. How have we arrived at such a predicament? The modern world's belief that nature does best with no human intervention is a key factor. For most of the past century, US society has blamed human activity as the major element in the decimation of America's natural environment. Indeed, the encroachment of commercial and residential developments on plant and animal habitat has greatly reduced California's natural landscape. However, ecological damage is also occurring away from urban centers and suburban sprawl. The condition of the land is not as it was when Spanish explorers first set foot in California. The "pristine wilderness" they found is now overgrown, unhealthy, and dying. Nearly one in three vertebrate species and one in ten native plant species in California is now in serious danger of extinction.[18] It is within this reality that California Native people are trying to rehabilitate their own tribal ecosystems as well as educate federal, state, county, and city governments and agencies about the traditional managements practices that were uniquely designed for California's diverse ecology.

In 2003, the Sycuan Band of the Kumeyaay Nation established Kumeyaay Community College (KCC) in an effort to educate its own people and others on the importance of traditional ecological knowledge and practices. The tribe has stated that its goal was to support Kumeyaay cultural identity, sovereignty, and self-determination, while meeting the educational needs of Native and non-Native students.[19] KCC offers classes in Kumeyaay culture, including language and history, along with courses in ethnobotany, ethnoecology, traditional plants/foods, and arts (including basketry). A primary focus of KCC is to strike a healthy balance between traditional Indigenous and Western systems of knowledge.

Richard Bugbee team-teaches the ethnobotany and ethnoecology classes with Michelle Garcia, an associate professor of biology from the nearby Cuyamaca College. Each approaches the subject with a different perspective that, when combined, provides students with a broad understanding of the subject. Students in these courses range from tribal members, state conservationists, and park rangers to students pursing degrees in horticulture and environmental science.

Similar educational and outreach activities are occurring elsewhere in California. Recently, the Environmental Department of the Pala Band of Mission Indians sponsored the "Creating Successful Partnerships with Native American Tribes" workshop. Among speakers at the workshop were tribal leaders and tribal environmental advocates, federal and state agency representatives, and academic faculty. The overall goals of the workshop were to build bridges between government agencies and tribes and to facilitate better understanding of tribal ecological management practices. Participants heard directly from tribal people and learned about the area's Native plant use firsthand.

The California Indian Basketweavers Association (CIBA), a statewide organization of Native basket weavers, is doing similar outreach activities. CIBA's members have been working to educate the public on the California Indian basketry tradition, advocate for increased access to traditional gathering sites, and oppose pesticide use on or near traditional gathering areas. For Indigenous basket makers, a full understanding of relevant plant materials is a necessity of their craft. Knowledge of plant habitats, growth and maturity of the plants, and proper harvesting techniques are essential to creating high-quality baskets.

Angela "Tangie" Bogner, is a Cahuilla basket maker who works as a community health representative for the Riverside–San Bernardino County Indian Health Service. She lives on the Cahuilla Indian Reservation and near her home grows a thick patch of juncus (*Juncus textilis*), one of the main plant materials used in the creation of southern California coiled baskets. The patch is well known among basket weavers, and they come from all over southern California to gather materials for their baskets. But Bogner says the juncus patch has changed over time. Drought and overgrowth have affected the health and quality of the plants. "One of the practices we used to have was to burn it . . . to help sustain it and help it grow," she says. "At one point, we were ready to burn it, but the Bureau of Indian Affairs stopped us and said we needed their approval, so we never did it." Today, government agency interference in traditional management practices is common. Even a tribal person with an administrative pass for gathering cultural material is not immune from harassment by rangers and Forest Service personnel. The questioning of identity and motive, and outright refusal to allow the gathering of materials in the forest, are common. Bogner and her husband, Sean, have sought out their own solution to the access issue: "We planted deer grass, sumac, and other basketry plants right here in our yard," she states. "That way, the rangers can't say anything." Plans for burning the juncus patch at Cahuilla continue to progress. The Cahuilla Tribe, CalFire, Riverside County Fire, BIA, and even CalTrans all have parts to play and must cooperate in order for this to be achieved. Proponents of the burning are busy educating the various public agencies about their practices, their tribal traditions, and the cultural importance of the management of their environment.

INTO THE FUTURE

A theme that consistently comes up when speaking with people knowledgeable about plants and traditional foods is a strong connection with Indigenous language. While many Indigenous California languages are endangered, numerous Native communities are teaching language classes as part of their efforts to reinvigorate their tribal customs and traditions. In many of these classes, the names of plants and the places they grow are prominent components of vocabulary lessons, as are songs and stories relating to plants.

Another theme mentioned by people who actively gather is the psychological benefits of plant gathering. Each may state it differently, but all express feeling gratitude, connection, and calming while out gathering foods and medicines in the environment. Many express pride in participating in an ancient and continuing practice that is something bigger than themselves and is part of the continuum of their culture.

Indeed, it is hard to speak about tribal sovereignty without acknowledging the roles that plants, traditional foods and medicines, and the environment have played in molding tribal identities. As we move through the twenty-first century, the current efforts of California's tribal communities seem to be on a trajectory to the restoration of cultural knowledge and of the health of both the community and its ecology.

NOTES

1. Alfred L. Kroeber, *Handbook of the Indians of California*. Bureau of American Ethnology Bulletin 78 (Washington, DC: BAE, 1925), 474.

2. Lowell J. Bean, *Mukat's People: The Cahuilla Indians of Southern California* (University of California Press, 1972), 165.

3. M. Kat Anderson, *Tending the Wild: Native American Knowledge and the Management of California's Natural Resources* (Berkeley: University of California Press, 2005), 4.

4. James Miller Guinn, *A History of California and an Extended History of Los Angeles and Environs* (Los Angeles: Historic Record Co. 1915), 39.

5. Pacifica Historical Society, *Portola Expedition, July 18, 1769, Diaries*, 2017, https://pacificahistory.wikispaces.com/Portola+Expedition+July+18%2C+1769+Diaries.

6. "Past and Present Acorn Use in Native California," Anthropology Museum, California State University, Sacramento, https://www.csus.edu/anth/museum/pdfs/Past%20and%20Present%20Acorn%20Use%20in%20Native%20California.pdf.

7. R. F. Heizer and M. A. Whipple, eds., *The California Indians: A Source Book* (Berkeley: University of California Press, 1971), 239.

8. USDA NRCS National Plant Data Center, "USDA NRCS Plant Guide: Holly Leaf Cherry," 2002–3, https://plants.usda.gov/plantguide/pdf/cs_pril.pdf.

9. Jan Timbrook, "Use of Wild Cherry Pits as Food by the California Indians," *Journal of Ethnobiology* 2, no. 2 (1982), 172.

10. Kristin L. Zouhar, *Pinus monophylla*, in *Fire Effects Information System* (Missoula, MT: US Department of Agriculture, Forest Service, Rocky Mountain Research Station, Fire Sciences Laboratory, 2001), http://www.fs.fed.us/database/feis/plants/tree/pinmon/all.html.

11. John C. Frémont, *The Exploring Expedition to the Rocky Mountains, Oregon and California; To Which Is Added a Description of the Physical Geography of California, with Recent Notices of the Gold Region from the Latest and Most Authentic Sources*, Project Gutenberg, http://www.mirrorservice.org/sites/gutenberg.org/9/2/9/9294/9294-h/9294-h.htm.

12. Anderson, *Tending the Wild*, 162.

13. Kent Lightfoot and Otis Parrish, *California Indians and Their Environment* (Berkeley and Los Angeles: University of California Press, 2009), 225.

14. Lowell Bean and Catherine Saubel, *Cahuilla Ethnobotanical Notes: Aboriginal Uses of Mesquite and Screwbean*. Archeological Survey Annual Report (Los Angeles: University of California, 1962–63), Page 193.

15. M. Kat Anderson and Michael J. Moratto, "Native American Land-Use Practices and Ecological Impacts," in *Sierra Nevada Ecosystem Project: Final Report to Congress*, vol. 2, Assessments and Scientific Basis for Management Options (Davis: University of California Centers for Water and Wildland Resources, 1996), 187.

16. Anderson and Moratto, "Native American Land-Use Practices," 187.

17. Rose Ramirez and Deborah Small, *The Ethnobotony Project: Contemporary Uses of Native Plants: Southern California and Northern Baja California* (Self-published, 2015).

18. Anderson, *Tending the Wild*, 328.

19. See Kumeyaay Community College, http://www.kumeyaaycommunitycollege.com/index.html.

DEVON G. PEÑA

12
ON INTIMACY WITH SOILS
Indigenous Agroecology and Biodynamics

Maíz y frijol seco; calabacita y calabaza—corn and dry beans; squash and pumpkin: these three companions, cultivated by Indigenous farmers across Turtle Island and Abya Yala for thousands of years, represent our collective biocultural heritage.[1] Alongside cilantro (coriander) and *alberjón* (sweet pea),[2] these seeds found their way into coffee cans and Mason jars that Margarita kept in a cupboard next to the old stovetop. I was ten when I asked my paternal grandmother why she kept these seeds: *¿Por qué mamá*? She replied: *Porque la semilla es la memoria de la planta de cómo vivir bien en este lugar.* "Because the seed is the plant's memory of how to live well in this place."

TIERRA SANA

To my *abuelita*, seeds had memory and agency. She thought of seeds in a manner akin to what some social theorists today call "vibrant matter."[3] Indigenous farmers long have understood how people, seeds, *and* soil merge capacities and grow together—we co-evolve.⁴ With a little help from human hands, pollinators, soil, and water, seeds enact an evolving memory of lived experiences in agroecosystems—when to sprout; how much water to absorb and how to survive drought; how to adapt to ultraviolet radiation; when to blossom; how many cobs and kernel rows to develop, or how many peas or beans per pod; how to respond to other plants and insects; the amount of protein, starches, sugars, and amino acids to store in leaf, root, grain head, rhizome, or tuber; when to mature as viable seed or rootstock; and so on.

Awareness of seed memory was a vital element in my grandmother's path to a relational sense of self through place attachment. I believe her care of seeds encouraged a *re-membering* of the body by means of affiliation with the land immediately around her. The simple and elegant practice of crafting a "land-connected self" may allow persons to develop a deeper, more relational sense of place. This connection can generate healing in the convivial spaces of home kitchen gardens, crop fields, and kitchens. A mindful kindness toward the soil defined what it meant for her to be a good human being in relation to others. She believed that if you want good corn and beans, you first need good seed and healthy soil.⁵ The same held for the soil of culture and for family and other social and ritual kinship relations like *comadrazgo* that were such a central part of our living social web back in the 1950s and early 1960s.⁶ Margarita's love of soil was grounded in direct lived experience. She did her best to care for the soil in our yard while raising ten children of her own and me, her firstborn and orphaned *nieto*. I'll never forget the joy of eating tender *nopalitos* cut from her beloved cactus fence on the east side of our home. She scrambled these with fresh eggs gathered from chickens living in a roost my uncles built in the backyard next to the cactus fence.

My childhood home was in Laredo, Texas, located on the Rio Grande where the border violently intrudes on local cultural, social, and economic life. The border is, as Gloria Anzaldúa says, "an open [colonial] wound." In this divided borderland, Margarita's emotional attachment to land avoided the shallow sentimentality so common in settler colonial ideals of "nature appreciation." She didn't care for overly manicured city parks or Lake Casa Blanca State Park. She preferred, as I did, the Tamaulipan mezquital—the mesquite, prickly pear, *huisache* (sweet acacia), and devil's claw habitat surrounding our ranchería in the barrio known as El Three Points.

I believe my grandmother eschewed binary constructs. What was important for her was for us to embrace daily labor as an act of re-creation with, within, and by courtesy of nature's amazing capacity for productivity. Margarita's intimacy with the land came from working directly with the soil in a regenerative and healing practice—she nurtured the soil to provide some beloved garden herbs, nopalitos, a few vegetables, corn, peas, and calabacitas. This produce did not by itself feed the family, of course, but it should not seem remarkable that Margarita achieved this sense of partnership with the soil despite being far removed from any ancestral lands. She refused to become a ghost of the primitive accumulation. Tending small patches of soil kept our sense of mindful respect for land, plants, animals, and water intact.

Margarita's relationship with seed and soil was an instance of mindfully embodied creativity. Her aim was to unleash the immediate use value of creative labor as an act of coevality with more-than-human beings and as a salve for healing colonial and diasporic wounds. She felt genuine sisterhood with other creatures and once sternly admonished one of her sons and me for hunting small birds with our BB guns. There were *remedios* for spiritual and physical ailments in the herb patch (*hortaliza*). Margarita's intimacies with soil, seed, and water are indicative of a worldview abiding by the principle of co-inhabiting with rather than dominating or developing a place. She practiced a niche-abiding presence based on respect for wildness—the Earth's self-willing agency in natural cycles of decay and renewal across seasons.[7] She gently imprinted these principles on anyone willing to pay attention.

Some years Margarita would instruct us to apply composted horse and goat manure to the backyard kitchen garden and have us spread it to other spots along the edible cactus fence or around the bountiful laurel bushes and twin sour orange trees in the front yard. For a time, we had a corral in the backyard with a rotating set of wild and trained mustangs, kept as part of an uncle's business. One of my chores was to tend that pile over the long, hot summer months all the way through *canícula* (the dog days). Rainfall is scarce in Laredo, so my grandmother would instruct me to sprinkle the pile with water and turn it over once every few weeks. Eventually, she would decide it was time to inspect the smelly, warm heap. Grabbing a clump of the dark compost from mid-pile to sniff, she used her nose and fingers to check its readiness for use. She wrinkled her nose to smell the clump and rubbed a bit of the dirt between her right thumb and left palm, feeling the texture. When she was satisfied the compost was ready, she would turn to us and declare, "*'sta listo*" (dropping the *e*). Sometimes she instructed my uncles to go fetch goat manure from the yard of a neighboring farmer named Amador. She would mix this with the horse poop. She described the concoction as *para la tierra sana* (for the healthy soil). Whatever signs she was smelling and feeling for remain a mystery to me to this day, but her technique worked: the milpa, the flowering plants and bushes along the edges of the house and yard, the sour orange trees and laurel bushes out front, and the edible cactus fence on the east side of our parcel—they all flourished despite the dusty desert conditions of our barrio.

Perennial flowering companion plants in Mexican clay pots lined the edges of Margarita's garden and were also distributed across the rest of the yard in beds, including in the circular margin borders under the sour orange trees out front. She loved geraniums and roses, so there were a lot of those in clay pots she tended on the smooth cement-slab front porch, nested with and overflowing alongside aloe vera (we called it *sabila*). *Tulipán* (tulip) was another favorite garden plant. Grandmother would sow a row of cilantro mixed with chives and Native aromatic plants like the *yauhtli* (Mexican marigold, *Tagetes lemmonii*), which I was to learn much later is also called *hierba de las nubes* (herb of the clouds).[8] This decorative plant is widely planted across greater Mexico alongside maize because it has a pleasant scent, attracts beneficial insects like ladybugs, and fixes primary nutrients essential for good soil and more nutrient-dense food and medicinal crops.

My grandmother's work made the soil dark and crumbly. It was filled with life—sleek wet, writhing earthworms; busy, shiny black and red-orange stinky ants; velvety red mites; and those strange roly-poly pill bugs. I remember the sweet yet slightly musty smell of the earth. What I now understand were white mycelium clumps, an indicator of healthy soil, would sometimes be exposed as she prepared the ground to sow seeds. Life sprung up as she gently weeded, irrigated, and watched for the sacred corn-bean-squash triad to sprout and grow amidst floral profusion. I believe she helped plants work together so that they could be healthier. It took decades before I fully appreciated Margarita's soil intimacies: diversity and a gentle hand are keys to agroecological and cultural resilience.

My abuelita's intimate relationship with seed saving and soil health during my childhood years had a lasting influence on my way of being in the world. In my adopted home today, in the Rio Grande headwaters bioregion of south central Colorado, her lessons continue to inform my farming practices on a 181-acre acequia farm located on the 1844 Mexican Sangre de Cristo land grant, in a high alpine desert valley surrounded by fourteen-thousand-foot peaks. This high valley was a vital stretch of historic spring hunting territory for the Caputa bands of the Ute Nation in what is now southern Colorado's San Luis Valley.[9] After the War with Mexico, the Caputa were violently driven out of their homeland by US Army troops stationed at Fort Massachusetts (Fort Garland) and forcibly resettled onto lands within the Southern Ute Reservation.[10]

In this chapter I offer some reflections grounded in my own intimacies with soil in a sacred landscape that is not my ancestral homeland. The Rio Culebra acequia commonwealth offers original instructions I seek to abide by as a displaced Indigenous person of German-Mexican and Mvskoke-Black-Irish heritage. Although my grandmother did not live to see us establish a farm school in this stretch of Ute high country, I wish to illustrate how the lessons she taught me resonate with Indigenous agroecological methods, or what modern advocates call biodynamic and permaculture principles.

These principles have deeper, more diverse roots than is acknowledged or understood by the most prominent modernist philosophers and groups associated with settler colonial sustainable agriculture movements and enterprises. This abbreviated missive is part of a larger project in which I seek to clarify how land ethics and soil conservation are political projects tied to struggles opposing the social, cultural, and environmental violence unleashed by settler colonial capitalist empires seeking control of the modes of human production and reproduction.[11] Indigenous resistance to enclosure and destruction of ancestral lands can proceed by our standing ground in solidarity with the ground we sow. As traditional Indigenous farmers, we can continue fulfilling our moral obligations to defend and regenerate this most vital of all the sources of right livelihoods by weaving together environmental wellness, spiritual integrity, cultural resilience, and community health. No matter where we are—for so many of us no longer inhabit ancestral ground—the work of regenerating soil health is a vital part of our decolonizing environmental justice and food autonomy work.

The first section of this chapter is a brief ethno-historiography highlighting some examples of Indigenous antecedents of modern biodynamic and permaculture principles. I include discussions of historic Mexica and contemporary Maya and P'urhépecha epistemic domains. I am interested in enunciating lessons from the study of ancestral and contemporary soil knowledge and stewardship practices as illustrative of Indigenous survivance.[12] The second section outlines eight principles I envision as germane to the elaboration of models of Indigenous agroecology. My articulation of this list is itself a decolonial act that seeks to disrupt and delink from the worldviews and politics enunciated by settler colonial sustainable agriculture advocates.

SURVIVANCE OF INDIGENOUS BIODYNAMICS

Rudolf Steiner (1861–1925) gets credit for "discovering" the principles of the dynamic biochemical qualities of soil, which he understood as a living organism or proper micro-ecosystem interacting with "cosmological" forces.[13] When Steiner was writing in the 1920s, it would have been difficult for him to know about the antecedent Indigenous sources of biodynamic practice. These sources of Indigenous knowledge were obscured to most of the world as a result of settler colonial acts of epistemic violence and deliberate erasures against multigenerational place-based cultures across the world.[14] Steiner's more modern interlocutors have deepened these acts of epistemic closure by presuming and fabricating the myth that biodynamics is somehow a uniquely European and neo-European (American and Australian) settler colonial modernist invention.[15]

A good contemporary example of deeper Indigenous sources—widely cited by scholars and soil tilth practitioners today—is *terra preta* (dark earth) in the long-duration rotation islands and agroforestry mosaics across the Amazon.[16] Other examples of resilient biodynamic agroecosystems include the famed *chinampas* (floating gardens) and *huertos familiares* (home kitchen gardens) of Mexico; the polyculture milpas of acequia farms in the upland valleys of the southern Rocky Mountains in Colorado and New Mexico; and the Three Sisters gardens of the Haudenosaunee nations of the northeastern United States and Canada. All these, and many more, are veritable storehouses and innovation hotspots of biocultural diversity and regeneration.[17] Deep soil knowledge, created over millennia by sustained Indigenous ecologies of place, is resurfacing across Turtle Island and Abya Yala as food sovereignty/autonomy projects take hold.[18] It is well past time for the biodynamics and permaculture discourses and movements to take notice of and reflect on the profound implications posed by the deeper Native sources across a broad range of agroecological knowledge, belief, and practice traditions.

A significant body of scientific research supports a relationship between soil quality and crop nutrient density, which in turn results in higher seed quality and preservation of agrobiodiversity.[19] There is also ample evidence of this understanding in the returning salience of interlinked fields of Indigenous knowledge of *ethnoedaphology* (Indigenous models of ecosystem processes in soil), *ethnopedology* (Indigenous classifications of soils), and *ethnobotany*, which I define as Indigenous knowledge of the soil biodynamic, nutritional, and medicinal properties of plants. This last epistemic triad embodies Indigenous understandings and framings of the relationships among soil quality, companion planting and polycultures, and medicinal and nutritional properties of food crops and their wild and intermediate relatives and companion plants.

The Dualistic Quality of Companion Plants

After Spanish forced entry into Mexico, and the unfolding of a veritable ecological and cultural apocalypse, a group of Catholic Franciscan friars coordinated what they viewed as a massive salvage ethnography project. This involved the co-production of hybrid codex manuscripts with Native scholars who were often described as "unworthy servants" of the king and queen in the frontispieces of the manuscripts. These manuscripts combined Mexica (Aztec) glyphs, numerology, and related symbolic forms alongside Latinized Nahuatl and archaic Castilian and Latin narratives and nomenclature as part of an effort to preserve Mesoamerican knowledge of the medicinal properties of native plants. The plants were selected solely based on their ethnomedical properties.

What interests me is how Indigenous farmers have long found many of the same plants to be important companions in polyculture milpa agroecosystems. The dual value of these plants in ethnobotanical *and* biodynamic terms is verified by past and existing Indigenous agroecological practices. The colonial manuscripts, and indeed most discourse in the biodynamic and permaculture movements, overlook the dualistic quality of Indigenous companion plants; simply put, these plants are medicine for the body and the land (soil) and other plants that co-inhabit the agroecological landscapes in biochemical feedback loops.

The earliest effort by a European mind to explicate the soil knowledge of Mesoamerican farmers was that of Bernardino de Sahagún (1499–1590), a Franciscan missionary and early colonial-period "salvage" ethnographer. His chronicles of the Culhua Mexica imply the early adoption of what we would today view as biodynamic practices. Sahagún mentions the Nahuatl term *quauhtlalli*, which he translates as "[rotten] wood soil." In volume 2, chapter 12, section 3 of his *General History of the Things of New Spain* he writes:

> Hay otra manera de tierra fértil, donde se hace muy bien el maíz y trigo, *Uamanla quauhtlalli*, que quiere decir, tierra que esta estercolada con maderos podridos, es suelta, amarilla, y hueca.
>
> > There is another sort of fertile soil, in which corn and wheat flourish very well, they call it *quauhtlalli*, which is to say, earth which has been manured with rotten wood, it is soft, rich, and golden. (Translation mine)

This is similar to the oft-cited European practice of *Hugelkultur*, a German word meaning "hill culture" or "mound culture," which is often invoked in settler colonial discourses to illustrate the ancient European roots of biodynamics and permaculture.

Similar evidence is found in the famous *Códice Badiano* of 1552, also known as "The Little Book of the Medicinal Herbs of the Indians."[20] This is one among several colonial-era codices that hint at the importance of companion planting and soil knowledge in Mesoamerican agroecology. The book has descriptions of more than 250 plants with 187 illustrations. One noteworthy example is a sketch of water nettle, or *Atzitzicaztli* (*Urtica chichicaztli*). The notes accompanying this lovely sketch describe how the juice of nettle, ground with salt and mixed with urine and milk, can be poured into the nostrils to stanch a nosebleed. Not mentioned are various biodynamic properties of this plant, which Native campesinos view as having a "strong spirit."[21] Indigenous and traditional smallholder farmers use water nettle because it is known to promote the formation of humus in the soil. There is some evidence that when it accompanies corn-bean-squash intercropping, it suppresses bindweed.[22] Three common companion plants recorded in de la Cruz–Badiano are still widely used in the milpa agroecosystems of Mexico and the acequia farms in parts of the Southwest: these are yarrow (*Achillea millefolium*), lamb's quarters (*Chenopodium album*), and stinging nettle (*Urtica dioca*). All three are recognized and valued for their medicinal properties in our bioregion today.[23] All three are also listed as ingredients in contemporary certified biodynamic concoctions added to compost.[24] Similar examples of this dual quality can be discerned among dozens of the companion plants known to have been cultivated since well before contact times.

One recent study of farmers at the *tiangüis* (popular open-air markets) in Mexico verifies the continuity of applied Indigenous knowledge of companion planting, highlighting its value in optimizing the availability and renewal of organic matter in soil. In the tiangüis study, Miguel Escalona Aguilar explains that farmers' soil tilth focuses on multicropping with companion plants. This knowledge is ancient *and* contemporary; it is not static; this knowledge responds to territorial shifts and environmental changes. The goal of regenerating soil quality is achieved through a combination of on-site inputs with different types of plants available as cover crops, since both the temporal availability of organic matter and its actively shifting composition are enriched by such diversity.[25]

Farmers practicing traditional polyculture view plant diversity as a fundamental method for sustaining good anthrosols; that is, human-influenced soils.[26] Our ancestral agroecologists understood the importance of protecting structural and species diversity in order to sustain what we today imagine as the chemical and biological constitution of good agricultural soils.

Epistemic Intimacies: Indigenous Soil Classification, Biodynamic Companions, and Ethics Instruction

The Mexica, like other Mesoamericans, developed a very sophisticated system of soil classification that, by some accounts, included more than sixty classes of soil.[27] Adopting knowledge from antecedent Mayan, Toltec, and other civilizations, the Culhua Mexica adapted sensible experience-based criteria for defining soils to determine their appropriateness for different uses, including being left alone in a state of relatively undisturbed wildness. Particular terms were adopted to identify the qualities of soils deemed arable or fertile: *atocpan* (lands with a deep topsoil horizon) and *atoctlalli* (lands composed of moist alluvium). It is unsurprising that these continue as categories designating soils that are appropriate for farming, as does the general category of *cuenchihu*, or *cuemitl*, meaning those soils classified as appropriate for being disturbed and supplemented or composted. This terminology conveys awareness of the potential for negative disturbance of soil as a result of misguided activities by farmers. It may also invoke ethical obligations to avoid abusive practices. The semiotics of Indigenous soil vernaculars suggest our soil knowledge remains connected to instruction in land-care ethics. This requires us to move beyond the settler colonial binaries associated with the ideology of human exceptionalism. The routine separation of the teaching of land ethics (soil intimacies) from instruction in soil science in the pedagogy of today's land grant universities illustrates a binary split in the Western paradigm and hints at why our survivance is so revolutionary.

Understanding the ethical dimensions of how people view and manage soil (ethnopedology) is only the beginning of a remapping of our Indigenous knowledge, belief, and practice complex. Yet these ethical concepts have not been explicitly addressed in the context of using companion plants to maintain and regenerate soil as a practice common to Indigenous agroecology. Based on my ongoing research, many of the 251 plants discussed in de la Cruz–Badiano for their medicinal properties were also valued, and are indeed still valued today, for their soil-enriching and allelopathic properties. A preliminary list of companion plants with dual ethnomedical and soil biodynamic/agroecological properties is presented in table 12.1. This list hints at the need for independent

Indigenous research on using these plants in biodynamic preparations to supplement compost and manure.

Beyond Deep History and the Mexica Fetish

Despite the importance of pre- and post-contact codices and testimonies, I wish to avoid giving the impression that the Mexica and Mayan civilizations and their heirs (contemporary Nahua and Maya communities) are the only sources of Indigenous knowledge of agroecology and specifically soil biodynamics. Antecedent and contemporary Indigenous communities across Mexico and the rest of the planet have soil TEK. We need not only to listen to but to enunciate these contemporary stories in acts of relational solidarity instead of ending inquiry with the clarification and celebration of ancient codices, as if Native soil knowledge were confined to some remote past.

In Mexico, a notable current form of Indigenous TEK is the soil classification–farming–ceremonial complex of P'urhépecha campesinos in Michoacán; I point to their historical and continued presence as peasant farmers in Colorado and other places north of the US-Mexico border.[28] These farmers continue to develop and transmit classification models and ecological knowledge across generations. They recognize five general classes of soil, or *echeriecha*, with two to five subtypes depending on location. The current soil classification system contains a total of seventeen soil subtypes.[29]

The highly nuanced criteria used to discern soil types (for example, based on gradations of color) and subtypes in the P'urhépecha model anticipate modern soil science by several hundred years. Unlike in modern research universities and land grant colleges, however, in the P'urhépecha community contemporary soil classification practices involve teaching and practical activities linked to collective and communal work organizations that are traditionally responsible for protecting and improving soil quality. Soil health is a matter of common property relations and moral obligations. P'urhépecha soil knowledge, alongside other domains of local ethnoecological knowledge (like agro-forestry), is shared and transmitted in conjunction with closely aligned ceremonial and agricultural calendar cycles. Soil conservation and regeneration practices and ethical instruction intersect and flow alongside the cycles of ceremonial life in the collective work of such place-making communities. Even in the internally colonized northern acequia communities, there are still vestiges of the ties among ceremony, agriculture, soil conservation, and irrigation. These ties are evident in the continued observance

TABLE 12.1
ETHNOMEDICAL AND BIODYNAMIC/AGROECOLOGICAL PROPERTIES OF SELECT COMPANION PLANTS IN TRADITIONAL MILPAS

Plant	Ethnomedical Values[a]	Soil Biodynamic/Agroecological Propwerties[b]
Chamomile[c] Manzanilla (*Chamaemelum nobile*)	Tea has calming effect; sulfur content has antifungal properties; hay fever relief; anti-inflammatory; combats muscle spasms, menstrual disorders, insomnia, ulcers, wounds, gastrointestinal disorders; prevents rheumatic pain; reduces swelling of hemorrhoids.	Roots loosen compacted earth so that other plants can find nutrients and water. Preparation of flower "tea" helps unblock plant sap and prevents stress from excess heat/cold; adds sulfur and calcium to compost mixes; high in potassium. Stabilizes nitrogen within compost and increases soil life to stimulate plant growth.
Datura Tolohua (*Datura innoxia*)	Analgesic; narcotic effects used in setting broken bones, making incisions, and relieving painful bruises and other injuries. Anti-angiogenesis properties.	Leaf hexane extract has insecticidal potential. Facilitates phytoextraction of heavy metals from soil (e.g., excess cadmium, chromium). Grows in alkaline soils. Should not be grown alongside potato plants, as it can be a center of viral infections.
Lamb's quarters Quelite (*Chenopodium album*)	Astringent; upset stomach relief; prevents scurvy. Tea used to treat diarrhea.	Is useful as "trap crop" for leaf miners. Hosts beet leaf hopper. Restores healthy nutrients to the soil (except that phosphorous may not be bioavailable to plants for many years through green manure).
Laurel; Mexican bay leaf Eca-patli (*Litsea glaucescens*)	Bruised together in frigid water with other herbs and stones, then soaked in a neck wrap, it can relieve sore throats and head and neck pains.	Creates a supportive microclimate for shade-tolerant pulses and vegetables.

Source: Plants listed in the Martín de la Cruz–Badiano Codex of 1552.

[a] As narrated in the de la Cruz–Badiano Codex and contemporary accounts such as William Gates, *An Aztec Herbal: The Classic Codex of 1552* (Mineola, NY: Dover, 1939).

[b] Based on my ongoing research with multiple sources and field measurements at the Almunyah de las Dos Acequias.

[c] Introduced species that Indigenous farmers and agroecologists widely accept as a naturalized companion plant.

TABLE 12.1 *(continued)*

ETHNOMEDICAL AND BIODYNAMIC/AGROECOLOGICAL PROPERTIES OF SELECT COMPANION PLANTS IN TRADITIONAL MILPAS

Plant	Ethnomedical Values[a]	Soil Biodynamic/ Agroecological Propwerties[b]
Marigold *Yauhtli* (*Tagetes lemmonii*)	Ointment can be used to soothe sunburns, warts, bites, acne, and ulcerations; and to heal wounds, dry skin, and blisters.	Protects other plants by releasing a chemical that deters potentially lethal root-infesting nematodes. Chemical also reduces chances of fungal, bacterial, viral, and insect problems; attracts beneficial insects.
Papalo *Papaloquelite* (*Porophyllum ruderale*)	Soothes coughs, headaches; reduces flatulence after consuming legumes (beans, lentils).	Cold-hardiness contributes to protective cover and supports soil moisture, prevents hardpan, and attracts beneficial insects.
Purslane *Verdolagas* (*Portulaca oleracea*)	Leaves are rubbed on insect or snake bites; soothes boils, sores, pain from beestings; contains more omega-3 fatty acids than any other leafy plant; high in vitamins A, C, E, and B.	Creates humid microclimate for nearby plants; deep roots can bring up moisture and nutrients other shallow-rooted plants (e.g., corn) cannot reach.
Water nettle *Atzitzicaztli* (*Urtica chichicaztli; U. dioica*)	Ground with salt and mixed with urine and milk to stanch bleeding.	Promotes formation of humus in soil; suppresses bindweed; can be sprayed on sick or stressed plants as liquid manure.
Yarrow *Tlalquequetzal* (*Achillea millefolium cv.*)	Flowers and leaves are eaten or made into a tea-like drink; fresh leaves are applied to stanch bleeding wounds; treats gastrointestinal problems and fevers (infections); lessens menstrual bleeding; promotes circulation.	Preparation enhances crops' ability to absorb potassium in a balanced manner. Enables soil to absorb and retain silicic acid. Aids in the formation of quality plant proteins.

of the feast day of San Isidro Labrador, the patron saint of farmers, in northern New Mexico and southern Colorado acequia communities.[30]

How Continued Primitive Accumulation Disrupts Soil Intimacies

We face the long-duration effects of settler colonial–capitalist enclosures (primitive accumulation) on the health of the land and the degradation of our soil knowledge. These conditions reduce our autonomous capacities to renew and replenish our agroecological systems, skills, and practices, including our ceremonies. Contemporary Maya communities embody the continuity of soil classification and restoration practices, but their evolving soil vernacular reflects the inescapable need for us to consider myriad political-economic and ecological problems posed by the history and continuing violence of enclosure and displacement. In Hocabá, Yucatán, Maya farmers today continue to classify soil types into eleven distinct subclasses.[31] Their soil terminology is influenced by the long-term effects of structural violence unleashed by settler colonial dispossession and displacement of Native farmers from ancestral lands. Many of the concepts used in the Hocabá Maya soil class schema today reflect the marginal condition of the lands these farmers have been forcibly displaced onto for their subsistence plots (*sembrados*). There are now constant references to different types of "thin" or "shallow" soils and soils that are "too rocky" or "too steep." Many of these concepts defining poor-quality soils were rarely enunciated by past generations of farmers. The evolving language of soil is therefore reflective of Indigenous peoples' active engagement with changing environmental conditions, most of which have been imposed on them by colonizers and usurpers who stole the more fertile ancestral lands. Imagine how climate change (the sixth mass extinction) and other ecological catastrophes will influence our changing soil and other land health vernaculars.

This brief ethnohistoriography suggests it is insufficient to acknowledge and celebrate the deep roots, methods, practices, and ethics of Indigenous soil knowledge. We need to understand the conditions of disruption and change as defined and constrained by the specific bioregional qualities and environmental histories experienced by each place-based culture. Despite epochs of resistance, our soil intimacies are tempered and transformed by the forces of settler colonial-capitalist dispossession and displacement. The lesson I draw from the Hocabá Maya is that our soil knowledge today includes signifiers of the degraded condition of the reduced and marginal landscapes and watersheds we have been forced to inhabit and cultivate.[32] The land and water are gravely wounded, and Indigenous linguistic shifts reflect this fact. It is in this subaltern realm where we may find especially poignant examples of hope for the resurgence of Indigenous biodynamic and related agroecological practices to heal the land, water, and people. This is not a lifestyle option or entrepreneurial opportunity in some newfangled farm-to-table touristic dystopia. Mapping our own soil-knowledge models across diverse bioregional landscapes remains an important task of our decolonization projects toward food autonomy by healing the soil and protecting our water to create la tierra sana.

THE RESURGENCE OF INDIGENOUS BIODYNAMIC AND PERMACULTURE PRINCIPLES

Indigenous principles of biodynamics and permaculture arose from place-based cultures and, in many places, associated diaspora communities who developed these over millennia *in and often forcibly out of place*. In the end, these principles are about a people's capacity for intimacy with more-than-human beings—the intimacy between people, land, water, the health of the soil, and all living beings sharing the biosphere. The integrated notion of soil-body health reflects principles of *relationality* that differ from Western epistemologies derived from the norms of *objectivity*.[33]

I encourage further discussion of the principles of Indigenous biodynamics and permaculture. How are these different from settler colonial principles? Underlying my opening contribution to this discussion is the idea that our ancestral lands have been appropriated and damaged. As a result, the land has suffered through the same conditions and effects of historical trauma and structural violence experienced by Indigenous peoples. From this location, our agroecological practices necessarily merge with the work of restoration ecology and the spiritual tasks of healing colonial wounds through our intimacies with soil, plants, animals, water, and each other. In this sense, the concept of permaculture is a more recent and limited Western analogue of the primordial fields of place-based, spiritually grounded Indigenous agroecologies. This immediately invokes an issue of social justice and a call for the restoration of Indigenous autonomy (including land and water rights claims), which makes our restoration efforts distinct from settler colonial movements whose starting point is sustainability and not Indigenous sovereignty.

The following eight principles can perhaps be viewed as a possible formulation of *relational practices* evident in multi-sourced Indigenous agroecologies:

1. *Spiritually grounded practical respect for extant landscape ecologies, including wounded lands*. As Indigenous farmers we engage ceremony

by working to repair and sustain the material and spiritual conditions of the social and cultural fabric of our communities and parent ecosystems. This is true of our respect for natural landscape qualities like contour intervals, as illustrated by *loma y bordo* [terracing] in milpas and huertos familiares [home kitchen gardens] across the upland zones of central and southern Mexico and as far north as New Mexico. The spiritual dimension involves acting with loving kindness toward the land and all our co-inhabitants, especially more-than-human beings. Recognizing the effects of the environmental violence of settler colonialism and capitalist enclosures, we perform actual service to the land by enriching and deepening soil horizons (removing rocks and adding tilth), actively repairing damage to "marginalized" lands, and protecting our native seeds and rootstocks while continuing to nurture their adaptability in a climatically chaotic world. Our spiritual practice thus becomes a moment in which we may embody "theory in the flesh," binding the well-being of our bodies to the well-being of the soil and other Earth life support systems as coeval partners.

2. *A preference for perennial and annual polycultures* and avoidance of monocrop systems. The principle of *agro-biomimicry* involves practices that create agroecosystems which mimic the environmental conditions of place by reproducing structural, species, and process diversity, including hundreds or even thousands of wild plants and relatives of cultivars, many with valued qualities as spiritual companions, as sources of medicinal substances, or as promoters of soil biodynamic qualities (see number 4). These plants are not weeds.

3. *Established patterns of multi-crop rotations with long-duration fallows.* Allowing soil to rest and regenerate tilth in order to sustain its biodynamic qualities is one focus of Indigenous agroecological practices. Rotations may often involve so-called green manure crops that fix nitrogen and facilitate regenerative processes in the soil (the practice of *frijol tapado* comes to mind). These practices abide by the extant bioregional qualities of each watershed and focus on "re-wilding" ecosystems and habitats that host our farming and gardening plots. These rotational cycles are often linked to ceremonial and community life event calendars. Recognizing the regenerative nature of their activities, Indigenous farmers produce a shifting mosaic of domesticated and wild plant associations that blur the boundaries between domestic crops,

other plants, and wild relatives and associated organisms. Many of us do not have enough land to practice long fallow periods, which is why intercropping and minimum tillage are important practices (see number 4).

4. *Intercropping with biodynamic and allelopathic companion plants*, including decorative flowers that attract pollinators and beneficial insects for natural pest control. Intercropping is also important for regenerative processes and further establishes the vital principle that diversity is the key to resilience in agroecosystems. This practice centers around a set of culturally important plants—the triad of corn-beans-squash plus the hundreds of companion plants present in traditional Indigenous polyculture milpas. Intercropping can occur within various resilient spatial strategies: linear rows, circular mounds, more random field dispersals, or multiple combinations and sequences thereof. Intercropping on top of mounded sheet layers is an important minimum tillage practice with the right cover crops (such as purslane).

5. *Classification (ethnopedology) and systemic care (ethnoedaphology) of soils.* Understanding the soil types of a given bioregion, and the native plants these favor, is an important step toward food sovereignty. We must work to support the emergence of appropriate place-based material and spiritual practices that include ethical instruction for healing the land and people. Soil is culture, and our elders' understanding of the land as a living organism is a central quality of the deep heritage we must support and teach to the next generations. We must also recognize soil classification as dynamic rather than static. Since anthrosols are transformed through biodynamic and regenerative practices, new soil classification models and ceremonies will surely emerge.

6. *Preparation and application of biodynamic soil treatments* (such as quauhtlalli). The uses of biodynamic preparations and soil treatments are especially vital for repairing damaged landscapes. They are also antecedent practices in our deepest heritage and right livelihoods. Beyond the corn-beans-squash triad is a vast field of knowledge of the allelopathic, medicinal, and biodynamic properties of hundreds of companion plants that can be mixed in potent, concentrated tinctures and compounds and applied to regenerate tired or damaged soils.

7. *Cognitive maps of frost, infiltration, and saturation topographies.* Each bioregion has qualities unique to the watershed—the biota, landforms,

local cultures, and environmental histories which run deeper than the exploits of settler colonial empires that remapped Native territories in acts of violent enclosure and displacement. As we "remap" our homelands, including the gardens, farms, wildlands, and watersheds of our bioregions, we can imagine various types of "storied landscapes," such as soils or areas susceptible to frost; areas where sub-irrigation water returns to in-stream flows or creates springs; or areas prone to the formation of marshy or wetland conditions. These maps will connect us to awareness of the niche-abiding patterns of co-inhabitation while nurturing habitat for wild plants with spiritual and medicinal qualities, since these often thrive in ecological niches that are part of this "storied" landscape.

8. *"Resilient co-inhabitation," an Indigenous analogue of adaptive management.* From the vantage point of Original Instructions, the biosphere itself provides the rules. Because these rules are subject to change, we require a constant set of adaptive practices in response to changes in the coupling of social and ecological systems.

Our Indigenous agroecological heritage dates back to well before the arrival of the currently fashionable, modern, and profitable advocates of biodynamics and permaculture. These principles are rooted in a deep agroecological heritage and patterns of ethical instruction that refuse to die out. All these and many more agroecological practices are results of long-established traditions of Indigenous knowledge, belief, and practice in the various centers of origin and diversification of native crop plants like maize. It is time for Indigenous water and land protectors to consider how these principles might inform their own autonomous projects toward strengthening food sovereignty, environmental wellness, and community health.

NOTES

1. *Maíz* is maize, or corn (*Zea mays*); *calabacita* typically refers to gray Mexican summer squash (*Cucurbita pepo*), both squash blossoms and tender squash. *Calabaza* may refer to any among a wide range of pumpkin varieties, generally but not exclusively *Cucurbita moschata* and including a vast range of northern winter varieties (*pepitas* are pumpkin seeds for roasting). *Frijol* can refer to any among thousands of unique local heirloom landrace varieties of the common bean (*Phaseolis vulgaris*); in our case, at home in Laredo, *pinto* is the more frequent name; in Colorado, *bolita*.

Among the Haudenosaunee, these three crops present an exceptional range of full-spectrum heirloom varieties with unique bioregional and distinctive landrace parent lines. They have therefore long been known as the "Three Sisters." In the Haudenosaunee creation story the Three Sisters are celebrated and enunciated as significant cultural memes grounding the people in a distinct place-based or bio-centric worldview. Beyond the Northeast, Indigenous maize growers and protectors today have widely adopted the term *Three Sisters* since the plants are almost universally viewed as a triad of sacred companions at the center of Indigenous agroecological practices across many nations.

On the centrality of maize in the civilizations of Mexico and Mesoamerica, see Roberto Cintli Rodríguez, *Our Sacred Maíz Is Our Mother: Indigeneity and Belonging in the Americas* (Tucson: University of Arizona Press, 2014); Devon G. Peña, Luz Calvo, Pancho McFarland, and Gabriel R. Valle, eds., *Mexican-Origin Foods, Foodways, and Social Movements: Decolonial Perspectives* (Fayetteville: University of Arkansas Press, 2017), 313–41; 375–82. Indigenous stories and scientific discourses long have celebrated these three companion plants for their medicinal, biodynamic, and allelopathic properties. Among the few studies that have examined all three dimensions, see Jane Mt. Pleasant, "The Three Sisters: Care for the Land and the People," in *Science and Native American Communities: Legacies of Pain*, edited by Keith James (Lincoln: University of Nebraska Press, 2001), 126–34; Mt. Pleasant, "The Iroquois Sustainers: Practices of a Long-Term Agriculture in the Northeast," *Northeast Indian Quarterly* 6, no. 1–2 (1989): 33–39. A more recent Western scientific study of these companion cultivars and their wild relatives can be found in Salvador Montes-Hernández, Laura C. Merrick, and Luis E. Eguiarte, "Maintenance of Spanish (*Cucurbita* spp.) Landrace Diversity by Farmers' Activities in Mexico, *Genetic Resources and Crop Evolution* 52, no. 6 (2005): 697–707. These authors note "the ancestors of maize . . . and common beans . . . have survived until the present, and they occur usually in small populations, grow in sympatry with the domesticated forms . . . two cultivated squash species, *C. argyrosperma* and *C. moschata*, and the wild type *C. argyrosperma* ssp. *sororia* also grow in sympatry (698).

2. Coriander (*Oriandrum sativum* L.), among other aromatic herbs and spices like cumin and clove; and sweet peas (principally *Pisum sativum* L.), among other legumes, were introduced to Mexican cuisine and milpa farming in the mid- to late sixteenth century by settler colonists from Spain. Research on the introduction of these crops to Mexico shows that most have Middle Eastern or North African origins rather than Iberian or European ones. See George W. Hendry, "The Source Literature of Early Plant Introduction into Spanish America, *Agricultural History* 8, no. 2 (1934): 64–71.

3. On the Latourian concept of vibrant matter, see Jane Bennett, *Vibrant Matter: A Political Ecology of Things* (Durham, NC: Duke University Press, 2010). Bennett uses the term "vibrant matter" to refer to more-than-human beings, including landscapes and larger-scaled biophysical Earth systems, as "actants" with agency; this is a concept borrowed from Bruno Latour. Bennett views these more-than-human actants as having an a priori capacity to create the very matter of the world sans and within humanity, leading to changing conditions in how human beings come to experience and perceive variant forms of "nature." In a related vein, Julie Cruikshank recounts the voices of three Native women in the Yukon who "portrayed glaciers as conscious and responsive to

humans. Glaciers, they insisted, are willful, sometimes capricious, easily excited by human intemperance but equally placated by quick-witted human responses"; see Cruikshank, *Do Glaciers Listen? Local Knowledge, Colonial Encounters, and Social Imagination* (Vancouver: University of British Columbia Press, 2005), 8. The parallel to my grandmother's soil knowledge is that I cannot speak of the condition of the soil before our presence, but I can address the condition of the soil now as a quality of my present relationship to the land as a co-actant. Soil is vibrant matter effecting changes in how I relate to the ground I sow.

4. This seems true among all the domesticated species selected to avoid "shattering," which are plants that rely on our sowing, harvesting, selection, and seed-saving to reproduce, as suggested by Carey Fowler and Pat Mooney, *Shattering: Food Politics and the Loss of Genetic Diversity* (Tucson: University of Arizona Press 1990), 14–18.

5. There is a growing scientific literature to support my grandmother's insight that soil health is associated with higher nutrient density in cultivars. For example, see Alexandria Bot and José Benites, "The Importance of Soil Organic Matter: Key to Drought-Resistant Soil and Sustained Food and Production," FAO Soils Bulletin 80 (Rome: Food and Agriculture Organization, 2005), http://www.fao.org/3/a-a0100e.pdf; Donald R. Davis, "Declining Fruit and Vegetable Nutrient Composition: What Is the Evidence?" *HortScience* 44, no. 1 (2009): 15–19; Margo Malone, "Improving the Nutrient Content of Agriculture Crops through Community Ecology," Syracuse University Honors Program Capstone Projects 953 (2016), https://surface.syr.edu/honors_capstone/953; also see the citations in note 19.

6. Ritual kin relations through co-parenthood, or *comadrazgo/compadrazgo*, encompass a wide range of productive and reproductive labor activities and networking practices related to midwifery and related healing traditions, the cultivation of food and medicinal crops, and the care of soil. See, esp., Paloma Martínez-Cruz, "Survivor Woman." in *Women and Knowledge in Mesoamerica: From East L.A. to Anahuac* (Tucson: University of Arizona Press, 2011), 73–95; Irene Lara, "Latina Health Activist-Healers Bridging Body and Spirit," *Women and Therapy* 31, no. 1 (2008): 21–40.

7. On the difference I draw between "wildness" and "wilderness" see the essays in Gavin Van Horn and John Hausdoerffer, eds., *Wildness: Relations of People and Place* (Chicago: University of Chicago Press, 2017). Wildness encompasses the principle of "self-willing" land, an idea that resonates with Bennett's concept of vibrant matter and the Indigenous epistemic frame of "original instructions"; see Devon G. Peña, "The Hummingbird and the Redcap," in *Wildness: Relations of People and Place*, 89–99, esp. 91–95.

8. Ana María L. Velasco Lozano and Debra Nagao, "Mitología y simbolismo de las flores," *Arqueología Mexicana* 78, no. 2 (2006): 28–35. Among the several biodynamic substances produced by the Mexican marigold is alpha-terthienyl, a phototoxin extracted from root biomass. It has been shown to be extremely insecticidal against mosquitoes without affecting beneficial organisms like the ostracod, caddisfly, and *Physa* sp.; Anuj Kumar, Florence V. Dunkel, Matthew J. Broughton, and Shobha Sriharan, "Effect of Root Extracts of Mexican Marigold, *Tagetes minuta* (Asterales: Asteraceae), on Six

Nontarget Aquatic Macroinvertebrates." *Environmental Entomology* 29, no. 2 (2000):140–49, https://academic.oup.com/ee/article/29/2/140/341923. While Kumar and colleagues' study did not address the activity of the phototoxin in soil, it would seem the marigold may repel some insects besides mosquitoes, since other sources note its effectiveness in the control of nematodes; see Maureen Gilmer, "Marigold in the Control of Root-Knot Nematodes," *Orange County Register* (June 10, 2015), https://www.ocregister.com/2015/06/10/plant-marigolds-to-stop-root-knot-nematodes/.

9. The people who settled our acequia community and land grant included *genízaros* and direct descendants of marriages and captives with Jicarilla Apache, Diné, and Pueblo ancestors.

10. The Southern Ute Reservation is located due south of Durango and Pagosa Springs, Colorado, and encompasses an area of montane, riparian, and high desert life zones bordering New Mexico. It is about 150 miles due west of the San Luis Valley and the acequia communities of the Culebra watershed (Sangre de Cristo grant).

11. Peña, "Sodbusters and the 'Native Gaze': Soil Governmentality and Indigenous Knowledge, in *Mexican-Origin Foods, Foodways, and Social Movements*, 343–64. My reference to the governmentality of soil alludes to the idea that the Soil Conservation Service, which is now the Natural Resources and Conservation Service, has a history of imposing a governmental regime that narrowly manages soil conservation programs to serve large-scale corporate farming interests while ignoring and erasing Indigenous needs, knowledge, belief, practice, and traditions. This regime, involving both the soil conservation paradigm and the subjects (farmers) of the managerial regime, largely reflects neoliberal capitalist values.

12. See Gerald Vizenor, *Manifest Manners: Narratives on Postindian Survivance* (Lincoln: University of Nebraska Press, 1999). Vizenor conceptualizes survivance as "an active sense of presence, the continuance of Native stories, not a mere reaction, or a survivable name. Native survivance stories are renunciations of dominance, tragedy and victimry" (vii).

13. For the standard account of the origins of biodynamics, see Jane Hodges Young, "The Ancient Practice of Biodynamic Farming," *Northbay Biz: Napa, Marin, Sonoma*, July 2017, https://goo.gl/fwhL58. All the Indigenous civilizations of North America shared a concern for cosmological forces as pertinent to material conditions and livelihoods and predate Steiner's and other modern Europeans' visions in this manner.

14. For environmental anthropology or ethnoecology (the study of Indigenous knowledge of ecosystems), Harold Conklin's 1954 dissertation research, conducted in collaboration with the Hanunóo agro-forestry farmers in the Philippines, finally challenged the colonial gaze of anthropology writ large by respectfully re-centering local truth claims as conceptual and discursive elements of the culture analyst's own interpretive narrative; see Harold Conklin, *The Relation of Hanunóo Culture to the Plant World* (PhD diss., Yale University, New Haven, CT, 1954).

15. I have in mind here the work of notable interlocutors and permaculture advocates, a large number of them white Australian settlers; see Bill Mollison, *Permaculture: A Designers' Manual* (Berkeley, CA: Ten Speed Press, 1997); David Holmgren,

Permaculture: Principles and Pathways beyond Sustainability (Hepburn, Victoria, Australia: Holmgren Design Services, 2002); and Looby Macnamara, *People and Permaculture* (East Meon, Hampshire, UK: Permanent Publications, 2012). None of these advocates acknowledges or seriously considers the deeper Indigenous (or Aboriginal) bases of the methods, materials, and practices they advocate. This seems like a cautionary example of cultural appropriation without attribution and a faulty lack of relational solidarity with Native sources as living communities. The same is true of their ethical discourses: Holmgren, following Mollison's original declaration, lists three main ethical principles for permaculture practice: "Care for the earth; care for people; set limits to consumption and reproduction and redistribute surplus" (1). Obviously, these principles are widely recognized Indigenous concepts, customs, and practices. The carelessness with which these white authors outline their ethics and practitioner claims is illustrative of a form of epistemic violence and exemplary of the type of settler colonialist mindset I object to, especially since these advocates benefit from abstracting and obscuring Indigenous agroecological knowledge through commercialization and "professionalization." Mollison briefly pays tribute to the "aboriginal tribespeople of Australia" since "for every scientific statement articulated on energy ... [they] have an equivalent statement on life" (2). I am working on an extended critique of settler colonialism and agroecology in assessing the prospects for "indigenizing" biodynamics and permaculture.

16. Bruno Glaser, Ludwig Haumaier, Georg Guggenberger, and Wolfgang Zech, "The 'Terra Preta' Phenomenon: A Model for Sustainable Agriculture in the Humid Tropics, *Naturwissenschaften* 88 (2001): 37–41, https://goo.gl/n82LH2.

17. D. Peña, *Mexican Americans and the Environment: Tierra y Vida* (Tucson: University of Arizona Press, 2005), 44–57, 81–87; Enrique Salmon, *Eating the Landscape* (Tucson: University of Arizona Press, 2013).

18. Examples of this resurgence include the work of groups like the Traditional Native American Farmers Association, Tesuque Farms Agricultural Initiative, and New Mexico Acequia Association in New Mexico; the Acequia Institute and the Sangre de Cristo Acequia Association in Colorado; Oglala Lakota Cultural and Economic Revitalization Initiative in South Dakota, in affiliation with the Permaculture Guild; the Indigenous Permaculture Convergence at the Woodbine Institute; permaculture design courses on dozens of Native American tribal reservations; numerous projects supported by the First Nations Development Institute; dozens of Indigenous projects like the Braiding the Sacred and Voices of Maize network of more than fifty Native corn growers and protectors; and Alianza Milpa.

19. An early work in this genre of scientific inquiry by Megan Ryan found that most organic farms are not biodynamic over the long term and have similar deficits to conventional farms. However, Ryan does not consider monocrop tendencies in industrial organics, which can create soil health deficits and eliminate wildlife habitat, or how polycultures can improve soil quality; see Megan Ryan, Is an Enhanced Soil Biological Community, Relative to Conventional Neighbours, a Consistent Feature of Alternative (Organic and Biodynamic) Agricultural Systems? *Biological Agriculture and Horticulture*

17, no. 2 (1999): 131–44. Another report found that the fertility of agricultural soils can be maintained over the long term only if plant nutrients removed are replaced in equivalent amounts and if added sources have a higher solubility than those present in the soil; see Holger Kirchmann Lars Bergström, Thomas Kätterer, and Olof Andrén, "Can Organic Crop Production Feed the World?" In *Organic Crop Production: Ambitions and Limitations,* edited by Kirchmann and Bergström (Dordrecht, Netherlands: Springer, 2009), 39–72. Results of one study support use of compost extracts as fertilizer substitutes or supplements, and provide testimonial reports on the growth-promoting effects of compost extracts and occasional superiority of biodynamic compost compared to untreated compost; see Jennifer R. Reeve, Lynne Carpenter-Boggs, John P. Reganold, and Will Brinton, "Influence of Biodynamic Preparations on Compost Development and Resultant Compost Extracts on Wheat Seedling Growth," *Bioresource Technology* 101 (2010): 5658–66, https://doi.org/10.1016/j.biortech.2010.01.144. A detailed study of tomato production examined issues related to healthier, more nutritious outcomes for humans rather than just soil and crop qualities. This meta-analysis of studies on the nutritive quality of tomatoes found that cropping production systems on organic or ecological farms that involved use of fertilizers such as compost, manure, and farmyard waste; minimal chemical pesticides; renewable, non-fossil-fuel inputs; and application of mulches and cover crops all had positive effects on tomato fruit quality and human health-related factors (e.g., density of absorbable lycopene factor). The authors conclude more studies and data are needed to understand how environmental conditions and production practices affect nutritive quality; how agroecological practices affect canopy characteristics and interfere with the effects of light, temperature, and carbon dioxide; and the role of genetic factors (decline of agrobiodiversity, e.g., heirloom alleles, is a serious problem). They also recommend assessing the effects of storage, distribution, and retailing practices (e.g., many packing houses spray supplemental ethylene gas to accelerate ripening). Finally, they point to cultural practices, like picking fruit before it is vine-ripened, which can affect taste and nutritive qualities. See Reza Ghorbani, Vashid Poozesh, and Surur Khorramdel, "Tomato Production for Human Health, Not Only for Food." in *Organic Fertilisation, Soil Quality, and Human Health,* edited by Eric Lichtfouse (Berlin: Springer, 2012), 187–226.

20. *Libellus de Medicinalibus Indorum Herbis* (or Aztec Herbal Manuscript) was composed in 1552 by the Nahua scholar Martín de la Cruz of an elite family from Tlatelolco and translated into Latin by Juan Badianus. The fact that this codex is named after the European translator rather than the Native Mexica author represents a continuation of the original act of epistemic violence that led to the production of the de la Cruz–Badiano, as I prefer to call it. The English translation I use here is William Gates, *An Aztec Herbal: The Classic Codex of 1552* (Mineola, NY: Dover, 1939 [2000]).

21. Rudolph Steiner observed that a compost prepared of stinging nettles is "like an infusion of intelligence for the soil" (100). See Steiner, *Agriculture* (1924 [1958]), published online in 2007, https://goo.gl/EJ2cho.

22. Evidence from experiments suggests some annual and perennial crops have shade effects on bindweed growth. These crops include companion plants in the milpa, such

as legumes (beans) and cereal grains like corn; nettle extract can be sprayed on bindweed; see Laurie Hodges, "Bindweed Identification and Control Options for Organic Production," *Historical Materials from University of Nebraska–Lincoln Extension* 48 (2003), http://digitalcommons.unl.edu/extensionhist/48.

23. For a comprehensive discussion of the medicinal herbs of the Southwest, see Michael Moore, *Los Remedios: Traditional Herbal Remedies of the Southwest* (Santa Fe: Museum of New Mexico Press, 2008). Introduced to Mexico by the Spaniards as a medicinal herb, chamomile (*Matricaria chamomilla*) is today another frequent companion plant in the acequia milpas of the southwestern United States. See Janmejai K. Srivastava, Eswar Shankar, and Sanjay Gupta, "Chamomile: An Herbal Medicine of the Past with a Bright Future, *Molecular Medicine Report* 3, no. 6 (2010): 895–901, https://www.ncbi.nlm.nih.gov/pmc/articles/PMC2995283/pdf/nihms250193.pdf for discussion of the biodynamic properties of chamomile associated with the presence of anti-inflammatory phytochemicals, including the flavonoids apigenin, quercetin, and patuletin. Among flavonoids, apigenin is the most promising compound for its medicinal properties; also see Sharol Tilgner, "Medicinal Uses of Herbs Used in Biodynamic Preparations" (Indian Trail, NC: Wise Acre Farms, 2000), goo.gl/RuAZnu. The three plants are all implicated in improving the uptake of essential micronutrients like zinc.

24. Steve Diver, "Biodynamic Farming and Compost Preparation," Alternative Farming Systems Guide *Appropriate Technology Transfer for Rural Advancement* (ATTRA), February 1999, http://www.demeter-usa.org/downloads/Demeter-Science-Biodynamic-Farming-&-Compost.pdf.

25. Miguel A. Escalona Aguilar, "Los tianguis y mercados locales de alimentos ecológicos en México: Su papel en el consumo, la producción y la conservación de la biodiversidad y cultura" (PhD diss., Servicio de Publicaciones de la Universidad de Córdoba, Spain, 2009), 51. See also Victor M. Toledo, "La racionalidad ecológica de la producción campesina," *Agroecología y Desarrollo* 5–6 (1992): n.p., http://www.clades.cl/revistas/5/rev5art3.htm; Victor. M. Toledo and Narciso Barrera-Bassols, *La memoria biocultural: La importancia ecológica de las sabidurías tradicionales* (Barcelona, Spain: Icaria Editorial, 2008), https://www.socla.co/wp-content/uploads/2014/memoria-biocultural.pdf; Moreno D. Escobar, "Valoración campesina de la diversidad del maíz: Estudio de caso de dos comunidades indígenas en Oaxaca, México" (PhD diss., Ciencias Ambientales, Universidad Autónoma de Barcelona, Spain, 2006); Miguel A. Altieri and Clara I. Nicholls, "Applying Agroecological Concepts to Development of Ecologically Based Pest Management Systems," in *Professional Societies and Ecologically Based Pest Management Systems* (Washington, DC: National Research Council, 2000), 14–19.

26. Miguel Altieri, *Agroecology: The Science of Sustainable Agriculture* (Boulder, CO: Westview Press, 1999).

27. Peña, "Sodbusters and the 'Native Gaze.'"

28. A former *mayordomo* (ditch rider) of the San Luis Peoples Ditch, the acequia that we use to irrigate our farm, has P'urhépecha campesino roots.

29. Pablo Alarcón-Cháires, *Etnoecología de los indígenas P'urhépecha: una guía para el análisis de la apropiación de la naturaleza* (Morelia, Michoacán, Mexico: Centro de

Investigaciones en Ecosistemas and UNAM, 2009), 53–54.

30. Sylvia Rodríguez, *Acequia: Water Sharing, Sanctity, and Place* (Santa Fe, NM: School for Advanced Research, 2007), 81–100.

31. Héctor Estrada-Medina, Francisco Bautista, Juan José María Jiménez-Osornio, et al., "Maya and WRB Soil Classification in Yucatan, Mexico: Differences and Similarities," *ISRN Soil Science* Article ID 634260 (2013), http://dx.doi.org/10.1155/2013/634260.

32. After observing the degraded condition of the soil on one neighbor's acequia farm, one of my mentors, Adelmo Kaber, remarked:

> Es puro garrero. Tenían un rancho muy bonito pero se lo robaron y ahora está aquí tratando de producir maíz pero es un garrero, no sirve la tierra, la debe de manejar con mas frijol, habas, y hasta alfalfa, para que le den fuerza a la tierra. Después, pue si bien, el maíz.
>
>> It is a pile of ragged clothes [*garrero*]. They had a beautiful ranch but it was stolen and today here he is trying to produce corn but it [the land] is a pile of ragged clothes, the land does not suffice, he should manage it with more beans, favas, and even alfalfa, so they can give life force to the land. Then, well okay, of course, corn. (Interview with the author, June 5, 2006, San Luis, Colorado)

33. For explication of the differences between relationality and objectivity as epistemological precepts, see Shawn Wilson, *Research Is Ceremony: Indigenous Research Methods* (Winnipeg, Manitoba: Fernwood, 2008).

DEVON A. MIHESUAH

NEPHI CRAIG

Life in Second Sight

I think that my perspective is a result of living on the reservation for the past ten years, working with our people, living with high mortality rates, and recognizing our vitality in light of perceived impending dooms. I suppose the pain you hear in my word choice is reflective of the strength in our journey toward resolve. I am just speaking on what I see, speaking on what I live. —Nephi Craig, 2017

Nephi Craig (Diné, White Mountain Apache) is notable in the culinary world, especially to those in the Indigenous foodie movement. He has cooked in eateries around the world—one being a five-star French restaurant—and founded the influential Native American Culinary Association (NACA) network. We see pictures of him cooking, working with fellow chefs, and gathering plants in newspapers and magazines such as *Newsweek, Huffington Post, Forbes, Zester Daily, Navajo Times, Indian Country Today, Tucson Weekly, MUNCHIES,* and *Food and Wine*.[1] We hear him on National Public Radio and dozens of other radio shows[2] and find him mentioned in conference poster presentations and food blogs. From 2011 to 2015, he explained his cooking methods and philosophy in his personal blog, *Apaches in the Kitchen*, a wealth-of-information site that is on hiatus as of late 2018 but will soon become active again.[3]

Despite the attention Nephi continues to garner, he has not sought out publicity, nor does he market himself beyond periodic posts on Facebook. He demonstrates his techniques at universities, food summits, and cooking schools, and he occasionally headlines upscale dinners. With his experience and amiable personality, he could work almost anywhere, but right now what holds interest for him is his homeland, and his desire is to serve the people who live there.

That homeland is the White Mountain Apache Reservation, covering 1.67 million acres in east-central Arizona. Multiple "Apache" groups, including Chiricahuas, Mescaleros, Lipans, Jicarillas, Coyoteros, and Kiowa-Apaches, once ranged a vast area known to Spaniards as Apachería that encompassed northern Mexico, eastern Arizona, New Mexico, southern Colorado and west Texas. Today's White Mountain Apaches are Western Apaches, a large group that also includes Cibecue, Tonto, and San Carlos bands. Almost fifty White Mountain Apaches served as scouts for the US Army in the 1870s and were instrumental in the subjugation of other Apaches, notably Goyanthlay (Geronimo) and his band of Chiricahuas. As a reward of sorts, the White Mountain Apaches were given a part of their homeland as their reservation.[4] Nephi Craig is enrolled as a White Mountain Apache and lives in his ancestral territory, the White Mountain Apache Reservation.

The 2,627 square miles of the reservation contains a spectrum of ecosystems, from the Salt River Canyon desert to grasslands, wetlands, and one of the largest ponderosa pine forests in the world. Dził Łigai (White Mountain), also known as Mount Baldy, stands to the east. The 11,409-foot mountain is a popular climbing destination, but because part of it stands on the reservation, nontribal members must request permission to hike to the summit. The beautiful golden-colored Apache trout, the Arizona state fish, lives only in the lakes and streams on tribal land, along with the wider-ranging rainbow, brook, and brown trout species. Catfish, bass, and sunfish inhabit waters at the lower elevations. The Apache-Sitgreaves National Forest is also home to bighorn sheep, mountain lions, javelinas (collared peccaries), turkeys, wolves, mule and white-tailed deer, antelopes, and black bears. This is the land of trophy elk, the spectacular ungulates whose antlers are prized by hunters wanting a high score on the Boone and Crockett Club "trophy scoring" scale.[5] Some guides charge more than $7,000 for a trophy elk guided hunt,[6] but permits to hunt trophy elk sell out almost immediately. Only tribal members can take deer, and outsiders can expect to pay upwards of $5,000 for a bear tag.[7]

It is indeed a beautiful environment, but the natural resources have become imperiled in recent decades. The reservation has been held in trust since 1871, yet federal mismanagement of surrounding lands has resulted in the loss of trees from diseases, insect infestation, inadequate forest thinning, and massive fires. Bark and spruce beetles have taken a toll on the tribe's forest.[8] The last time I came through this area was more than fifteen years ago, on my way to skijor and race sled dogs at Hannagan Meadow outside Greer in the eastern White Mountains. The trails wound through quiet and cold forest, with steep climbs, fast downhills, and sprints across the open meadows. I recall seeing close-up that various destructive beetles had started to encroach on the landscape, and my colleagues in the School of Forestry at Northern Arizona University—where I worked at that time—were seeking to restore and protect those western forests.

Despite the degraded health of the forests, the land still nourishes a variety of edible plants. A wealth of flora grows in the Apache-Sitgreaves National Forest: dandelions, cottonwood buds, piñon pines, acorns, walnuts, Indian rice grass, sunflowers, cholla cactus fruit, western juniper, sumac, elderberries, ground cherries, century plants (agave), mesquites, locust and kidney beans, milkweed, water chestnuts, pigweed, twin crest onions, water parsnips, and yucca pods.[9] White Mountain Apaches historically gathered these foods; hunted game; and farmed corn, melons, squashes, and beans. They cultivated wheat by at least 1900. Nephi's family continued the farming tradition. Joseph C. Ivins, Nephi's great-grandfather, was born in 1899 on this land. Ivins grew a garden with apples, pears, apricots, plums, roses, corn, beans, and squashes. Some of Nephi's early memories of growing up on the reservation were in that garden. "I was a small kid and I did little things like help water the garden," he says, "but it was mostly a great place for adventure back then."

Growing and cooking food stayed on his mind, and after high school, Nephi attended a culinary program at Scottsdale Community College. After earning a cook's certificate in 1999, he worked at the Country Club at D.C. Ranch, then received a position at the renowned Mary Elaine's at the Phoenician, a Five-star French restaurant in Scottsdale Arizona. Until its closure in 2008, Mary Elaine's was elegant, opulent, and expensive, and employed chefs honored by the prestigious James Beard Foundation. The place had a wine inventory of more than 40,000 bottles and five sommeliers to offer advice; sold Iranian Gold Asetra caviar at $500 a portion; and provided purse stools.[10] This was not a venue for novice cooks. The dishes required precision, perfection, and dedication, and everyone from management to the head chef, maître 'd, and servers had to be at the top of their game for each service. Working at such a high-level gourmet venue is demanding and exhausting, often resulting in yin-yang, love-hate feelings about the food industry, most colorfully described by the late food adventurer Anthony Bourdain. Nephi felt the pressure, yet the culinary skills he garnered and lessons learned in navigating the stressful world of haute cuisine at Mary Elaine's were invaluable to his growing confidence—and reputation—as a chef.

After his stint at Mary Elaine's, Craig traveled abroad, including to Germany, São Paulo, and London. In Japan he taught about Native foodways and, conversely, came to appreciate the precision and rituals involved in Japanese food preparation and eating. At the age of twenty-nine, Nephi became head chef at the Summit Restaurant at the Sunrise Park Resort, an eight hundred–acre ski resort owned by the tribe. It is a destination for skiers in winter, and in warmer months offers hiking, fishing, zip lining, lift rides, 3-D archery, and mountain biking. Restaurant owners want recognition, and the Sunrise Park Resort created an enviable advertisement of its Summit Restaurant as "home to world renowned Chef, Nephi Craig." For nine years, he made standard food fare, but Thursday through Sunday, Nephi and his fourteen-member White Mountain Apache culinary team cooked what he really wanted, a unique menu featuring traditional and local ingredients. From that hub, he received invitations to demonstrate his expertise at events such as the Phoenix Indian Center's Silver and Turquoise Ball annual black tie gala for five hundred guests at the Talking Stick Resort in Scottsdale. During the live auction he offered a private dinner for ten, but so many bids came in that he offered a second private dinner. While working at the Summit restaurant, Nephi began to acknowledge to himself what was important to him. Except for the nights when he prepared Indigenous dishes, the majority of the meals he created were mainly for non-Native patrons and were nothing like what his ancestors consumed. What he was actually concerned about were the serious problems among members of his tribe: poverty, racism, poor health because of inadequate food, and emotional distress because of colonialism. He performed his job, but continued to consider how he could use his knowledge and talents to assist his tribe.

Nephi already had put some of his thoughts into action after he graduated from culinary school. By that time, a few professionally trained Indigenous chefs had emerged around the country, enough for him to take the initiative to found NACA. NACA's initial goal was to serve as a grassroots network of Indigenous chefs who would connect, strategize, and collaborate in the spirit of ceremony around cooking, using Indigenous ingredients as the foundations of their dishes. He hoped to organize symposiums with cooking demonstrations, discussions about food and culture, and a chef apprentice program, among many other initiatives. "NACA is propelled by those who believe in it," he says. "If a person couldn't participate, they might help in other ways, such as donating time, skills, tools, and food." One of NACA's events was the Food Symposium and Celebration of Basketry and Native Foods Festival at the Arizona-Sonora Desert Museum in Tucson in November 2015. Throughout the symposium, NACA chefs prepared meals, but the highlight was the final evening's dinner and reception, with a five-course meal created by chefs Nephi, Lois Ellen Frank (Kiowa), Loretta Oden (Citizen Band Potawatomi), and Walter Whitewater (Navajo), assisted by handpicked sous chefs. The dishes featured Indigenous ingredients, including Whitewater's nopal cactus pad salad, Frank's elk tenderloin, and Oden's seared achiote duck breast.[11]

Nephi prepared a striking and unexpectedly earthy dessert composed of Western Apache seed-mix fritters, honey-braised butternut squash, piñon cloud, chocolate, and amaranth, adorned with tiny citrus marigold blossoms (see illustration). Because I thrive on nuts, seeds, and dried fruits and do not crave sweets, this plate fascinated me. Afterwards, I wrote on my *Indigenous Eating* Facebook page with regard to Nephi's plate: "This dish was particularly interesting, because unlike, say, a bowl of ice cream, every bite of this dish was different. On the plane ride home today I tried to think of how to describe it. Not exactly thick or condensed, but it was very filling. This combination is maybe better described as intricately eclectic. It took some real skill to put this together."

While Apaches and Navajos historically did not make piñon whipped cream, nor have the wherewithal or reason to construct such an intricate dessert, Nephi's ancestors did depend on the nutritious ingredients and did carry the calorie-rich seed mix when they traveled. Despite the artfully precise placement of each ingredient and of the marigolds, the dish was not pretentious and it was not Nephi's intent to dazzle. Rather, the dish held significant cultural and personal meaning.

In conceptualizing the dish, Nephi took into consideration the four sacred mountains of Dinetah, the Navajo homeland—to the east, Tsisnaasjini' (Mount Blanca); to the south, Tsoodzil (Mount Taylor); to the west, Doko'oosliid (San Francisco Peaks); and to the north, Dibé Nitsaa (Mount Hesperus)—so arranged the components as if on a compass. The empty area at the top right represents ideas, the conception of the dish. Looking south, the plate appears empty, but that space is actually the planning stage that has not yet taken form. The dish begins to take shape at the southwest (lower left) of the plate, and as the eye moves up from south to north, the nutrient-dense result is realized, what Nephi calls the Indigenous sensory experience formulated by ancestral memory.

Nephi Craig's dessert composed of Western Apache seed-mix fritters, honey-braised butternut squash, piñon cloud, chocolate, and amaranth, adorned with citrus marigold blossoms, 2015. *Photograph courtesy of Nephi Craig.*

It is ironic that such a symbolic food prepared with so much reverence manifested as a dessert for a crowd of non-Apaches at an expensive food gathering. This combination of ingredients and special plating satisfies those with discriminating palates because it tastes good and looks pretty, but how many diners (and chefs who copy the plating) understand what it signifies? Perhaps Nephi was not aware of it yet, but that dish represented what he really wanted to do. Despite his successes and growing fame, he remained emotionally and spiritually unsatisfied, in large part because of the restaurant industry.

Any chef or business that wants to thrive must attract and keep customers satisfied by consistently serving creative, delicious dishes concocted of high-quality ingredients. That is a requirement of any eatery. However, the more the media profiles and celebrates a chef, the more that chef must live up to a created image. To do so, the chef must win approval from mainstream society.

But this goal can put Native chefs in an awkward position because it might entail neglecting Native values of being primarily concerned with providing for the community, sharing what one has, and promoting Indigenous foodways among one's tribe. Nephi was capable of producing outstanding and unique dishes at the Summit Restaurant, but unlike most chefs, he did not care about a Michelin Star or James Beard Award. "The standards of the Michelin Star are not the standards in traditional Native communities," he says. "It's not our goal to get attention."

Even if an Indigenous haute cuisine restaurant were established in Indian Country, whether the place could depend on repeat Native diners remained a question. An Indigenous chef might play on the current trendy brand of "traditional Native American food" for the short term, but the diners who stand to appreciate that cuisine the most are not the ones who can afford to purchase it.[12] Restaurant meals, especially those at high-end places such as Mary Elaine's, are out of financial reach for most Natives. In many tribal communities, meals consist of ingredients taken from the federal Food Distribution Program. These USDA commodities, or "commods," consist of the basics: canned fruits, vegetables, and meats; cheese blocks; dried and canned milk; pasta; and cereals. There are some seasonal items, such as canned pumpkin and cranberry sauce, wild rice, and frozen bison.[13] Those who do not qualify for commodities are forced to purchase what they can afford. The problem is that tribal lands are often "food deserts" with few stores that offer inferior items.

Merely meeting the goal of creating beautiful dishes is "short-sighted," Nephi says. Pleasing the general public contrasts with what he now espouses. His impetus for serving Native foods "is about prevention, recovery, and maintenance," not seeking rave reviews from food critics. He has little interest in what he calls "The Chefy Phenomenon." The rise of social media, food photography, and food blogs—combined with the public's hunger for exciting images, bad-boy chefs, and scintillating memoirs—push reporters and writers to find new ways to keep readers' attention. More than ever, reporters highlight people they think are the authoritative voices on "Native American food." The irony is not lost on Craig. "Native food is not a trend," he said to a *Huffington Post* writer in 2012,[14] and he has repeated that sentiment in various ways since. Indeed, thousands of Native people have been cultivating, hunting, gathering, and consuming their traditional foods for their entire lives and are at the forefront of the political, economic, and social realities that drive the "Indigenous food sovereignty movement." Yet current popular writings about Native foods give the impression that no one had really thought much about traditional Indigenous ways of eating or Indigenous health issues prior to 2012.

Nephi knows this. "A little more than half of the time," he says, "when a writer reaches out to me and we talk, I am often the first Native American chef that they have ever spoken to, and what I say can always be flipped to echo themes of colonial valor. But since colonialism is cunning, the reality could be that I am being interviewed out of colonial sympathy and my truths that will be heard will always be determined by the gatekeepers of the press." Non-Native writers seem to have a fascination for Indigenous chefs. One reason is that Indigenous chefs represent something shiny and new. Reporters are always on the lookout for the next food movement or for what a new story might do for their careers. Most food writers, however, possess little to no knowledge about Indigenous cultures. Indigenous, or "traditional," food has now become part of the food world lingo, and the writing is predictable. Articles about Indigenous chefs contain the same themes and tired food clichés. Many tell us that the "leader" of the revitalization of Indigenous foods is the subject of the story. Oftentimes writers who do not know much about Indigenous peoples resort to stereotypes and romanticism.

The explosion of food competition and instruction shows the revitalization of what Rosalind Coward called "food porn"[15] and *Time* magazine referred to as the "cult of celebrity chefs," has nurtured a flourishing interest in beautiful and complex food made by equally intriguing people. Non-Native chefs and entrepreneurs—including Francis Ford Coppola, who opened his restaurant Werowocomoco in the Bay Area—have also pounced on the Indigenous food movement. To the myriad Natives who have grown up eating their traditional foods, the idea that these are somehow new trends in the culinary landscape is absurd. Grinding corn, preparing acorns, making tools, stalking game, and butchering takes time. During the appropriate time of year, some tribes foraged for wild fruits, roots, and nuts, often expending much energy looking for them. The people had to plan because foods were often hard to come by, and what they found needed to be dried and stored for the cold months. There also was ceremony associated with planting, harvesting, hunting, and consuming. That type of reverence for food is not found in a commercial, for-profit food enterprise. Nephi finds the trend amusing: "Falsehoods of luxury built on the appropriation of Indigenous foodways is the equivalent of imperial slumming and often comical from an Indigenous person's perspective," he says.[16] Further, "White Mountain Apache cuisine can only be produced in this region. It may be reproduced in other areas of the world, but in my opinion, it must be created here in the White Mountains and prepared by White Mountain Apaches."[17]

Craig still accepts interviews, but knows the probable result: "I get annoyed when writers (Native and non-Native) try to make [Indigenous foodways] sound noble, romantic, or overly spiritual. Phrases like 'saving Native foods from extinction,' 'a new cuisine from ancient methods,' 'living in two worlds,' 'honoring Native people and food,' 'modernizing ancient cuisine'... blah, blah, blah—these phrases are for the comfort of the wider audience that has already demonstrated they have no interest in the real people deep in the plight of surviving colonialism and colonial violence or, essentially, it is to keep the larger society/readers comfortable while continuing to dismiss, minimize, and deny the truth of our current Indigenous reality. Larger society wants our culture, but not our struggle. Terms like 'oppression,' 'racism,' 'decolonization,' 'genocide,' 'murder,' 'deliberate violence(s),' 'land theft,' and many of the other culinary truths are not mentioned." Indeed, scholars have written about food history and the political, economic, social, and environmental challenges to Indigenous food sovereignty for decades, yet reporters have no interest in these issues. For food writers, it appears that only those with a connection to a legitimate restaurant, or those who have created a brand, have food expertise.[18]

Even though Nephi has received a plethora of publicity and has been heralded as an Indigenous master chef for years, people have not flocked to the Southwest to taste his food. We gave that fact some thought and came up with possible reasons why, the most likely being his location. Whiteriver is at least a three-hour drive from Phoenix Sky Harbor airport, almost four from Tucson International Airport, and not everyone is willing to jump on the tiny commuter plane that flies between Phoenix and Show Low (thirty-five miles from Whiteriver). The other reason we considered is his cultural accessibility: Southwestern tribes tend to adhere to traditions, they live on tribal lands, and they look more phenotypically Native than tribes that intermarried with whites over multiple generations. The harsh effects of colonialism have hit Nephi's community hard, and one can see evidence of rough living on a drive down Arizona Highway 73 through Whiteriver. Combine those factors with the most prominent one—a culture that is not always open to sharing tribal traditions and that desires to keep Indigenous food knowledge at home—and you end up with a mysterious, perhaps intimidating, culinary presence.

This reality does not bother Nephi. After all, why should Indigenous people share their cuisine with the world? Which foods were here when Columbus stumbled on the shores of San Salvador in 1492 is no secret. There are plenty of books and articles about flora and fauna of this hemisphere pre- and post-contact. So what, exactly, is the purpose of putting these foods in headlines now—except to sell them, or oneself? Population reduction, loss of land, racism, and abrogation of treaty rights are only a few results of ongoing colonialism. Diabetes, obesity, and high blood pressure are the results of removal, relocation, poverty, cultural loss, and resource depletion. If these are the struggles facing Indigenous peoples of North, Central, and South America, then a return to Indigenous ways of eating (and the complicated steps doing so entails) is paramount. How will showcasing traditional foods to non-Indigenous people assist tribes in dealing with the myriad issues they must contend with every day? Perhaps if Indigenous food activists insist on publicizing the tough issues, then we might gather more allies. Even then, Nephi is skeptical: "I know writers who, I now recognize, just used our stories as stepping stones, to contribute to their body of works, to position themselves as politically correct 'allies,' and chose not to include our Indigenous truths. When you read through any Native chef's interview in our generation, you can often detect these notes/themes."

"You want real Indigenous food advocacy?" he asks. "Then speak from a point of community, mortality, healing, hurt, pain, violence, suicide, spiritual malady, unnatural death, and recovery. Come to ground zero and speak about truths while offering solutions—not just speaking about Native foods from a disconnected professional level with a message that is designed to appeal to the colonial culinary elite (Native and non-Native) for the sake of making a name for one's self, company, or organization. Not many chefs, cooks, scholars, and publications are willing to do this, possibly because it is painful, requires accountability and authenticity, and does not sell."

Indeed, those kinds of stories do not appeal to mainstream readers, nor to most foodies who subscribe to *Bon Appétit*. They know little if anything about Indians. The realities of poverty, drug and alcohol abuse, and severe food-related illnesses do not play well on social media. What is popular are photographs of beautifully prepared dishes, backyard gardens, braids of corn and onions, and Indigenous people wearing chef's jackets.

Nephi is uncompromising in wanting to focus on the issues in his community, and on what he feels is his responsibility to his home and the larger struggle in "this colonial reality." He does not want to be critical; rather, he admits he "has to remember that some of us don't even recognize the layers of oppression that we exist under, so in speaking with writers, I do my best to be sensitive to that as I speak." Now he lives "life in second sight," a phrase his wife, Jandi, uses to describe living a changed life. Nephi applies that concept to his sobriety and to what he refers to as "a life beyond the threshold of just understanding decolonization and applying decolonization principles for a life in reconstruction." Life in second sight is "a life-saving paradigm shift that speaks to humility, hard work, and dignified resurgence. This also refers to my life in second sight as a husband and father that just happens to be a chef, because I no longer see the practice and purposing of culinary arts the way I used to." Nephi also views life in second sight as activism. Not everyone can be out on the front lines, Nephi argues. "There is a difference between activism and home activism," he explains. Much of the crucial, difficult, and time-consuming political, social, and economic work that has to be done in order for people to thrive must be done at home. Grassroots activism is often personal, is not always exciting, and occurs on what Nephi calls "a quieter level" than what we see on the news and in papers. "We lose good people to the romanticism of the front lines," he observes.

Nephi has always been cognizant of the serious problems stemming from historical trauma on the reservation, including alcoholism, drug abuse, and emotional distress. Whiteriver has a population of around 4,100, with 96 percent being Native. Within the community, which Craig calls "ground zero," is a hotbed of serious socioeconomic problems. Indeed, historical trauma still lingers—not just at Whiteriver but on reservations across the United States—and Nephi sees the effects firsthand. It might be easy to fall prey to colonial violence, and for several years he struggled with alcohol abuse, and admits that he carries a fear of lapsing back into old habits. However, his reconnection to tribal ancestral knowledge gave him the self-control to stop drinking almost eight years ago. And at Whiteriver he can stay sober, although he freely admits it is not easy. "This is life-and-death-blood-and-guts-life-saving-realization-and-practice." He sees hope. Despite the dysfunction on the reservation, ceremonies are still intact, the language is spoken, and there is a community dynamic. Recently, there has been a resurgence of agriculture and interest in foodways.

He believes there is a need to repurpose what cooks do, although he is not yet entirely sure what that entails. After his success at the Arizona-Sonora Desert Museum, Craig continued to ponder how to "shift perceptions of the white tablecloth" and to instruct fellow Apaches about the connections among food, culture, and the landscape. His marriage to Jandi in June 2014 solidified his determination to find solutions to problems on the reservation. He seems to be in perpetual motion, determined to spread his enthusiasm about cooking.

He spends much time educating and cooking for Natives around the country, such as at the Annual Apache Harvest Festival at Canyon Day, Arizona, and the Phoenix Indian School Visitor Center grand opening.

He inspires the young cooks at the Arrowhead Café and Marketplace that opened in the summer of 2015 at Fort Apache.[19] He gives nutritional advice to young participants in the Ballet White Mountains "Glam Camp," held at the dance studio his elder daughter attends.[20] "I also enjoy sharing the similar principles of disciplined training that exist for professional cooks and professional ballet dancers," Nephi says. "Principles like repetition, balance, consistency, commitment, discipline, grace, determination, athleticism, health, stamina, etc. I call us ninjas. It is something that I would have probably never been involved in if it were not for my daughter Kaia, for which I am grateful." His wife, Jandi, is also an indigenous activist, an engaged member of the Continental Network of Indigenous Women of the Americas, North Region, who in 2017 traveled to Geneva as a United Nations Office of the High Commissioner for Human Rights (OHCHR) Fellow.[21]

His daughter Kaia and son Ari, a skateboarder and budding guitarist, have always grounded him. The births of another daughter, Tawny Cassia Craig, in early 2017, and a son, Rowley Nephi Craig, in late 2018, have further inspired Nephi, and as he grows older, he pays closer heed to the lessons of his ancestors. Nephi's Diné grandfather, Bob E. Craig, fought at Iwo Jima and served as a Navajo Code Talker. Bob's son (Nephi's father), Vincent Craig, was a well-known comic, singer, and creator of the comic superhero Mutton Man, a character who gained powers after consuming uranium-contaminated mutton. Vincent had a colorful history as a police officer for the Navajo Nation, Salt River Pima-Maricopa Tribe, and White Mountain Apache. He served as a sergeant and sharpshooter in the Marine Corps and built Huey helicopters. Vincent was instrumental in influencing Nephi's attitudes about cooking, from the Navajo philosophy of planning to the strategy of focusing on one's strengths in order to heal the trauma of colonialism. Vincent Craig died of cancer in 2010 at age fifty-nine. Nephi's mother, Mariddie Ann Johnson Craig, is White Mountain Apache and has served as a tribal council member. His father's mother, Nancy Mariano, was the school cook and butcher, and Nephi recalls that she "always cooked with love for all of us. She passed within the year my dad did. She was born in the eastern part of the Navajo Nation, in New Mexico. I have many memories of visiting Dinetah with my family to see all my cousins. She had an adobe oven in the backyard. It was always such a warm feeling to be there as a kid. Even into my early adulthood, it was always welcoming."

At the heart of many Native rehabilitation programs is a return to cultural pride. Indigenous knowledge is place-based and focused on how to survive and how to behave as a member of the group. Elders teach the younger generation about weather patterns, animal behavior, and seed saving. Prior to colonization, tribes made their own decisions about how to utilize resources, and their priority was to protect the Natural World that provided them with what they needed to stay physically and mentally healthy, what Nephi calls "Ancestral Health." Food is one of the most basic human needs, and for Natives, food intertwines with culture. Food has meaning beyond just sustenance. Ceremonies associated with food—from planting, cultivating, and harvesting plants to hunters giving thanks for what they kill—are inexorably tied to spirituality and identity. Stories and symbolism teach about the cycles of life. Among Navajos, for example, a spiritual entity is Changing Woman. She represents hope and the promise of life, and like the corn Navajos cultivate, Changing Woman is young in the spring, bears her harvest in summer, grows old in the fall, and dies in winter. Because she is self-renewing, she is reborn each spring.

Nephi's goal is to assist those on his reservation to "decolonize" and change their lives for the better by adopting Indigenous frameworks of thought. The way to do that is to be introduced to, or be reminded of, cultural values and traditions.[22] This is the same strategy used at Indigenous alcohol and drug rehabilitation centers, but Craig focuses on what he calls "restorative indigenous food practices." After all, hunting, gathering, planting, harvesting, and saving food for winter required knowledge of all aspects of the Natural World and intratribal cooperation.

The nine hundred–acre Ndée Bikíyaa (The People's Farm) is south of the town of Whiteriver. Under the direction of the Water Resources Department and farmer Clayton Harvey, the farm provides crops for the tribe's farmers' market and contains a greenhouse, two large hoop houses (cold frames), and a community education center. On this bright and warm October morning, Emily Maheux talks to eight outpatients from the Rainbow Treatment Center about corn and beans and cross-pollination. The Rainbow Treatment Center is a long-term residential treatment center in Whiteriver for Apaches dealing with substance abuse, and the men and women who receive treatment here are those whom Nephi hopes to help heal. On this day, male clients are at the farm. This session at Ndée Bikíyaa is part of the vision Nephi has had since 2011, and he has never relented from his mantras: "Be the change we want to see" and "make thought into reality." Phase 1 is residential treatment. Phase 2 consists of outpatient programs, including the Recovery Training Program, Working 2 Wellness, and Our Scholarship Program. The session at the farm is part of this second phase.

We sit under the shade of a small arbor; ears of corn hang from the roof. Hundreds of ears are piled on the ground, and the tubs are full of kernels. "Look around," Nephi says, motioning at the farm. "The Creator is in the soil, the plants, and the whole environment. Our Indigenous knowledge comes from across the cosmos. Lessons are everywhere. Inside the plants." Everyone listens intently. "This seed," Nephi says as he pulls out a kernel from a dry cob, "is like an indigenous microchip. All the data we need to recover are inside this seed."

Indeed, when one considers how corn has sustained those who live in what are now known as North, Central, and South America for millennia, those little seeds do contain Indigenous information about history, sustenance, cultural connection, and spirituality. I felt sad at that moment, assuming that the men under the arbor had not been taught about corn, about beans, about traditional Apache foodways, or about ceremonies crucial to sustenance and thanksgiving. Perhaps this cultural disconnect had led to their frustrations and addictions.

For the next half hour, everyone remained engrossed in shucking corn and removing kernels off the cobs, then they dispersed to look for bean pods. One of the farm's workers, Shalitha Peaches, dislodged kernels by kneading a burlap bag of dried corncobs with her feet as intently if she were working out on a Stairmaster. An optimist, she was clearly happy in her comfort zone at the farm as she talked enthusiastically about her goal of becoming an ethnobotanist: "I learned much about Apache foods from my grandparents," she said, clearly proud of them. She had recently lost weight by paying close attention to her diet, and planned to continue that lifestyle. "I just live down the road and want to start riding my bike to work." I asked about bicycle safety on the lonely road, and she said she takes the back routes. We talked about elk jumping the tall fence surrounding the farm, the types of pollinators that visit the plants, and monies just awarded Ndée Bikíyaa for construction of larger hoop houses.

After lunch, we returned to the former Chevron gas station that sits at the junction of Arizona Highway 73 and East 57th Street north of Whiteriver. This boarded-up building with dilapidated gas pumps will soon be transformed into Café Gozhóó, an enterprise that will operate as part of the Nutritional Recovery Department at the Rainbow Treatment Center and will provide the actual work experience and professional development that constitutes Phase 3. There are plenty of remote but "essential" restaurants on TripAdvisor lists. Café Gozhóó could eventually become such a destination, but that is not Nephi's goal. Workers at Café Gozhóó are people whom Nephi says are "actively working to change their lives and are in an important phase of transformation." With Nephi as the nutritional recovery program coordinator and executive chef, Café Gozhóó will offer fundamental, disciplined training that will prepare individuals to gain employment in other food establishments. He will teach all the workers the basics of kitchen protocol and working as part of a kitchen team. He cannot predict how long each worker will stay with him, but he hopes they all will do so until they heal spiritually and emotionally and can use their knowledge to find satisfying jobs as a part of their support system in their recovery. "Through investing in our people and highlighting our inherent Ancestral Intelligence," says Nephi, "we will cultivate skilled and conscious individuals equipped to change their lives." The foundational premise is the "nuts and bolts" of the reconstruction of a life in recovery. And that includes Nephi himself. "Addictions and violence are physical manifestations of historical trauma in various forms of detachment, including detachment from Indigenous foodways, parenting, landscape, identity," he says. "Café Gozhóó will provide an opportunity for individuals to reconnect with Indigenous foods, Indigenous agriculture, and emotional intelligence, which are all ingrained in our Western Apache lifeways. The word 'Gozhóó' in the Apache language has many similar meanings and applications, and can mean 'beauty,' 'balance,' 'harmony,' 'truth,' 'happy,' just to name a few. All the principles to live a life in perpetual recovery exist within that Apache word. To me, this tells us that our language and culture contain all the intelligence to recover from anything. Café Gozhóó is an action-oriented response to historical trauma."

The café is an ambitious project, yet Nephi visualizes exactly how it will be designed and what its purpose will be. Because it is a café, it will serve dishes made from foods produced at Ndée Bikíyaa and from the flora and fauna of Western Apachería, but he will also use non-Indigenous ingredients when the occasion calls for them. He plans to provide coffee products, arts and crafts, informative literature, and gasoline and auto service dispensed by old-fashioned gas jockeys out front.

The building still has to be redesigned to fit Nephi's vision, but for the moment, the structure has electricity and water and is very clean. On this day, Nephi and chef de cuisine Kristopher Bergen will prepare a dish while Nephi executes his lesson plan. Bergen, an energetic army veteran, is busy chopping

The Café Gozhóó logo incorporates two crossed arrows that represent the people's strength. The *tus* (water jug) at the top represents traditional food technology, ancestral knowledge, and the sacredness of water. The coffee cup is a familiar household item. The acorn at the bottom symbolizes historical health, wild foods, and the Western Apaches' favorite flavor. The coffeepot is a familiar cultural and ceremonial item.

onions and squash, and setting up plates, burners, and cutlery. He answers every request from Nephi with a quick "Yes, Chef," as he bustles around the makeshift kitchen.

Watching Nephi cook, and tasting his dishes, are always highlights of these outpatient sessions, but the real goal is for participants to think about accountability, responsibility, humility, and honesty. Nephi is confident and authoritative, but he does not say "me" and "them," or "I" and "they." Instead, he says, "We will learn about food and foodways together." He explained to me earlier that this type of session "is not complete immersion yet, because I am a small part of the larger treatment plan, but connected to all parts of the treatment plan because it is food. For the moments we work together, we focus on the sensory experience of tasting/learning and hope to relate that to the sensory experience of recovery. When we staff and open Cafe Gozhóó, it might be better described as 'immersion therapy.'"

Nephi talks as he cooks on two small table burners. The principles of flavor, he explains, are hot, sour, salty, sweet, umami, and bitter. The caramelized zucchini is the sweet, spiced with a dash of hot chili flakes and lemon juice. Nephi previously explained his strategy: "I'm currently teaching fundamentals to help individuals cook at home first. Sautéing, roasting, boiling, simmering, steaming, pickling, and we focus on seasoning. I teach some of the history behind techniques and linguistic origins of some terms like 'sauté' being a French word meaning 'to jump.' In that example, we ask what is the Apache language for 'to jump,' then talk about how when we use our own language, we gain ownership of techniques."

"Recovery is about transformation," he tells the group around the table. Through cooking even simple dishes, one can see, smell, and taste changes in addition to feeling history. As Nephi swirls and tosses the corn and acorns, he muses about how many Indigenous people have done the same thing for thousands of years. Granted, the ancestors probably used a clay pot, but the concept of parching is the same. "This combination of foods has survived," he says. "Apaches have always eaten it because they depended on it." The cultural connection is undeniable. As he showed me a few years earlier via his beautiful dessert, such high-calorie and nutritionally dense foods were crucial for survival. Nephi reminds the group that "thinking ahead" means drying food for hard times and winter. Like other tribes, Apaches used seeds and nuts as travel food.

Parched corn and acorns are a bit bitter, and not everyone likes the taste. Nephi is not deterred, saying, "Modern eaters are used to processed foods." I wondered if the men who were at the café that day would prefer the fancy dessert. As I compared what he was cooking there with what he presented at the elaborate Desert Museum dinner, I concluded that these men might not appreciate the intricate dessert that Nephi prepares on occasion. What he made that day was realistic and humble, and made perfect sense.

"What does the taste and smell of parched corn remind you of?' he asked each man after he had a taste. "Grinding corn with my grandmother," answered one man. "Eating after herding cattle," said another. Two recalled eating parched corn and acorns at sunrise dances. This exercise revealed that these men had cultural food memories to draw on for their path forward to recovery.

The last component was a green salad tossed with tomato confit and garnished with sunflower seeds, dried pumpkin, and hibiscus. "This is hardcore Apache cooking," Nephi said. "The food is grown by Apaches on an Apache farm, cultivated by Apaches, cooked by Apaches and eaten by Apaches." Everything is consumed except for a small pile of tomatoes left behind by one eater suffering from allergies.

Nephi continued the session for another hour. Everyone took notes. He managed to present enough difficult questions and truisms to fill pages. "If it is socially acceptable for Apaches to be drunk and faded and asking for handouts at the local grocery store," he asks, "then why is it not socially acceptable or cool to promote healing?" The group mulled over that one. One last comment compared hard life experiences to composting food scraps into useful, nourishing soil: "Don't transmit pain," he said. "Transform it."

After the men depart for the Rainbow Treatment Center with their bags of parched corn and acorns, Nephi smiles and is ready to share some good news from the morning planning meeting. Refurbishment for the structure that will become Café Gozhóó is underway, and should be completed in 2019. He is calm and satisfied, yet still a bit apprehensive. We walk outside to enjoy the quiet afternoon. Only a few cars zoom past. Leaves on some of the trees are changing color. In a few hours, the air will become chilly, but for now, we soak up the bright sun's warmth.

"Not everyone will stay sober," he says. "But my hope is that this type of treatment is therapeutic and will give them hope." Nephi calls his personal recovery "life changing," allowed by the sacred mountain Dził Łigai. "The stress will always be present," he says. "But there is nothing to compare to working in my own community [rather than] working in a big city. I am where I need to be—with my family. I am enjoying being a parent to my teenaged kids, my toddler daughter, and infant son. Much of what I have learned and developed could have never been created in a large city."

The therapy Nephi has created for his fellow Apaches has strengthened him. "I spent nineteen years getting to where I am," he says. The key, Nephi believes, "is that you don't let the negative side get to you."

NOTES

1. See Paul Wachter, "Nephi Craig, Farm to Table Food, and the Movement to Rediscover Native American Cooking," *Newsweek*, August 23, 2013, http://www.newsweek.com/2013/08/23/nephi-craig-farm-table-food-and-movement

-rediscover-native-american-cooking-237856.html; *Huffington Post*, August 18, 2012; "The A List," *Forbes*, October 2013; "How Native Americans Are Rescuing Our Food Culture," *Zester Daily*, October 18, 2013; *Navajo Times*, October 10, 2013; "In the Kitchen with Chef Nephi Craig," *Indian Country Today*, September 12, 2016; "Native Noshing," *Tucson Weekly*, November 12, 2015; "How to Decolonize Your Thanksgiving Dinner," *MUNCHIES*, November 18, 2016; and Jody Eddy, "Will Native American Cuisine Ever Get Its Due?" *Food and Wine*, March 31, 2015.

2. "'Going Green' Is Really 'Going Native': Western Apache Chef Nephi Craig," *NPR*, April 4, 2016.

3. *Apaches in the Kitchen*, http://apachesinthekitchen.blogspot.com/.

4. See Donald E. Worcester, *The Apaches: Eagles of the Southwest* (Norman: University of Oklahoma Press, 1979); Worcester, "The Apaches in the History of the Southwest," *New Mexico Historical Review* 50, no. 1 (January 1, 1975): 25–43. Worcester was my PhD advisor at Texas Christian University.

5. Boone and Crockett Club, "Scoring Your Trophy," http://www.boone-crockett.org/bgRecords/ScoringYourTrophy.asp?area=bgRecords&ID=416327E9&se=1.

6. Stephen Kent Amerman, "This Is Our Land: The White Mountain Apache Trophy Elk Hunt and Tribal Sovereignty," *Journal of Arizona History* 43, no. 2 (2002), 133–52, https://www.jstor.org/stable/41696697?seq=1#page_scan_tab_contents.

7. "The Hunting Report, Hunt-Planning Package; White Mountain Apache Reservation Bear Hunts," 2012.

8. Brandi Buchman, "Apache Tribe Blames US for Forest Ailments," *Courthouse News Service*, March 21, 2017, https://www.courthousenews.com/apache-tribe-blames-us-forest-ailments/.

9. Albert B. Reagan, "Plants Used by the White Mountain Apache Indians of Arizona," *Wisconsin Archeologist* 8, no. 4 (1929): 143–61. See also Ian W. Record, *Big Sycamore Stands Alone: The Western Apaches, Aravaipa, and the Struggle for Place* (Norman: University of Oklahoma Press, 2008), 38–42; 109–16.

10. "Bring Back Mary Elaine's! And Get Me a Purse Stool!" *Phoenix New Times*, July 30, 2012.

11. Arizona-Sonora Desert Museum, "Taste of Native Cuisine Dinner & Reception," November 13, 2015, https://desertmuseum.org/visit/basketry/dinner.php.

12. I discuss this issue in Mihesuah, "Indigenous Health Initiatives, Frybread, and the Marketing of Non-Traditional "Traditional" American Indian Foods," *Native American and Indigenous Studies* 3, no. 2 (Fall 2016): 45–69.

13. USDA Food and Nutrition Service, "Food Distribution Program on Indian Reservations," https://www.fns.usda.gov/fdpir/food-distribution-program-indian-reservations-fdpir.

14. *Huffington Post*, August 18, 2012.

15. Rosalind Coward, *Female Desire: Women's Sexuality Today* (London: Paladin, 1984), 103.

16. "Interception," entry in *Apaches in the Kitchen* blog, November 5, 2014, https://apachesinthekitchen.blogspot.com/2014/11/.

17. "Sense of Direction," *Apaches in the Kitchen* blog, September 26, 2011, http://apachesinthekitchen.blogspot.com/2011/09/sense-of-direction.html.

18. Nephi Craig, personal communications, various dates.

19. The café is largely run by young members of the tribe. Barclays Bank, Johns Hopkins Center for American Indian Health, and the White Mountain Apache Tribe collaborated to form the Arrowhead Business Group that backs the café. Karen Warnick, "Apache Youth Run Arrowhead Café," *White Mountain Independent*, September 2, 2015, https://www.wmicentral.com/news/latest_news/apache-youth-run-arrowhead-caf/image_c2d78e66-5052-11e5-8528-3f7525aa4fef.html.

20. Barbara Bruce, "First-Ever Glam Camp Builds Beauty, Confidence," *White Mountain Independent*, May 15, 2015, https://www.wmicentral.com/news_premium/first-ever-glam-camp-builds-beauty-confidence/article_f4fd6a02-fa97-11e4-88d8-c3450d1d5309.html.

21. Carmen Herrera, Marie Léger, Janneth Lozano, Martha Mendoza, Ana Manuela Ochoa Arias, Joanne Ottereyes, Laura Ramos, Sofía Robles, Natalia Sarapura, and Julia Suárez, *Indigenous Women of the Americas: Methodological and Conceptual Guidelines to Confront Situations of Multiple Discrimination*, https://www.forestpeoples.org/sites/fpp/files/publication/2014/02/iw-complete-english-new-photos.pdf.

22. Part of the Fort Apache Historic Park is Nohwike' Bagowa (White Mountain Apache Cultural Center and Museum), which features Ndee Bike' (Footprints of the Apache) a multimedia chronicle of the White Mountain Apache creation story.

KYLE POWYS WHYTE

14
INDIGENOUS CLIMATE JUSTICE AND FOOD SOVEREIGNTY
Food, Climate, Continuance

One of the distinctive aspects of Indigenous climate justice movements is their focus on food as a vector for understanding the relationship between environmental change and justice. I have worked with nearly twenty tribes and First Nations in North America on climate change planning and have connected to many more through events and educational programs I have organized with numerous colleagues and partners. I am a coauthor of several recent reports of the United States Global Change Research Program (USGCRP) and the US Forest Service that review the available empirical research and testimonies on climate change trends and issues. In all this work, Indigenous peoples' emphasis on food is significant.

However, the word "food" in the English language often connotes something far simpler than what many Indigenous persons I know seek to invoke when they say "food justice," "food sovereignty," or "first foods." In reality, and depending on the peoples, food actually refers to different conceptions of collective self-determination that integrate ecological, cultural, social, and political dimensions of our lives as members—or relatives or kin—of our Indigenous societies, communities, and nations. And, in many Indigenous worlds everywhere, societal membership and kinship are not confined to the human species; or, in many cultures, there is not even a privileging or concept of the human in the first place.

In this chapter, I will visit my previous experiences and work in Turtle Island/North America and share some of my own reflections on how I have tried to make sense of the complex meanings of food. The way I have thought about it, quite broadly, is that Indigenous food sovereignty aspires to increase the degree of what I call the collective continuance of our peoples. "Collective continuance" refers to a way of understanding our places in the world in which our peoples' capacities to adapt to changes, whether natural or human caused, are based importantly on qualities of our relationships and responsibilities to human and nonhuman relatives.

Today's climate change ordeal represents a particular kind of challenge for Indigenous collective continuance. For changes such as sea-level rise or increases in severe weather are made all the more difficult to endure owing to the conditions of domination our peoples face from nation-states, corporations, and widespread discrimination against our sovereignty, cultural integrity, economic vitality, and aspirations. The relationship between Indigenous foods and climate change is a powerful vector for understanding climate justice in relation to our collective continuance as peoples.

OBSERVING CLIMATE CHANGE

"Climate change" refers to the idea that many Earth observers, including Indigenous elders, scientists, and knowledge keepers have been tracking a spike in the rise in global average temperature in the latter part of the twentieth century and into the twenty-first century. Climate scientists of different cultures and heritages, working in government, academic, and other research organizations, have also begun to track this change through their measurements of contemporary and historical temperature trends. The spike is significant, for rising temperatures can potentially destabilize the climate system, altering the patterns of ecological change expected by humans.

Many human societies are not prepared to absorb the impacts of climate destabilization without incurring problems of economic decline, cultural loss, environmental destruction, social unrest, and political conflict. These problems affect humans' quality of life and their self-determination or sovereignty—whether as individuals or as members of political, social, and cultural groups. Of course, some types of climate destabilization engender opportunities. For example, Greenland will experience a longer agricultural growing season, though increases in invasive species will perhaps become a new challenge. I also recently saw a story about how Greenland will have more economic opportunities in tourism for visitors interested in witnessing glaciers melting away.

The recent climate change is sometimes called "anthropogenic," because we know that much of the temperature spike arises from human capitalist, colonial, and industrial collective actions since the nineteenth century that have increased concentrations of greenhouse gases in the atmosphere or changed local temperature conditions (such as the heat island effect in urban areas). These activities include burning fossil fuels (coal and oil industries, among others) or deforesting landscapes (for commercial agriculture, urbanization, recreation, among other reasons). Greenhouse gases trap heat on the earth, so that their increasing concentration is associated with rises in the average temperature. Scientists often measure human industrial contributions to climate change through comparing temperature fluctuations and atmospheric carbon concentrations.

Many Indigenous persons I know have had the sense for many decades that recent environmental change is probably not entirely natural. Whether through local experience or hearing from Indigenous persons living elsewhere, people I know often tell me that the massive, rapid environmental transformations of colonialism, capitalism, and imperialism are to blame. It is not uncommon to hear Indigenous persons in societies, communities, and nations across Turtle Island highlighting the connection that the United States and Canada made Indigenous peoples suffer in order to usher in the natural resource extraction, water and air pollution, and deforestation that are now driving the spike in global average temperature.

Sheila Watt-Cloutier is among the major voices articulating this way of thinking about climate change:

> Inuit culture is based on the ice, the snow, and the cold.... Therefore when the climate changes and/or warms... [t]hen our right to culture, our right to educate our children on the land, our right to safety, our right to health all become impacted by these rapid changes. In essence our Right to exist as Inuit as we know it is impacted.... We are a very adaptable people

and yet others tend to think that it is [due to] our inability to adapt to the modern world that we are facing these challenges of social and health related issues. Not true.... It is the speed and intensity in which change has occurred and continues to occur that is a big factor [in] why we are having trouble with adapting to certain situations. Climate change is yet another rapid assault on our way of life. It cannot be separated from the first waves of changes and assaults at the very core of the human spirit that has come our way.[1]

For Watt-Cloutier, climate change cannot be unraveled from the colonial invasion, industrialization, and capitalist expansion that continue to occur at the expense of Indigenous peoples' self-determination and wellness. Candis Callison writes that "Watt-Cloutier assigns moral meanings and not just an explanation of physical mechanisms when it comes to climate change. In so doing, she underscores [climate change as] "industrialization gone terribly wrong" given its connections to colonial and other forms of anti-Indigenous power. Callison emphasizes how Watt-Cloutier "sees connection between culture, environment, and community survival" in relation to climate change.[2]

STORYING CLIMATE CHANGE

Even before the twentieth century, what "climate change" refers to in English was not a new or profound concept for many Indigenous peoples. Indeed, many Indigenous peoples have ancient traditions of governance and political philosophies that seek to understand how peoples should be organized to be most adaptive to seasonal environmental changes and to keep track of trends in interannual variations in environmental conditions. I have had the chance to experience and learn from Indigenous calendars, seasonal rounds, and almanacs throughout the world. More recently, in collaboration with Jared Talley, I have reviewed large, often forgotten literatures on Indigenous adaptive governance, from Anishinaabe seasonal rounds to Lakota winter counts, to Nuu-chah-Nulth potlatch traditions, to Aztec and Mayan calendars.

Each example I have seen usually involves very detailed understandings of how societal institutions—from ceremonies to harvesting and food storage, to political leadership, to gender relations—should be designed to support resilience in the face of inevitable annual seasonal changes and interannual trends. Significantly, Indigenous peoples used very different concepts to understand what is today called "climate." Outside Indigenous traditions, "climate" is often taken to be an abstract notion tied to Earth-scale changes that are, in a sense, invisible to humans. Indigenous calendars or seasonal rounds, quite differently, track environmental change through the actions of humans and nonhumans—actions situated within ceremonial, ethical, political, conservational, and other types of responsibilities that work to create accountability across human and nonhuman members of a society.

We know historically that Indigenous peoples learned to adapt to a certain range of changes, but that many also had to experience more disruptive shocks that are hard to recover from without incurring harms, such as malnutrition, painful geographic displacement, disease, and violent conflict. The transatlantic fur trade of the sixteenth through nineteenth centuries is one case. Michael Witgen, Susan Sleeper-Smith, and other historians have shown how Anishinaabe/Neshnabé peoples responded in ways that strengthened their sovereignty and well-being while also being harmed severely over time by environmental degradation, racism, violence, patriarchy, economic marginalization, and forced displacement.[3]

Fast-forward to the present: numerous Indigenous testimonies and scientific reports claim that Indigenous peoples are among the populations that face the most risks and will suffer the most harms from anthropogenic climate change. Importantly, impacts on food are perhaps the most commonly discussed topics. I contributed to a health-focused USGCRP report that discusses climate risks to Indigenous peoples' "food safety and security," including decreasing access to berries (Arctic), moose and wild rice (Great Lakes), water for farming (Southwest), and shellfish and fish (coastal regions), among other examples.[4] Damage to infrastructure from climate change threatens food storage and preparation.

Critically, the report discusses how loss of access to traditional foods threatens economic vitality, cultural integrity, and political sovereignty. I interpret the report as tying these risks and harms to US colonialism, in terms of how land dispossession and failed implementation of treaty rights affect peoples' capacity to steward the environment. Moreover, the risks and harms are particularly challenging in contexts where Indigenous nations, societies, and communities are already negotiating historical and personal trauma, alcohol abuse, suicide, high infant and child mortality, diabetes, and unsafe levels of exposure to pollutants or toxic substances.[5]

Specific Indigenous groups and Indigenous initiatives have been at the forefront of calling out climate change risks. The Northwest Treaty Tribes have launched an initiative called Treaty Rights at Risk.[6] The initiative discusses how the US federal government continues to violate treaty rights by endorsing activities that permit US settlers to establish their own ways of life in western Washington at the expense of Indigenous peoples' ways of life, which heavily involve salmon but also other treaty-protected plants and animals. Dams, intensive agriculture, urban development, industrial pollution, and other land-use practices, including recreational activities, are among the actions that advance settlers' aspirations at the same time that they degrade the habitats of salmon and other species.

Warming waters, ocean acidification, and other climate change effects exacerbate these threats to salmon and shellfish populations. The initiative references how closely salmon, for example, is related to economic vitality and cultural integrity. One ceremony, often referred to as a potlatch ceremony, involves people giving away their wealth in salmon to others, accompanied by intertribal treaty talks and cultural performances.[7] The late Billy Frank Jr. stated "as the salmon disappear, our tribal cultures, communities and economies are threatened as never before.... Some of our tribes have lost even their most basic ceremonial and subsistence fisheries—the cornerstone of Tribal life."[8]

The Mohawk Nation, which straddles the US-Canadian border along the St. Lawrence River, has a long track record of working to clean up massive industrial pollution in that region.[9] The St. Regis Mohawk Tribe is a leader in climate change planning thanks to the work of Mary Arquette and Angela Benedict, among others. One of the issues their climate change plan focuses on is the Three Sisters, "the corn, beans and squash" that supported survival and "sustained life" and continues to be important to Mohawk people today. The plan states that "The Three Sisters are the foundation of the Haudenosaunee culture. Ceremonies are to be concluded with the Three Sisters as an offering to the spirit beings."[10]

Though agriculture centered around the Three Sisters had long been a major economic activity for Mohawk people, the rise in industrial agricultural and rampant pollution from General Motors and Alcoa facilities weakened the production of abundant, nutritious, and culturally significant foods.[11] Relative to climate change, the plan states how the Three Sisters are further threatened by recent hailstorms; higher temperatures; droughts; and increases in pests, weeds, and diseases. According to the plan document, these changes affect not only food production but also cultural integrity.[12]

The Karuk Tribe in what is now called California has also been a leader in preparing for climate change, having recently issued its own climate change plan.[13] The Karuk plan emphasizes a range of species as historically and contemporaneously important to them for nutritional, economic, and cultural reasons—including salmon, deer, and a variety of nuts and berries. Karuk people relate to many of these species spiritually by understanding themselves as bound together through reciprocal responsibilities. In addition to honoring the environment through ceremonies, humans engaged in burning practices to maintain habitats, among other stewardship practices. US colonialism in California degraded the environment, outlawed practices such as burning, and dispossessed Karuk people of large portions of their lands and water.

Today, the Karuk plan states, "From a Karuk perspective, continuance of these traditional lifeways and practices is essential not only for food, but for the maintenance of traditional knowledge, cultural and tribal identity, pride, self-respect, and above all, basic human dignity."[14] As climate change creates further threats to the maintenance and resurgence of these relationships and practices, the Karuk people have created a knowledge sovereignty plan that seeks to strengthen both traditional knowledge and also their self-determined use and application of scientific techniques from Western institutions.[15]

While these examples focus on Turtle Island/North America, Indigenous peoples globally very much emphasize food in their climate change work. The Tebtebba organization's guide on climate change and Indigenous peoples discusses a range of issues and consequences on Indigenous peoples living in different regions of Asia, Africa, the Pacific, Central and South America, and the Caribbean. For example, the recent climate change report discusses how "massive floods, strong hurricanes, cyclones and typhoons, and storm surges" affect "forests, agricultural lands, crops, livestock, marine, and coastal resources." The guide references a range of food insecurity issues related to climate change effects on freshwater, coral reefs and mangroves, spawning beds for fish, and agricultural lands. It cites the "disappearance of plant and animal species that have sustained indigenous peoples as subsistence food sources or as essential to their ceremonial life" and how "traditional livelihoods ranging from rotational agriculture, hunting and gathering, pastoralism, high montane livestock and agriculture . . . rain fed agriculture, bird migrations used to guide hunters and mark agricultural seasons" are being lost, along with "availability of drinking water."[16]

Again, here we see food as a vector for discussing climate change. United Nations Special Rapporteur on the Rights of Indigenous Peoples Victoria Tauli-Corpuz states, "Climate change poses a serious threat to the rights of self-determination and development ... as well as the rights to food, water, land, territories, resources, and traditional livelihoods and cultures."[17]

TOWARD COLLECTIVE CONTINUANCE

I have been part of organizing educational programs on planning for climate change in my collaborations with the College of Menominee Nation Sustainable Development Institute, the Indigenous Planning and Design Institute, the Affiliated Tribes of Northwest Indians, and the United Southern and Eastern Tribes. I have worked with great colleagues, including Marie Schaefer, Chris Caldwell, Michaela Shirley, Mariaelena Huambachano, Ted Jojola, Tania Wolfgramm, Paulette Blanchard, Wikuki Kingi, and Don Sampson. I continue to be inspired by the work of many great tribal delegates, including staff, leaders, and Indigenous students who participated.

At each of these climate change programs, organized into a weeklong camp, thirty to fifty participants engage in activities, games, and experiences that open dialogue, creative expression, and concrete planning on climate change. Though the general information pertaining to specific nations, societies, and communities exchanged in these educational programs is not really for public consumption, I will share that food and medicine were among the most heavily covered topics in relation to climate change. Food was almost always discussed in a complex context of its connections to treaty rights and political sovereignty, gender and intergenerational relationships, ceremonial protocols (including eating and food-sharing protocols), kinship relationships, and economics, among other relationships.

Unlike scientific reports on climate change, which tend to reference temperature and precipitation, Indigenous accounts of climate change often focus on foods that people relate to in diverse ways. Perhaps this arises from our continuing to acknowledge and remember our heritages as peoples who always organized themselves to respond to seasonal, climatic, and other environmental changes. "Foods" is not an adequate word. If I had to write it broadly, I would say that I and the Indigenous peoples I have worked with are referring to food systems. At one level, food systems are simply about how nations, societies, or communities cultivate and tend, produce, distribute, and consume their own foods; recirculate refuse; and acquire trusted foods, ingredients, and technologies from others. At another level, food systems are collective capacities that are part of the production and exercise of certain valuable goods, including political self-determination, living a healthy lifestyle, and expressing spirituality and cultural uniqueness.

Food-related collective capacities facilitate the avoidance of preventable harms, including political powerlessness, undernourishment, and depression. In the Treaty Rights at Risk Initiative, the relationship between salmon and potlatch ceremonies reflects at once the sovereignty of Northwest tribes, their ideas of just and effective diplomacy, the values of generosity and accountability they want to pass on to future generations, their ceremonial and cultural lives, their economies, their history, and the environmental conditions required for their people to flourish.[18]

In each of the aforementioned cases, collective capacities revolving around food involve deep connections between human relationships or institutions (such as salmon giveaways, ceremonies for the Three Sisters) and ecosystems (such as burning practices, seasonal calendars, rain-fed agriculture). Collective capacity is also valuable as a mechanism for facilitating adaptation to changes that arise from environmental and human sources. Indigenous environmental studies scholars, such as Ronald Trosper and Robin Kimmerer, show that buffering environmental change is a key role that food systems play as collective capacities.[19]

Part of the importance of these scholars' work is that they describe food systems as systems of reciprocal responsibilities that connect humans and nonhumans culturally, spiritually, economically, and nutritionally. Marlene Atleo and Ben Colombi show that Indigenous peoples' sustainability or resilience involves ways of gathering and passing on different forms of knowledge, land-based educational practices, traditions of leadership and diplomacy, and methods of visioning.[20] Both scholars write about these ways in close relationship to food and food-based activities, such as salmon harvesting, processing, and consumption and the stewardship of salmon habitats.

Thus, when we refer to Indigenous "ecologies" or Indigenous "ecology," as in the work of Dennis Martinez and colleagues,[21] we are oftentimes referring to relationships among interacting humans, nonhuman beings (such as animals and plants), other entities (spiritual, inanimate, and so on), and landscapes (for example, climate regions and boreal zones) that are conceptualized as operating purposefully to facilitate a collective's (such as an Indigenous people's) adaptation to changes.[22] Here, "relationships" refer to institutions that range from interpersonal norms to skill sets, to decision-making protocols, to social hierarchies or social fluidities.

To be in one of these relationships is to have responsibilities toward the other participants in the relationship. Responsibilities refer to the reciprocal (though not necessarily equal) attitudes and patterns of behavior that are expected by and of various parties or relatives by virtue of the different contributions that each may be understood to be accountable for.

Food systems, as collective capacities, promote collective continuance by facilitating at least three types of responsibilities: those that support (1) the means of advancing robust cultural and social ways of life, (2) peaceful diplomacy and emboldened resistance, and (3) the societal decision-making protocols required for evaluating high-stakes decisions. All of these responsibilities can flourish if they are organized in ways that facilitate carrying out the contributions associated with being in specific relationships.

The types of relationships just described are in some cases very hard to replace given their connection to particular ecologies. For example, the desire to discharge reciprocal relationships of mutual giving in societies with giveaway ceremonies is very closely connected to persons' experiences with the cultural and spiritual value of salmon and the fact that those persons have been gaining specific knowledge about salmon since time immemorial. If salmon and salmon habitat disappear very rapidly, then those responsibilities are also endangered because they are so closely associated with the experience and knowledge of salmon. There are many qualities or characteristics of relationships that cause the relationships to facilitate responsibilities and that are hard to replace. "Qualities" in this context are properties of relationships that make it possible for the discharge of a responsibility to have wide societal impact. I will discuss one quality here: trustworthiness.

Trustworthy relationships are built in situations where the parties have reasons to feel that they all have one another's best interests at heart, both in a sense of confidence that individual participants in the relationship (human or nonhuman) will do what they are supposed to do and that they are motivated to do so out of general concern for everyone's well-being.[23] The Karuk food sovereignty and climate change planning efforts are largely about rebuilding systems of accountability in the tribe that restore trust in their own knowledge systems, experts, and educational processes, and about opening people up to developing more reciprocal responsibilities with nonhumans.

This relationship building involves rekindling language too, as being able to speak and understand one's Indigenous language supports trust in the society—especially in situations where a colonial language is imposed as a mechanism of power. Here, trustworthiness also includes reliability and accountability, which many philosophers distinguish from trust (I treat these concepts as equivalent in this chapter). For example, individuals are more likely to discharge their responsibilities if they know the other parties have a reliable track record of fulfilling their reciprocal responsibilities in return and they have been sufficiently trained or educated to do so. Trustworthiness also reaches into the nonhuman world.

"The Three Sisters" refers to kinship relationships among humans, plants and related species, and aspects of the environment. The Three Sisters, to me, suggests a long-standing building of trust with plants because it is understood that when the plants grow that way, they take humans' best interest to heart and provide nutrition and ceremonial vibrancy.[24] This is a reliable ecological relationship that humans are responsible for maintaining. The spiritual relationships between humans and salmon is similarly based on a complex relationship of trust and reliability. A salmon's sacrifice is considered a gift.[25]

Trustworthiness in relationships is a characteristic that serves a people's or society's collective continuance in at least the following ways. Trustworthiness and reliability facilitate the capacity to adapt to change without incurring preventable harms and with avoiding some harms altogether. For example, being able to trust where one's food comes from allows one to make healthy food choices. Trust also facilitates education across generations, as children and learners of all ages can get the mentorship they need to be responsible stewards. Trust means that there are people who will not falter in their responsibilities to care for plants, animals, fish, and insects, ensuring in return that these nonhumans will continue to provide gifts back. Trust provides for meaningfulness in life tied to the camaraderie that promotes cultural and social flourishing, which enables people to work together when conditions get tough instead of opposing one another.

Trust and reliability, if lost, are hard to replace, given that it can take many years to build the emotional bonds of trust or for someone to develop the needed track record to demonstrate his or her reliability. So it should not be surprising that Indigenous peoples discuss climate change as a threat. For if damage or risks pertaining to foods occur too rapidly, they will have ripple effects across politics, economics, culture, and other dimensions of life because food is central to collective continuance. While trustworthiness can support Indigenous peoples' resilience to change, climate-related threats (among other threats) can nonetheless undermine the exercise of responsibilities which are associated with trustworthy relationships that have been cultivated for years and generations. By calling food central to collective continuance, I am referring to how food-based practices are associated with responsibilities that bear qualities like trust.

Although Indigenous peoples have always adapted and changed over time, well before the emergence of the United States or Canada, it is also the case that many Indigenous peoples are concerned because their degree of collective continuance is lessened by dominant societies (nation-states, racist cultural practices, and so on) and organizations (such as corporations and educational institutions), which make certain climate change consequences more severe. This issue is called climate justice, a topic I turn to in the concluding section.

CONNECTING CLIMATE JUSTICE, COLLECTIVE CONTINUANCE, AND FOOD SOVEREIGNTY

Being from a relocation tribe, the Citizen Potawatomi Nation, I have long been concerned with understanding what harms and risks our people became susceptible to because we were forced to move from the Great Lakes region to Kansas and Oklahoma in the nineteenth century. One issue that is always brought up is that being relocated drastically increased our vulnerability to harms and risks that previously were not an issue for us—from restrictions in the diversity of our diets to dependence on fewer sources of financial stability and sustenance to having to adjust rapidly to new property and farming systems. The more I engaged with current Indigenous issues surrounding climate change, the more I also saw these vicious patterns occurring as well.

For example, shifts in salmon abundance historically were handled through diplomatic protocols in which groups with plenty supported those harmed by shortages. Tribes attempting to relocate due to sea-level rise are vulnerable because as a product of US colonialism, they are now confined to small areas of permanent residence they cannot leave. Historically, they had a larger range of movement that would offer greater adaptive options for sea-level rise in a particular location. Tribes facing loss of a traditional species due to warming climate or changing precipitation are more vulnerable owing to massive prior levels of deforestation, water pollution, and land dispossession.[26]

For this reason, vulnerability to climate change is part and parcel with the terraforming of colonialism. "Indigenous climate justice" refers to the idea that these harms are not accidental. Oftentimes the climate change consequences themselves are caused by the perpetuation of certain policies toward Indigenous peoples that restrain their ability to adapt. In the case of treaty rights, the United States disrespected the terms of treaties with Indigenous peoples in order to make way for many of the industries and lifestyles that have brought about the current rise in global average temperature. Heather Davis and Zoe Todd refer to the ecological footprint and the ripple effect of resulting traumas on Indigenous peoples as a "seismic" impact.[27]

At the same time, as the Treaty Rights at Risk Initiative shows, the United States still does not honor the terms of its treaties, which exacerbates climate-related harms and risks, as well as other activities that degrade habitats. Moreover, the entire situation is ironic given that Indigenous peoples did the least among US residents to cause anthropogenic climate change. Indigenous peoples today widely reject the current energy policies and consumer lifestyles in the United States and Canada. And Indigenous ancestors also rejected the capitalist, industrial, and colonial projects that imposed the economic, infrastructural, and landscape-scale conditions required for making these lifestyles possible in the first place. So climate justice is a vicious spiral.

Indigenous foods represent one of the most profound expressions of Indigenous climate justice. Indigenous food sovereignty brings out the idea of collective continuance, where what is at stake is far more than the loss of a particular nutrient or flavor. Rather, what is at stake is Indigenous collective capacities to adapt to change without incurring preventable harms or risks. Indigenous collective capacities are about retaining options for living good lives that are tied to responsibilities with relational qualities like trust. Whether an Indigenous environmental movement is explicitly about food or climate change does not matter. Food sovereignty, insofar as climate relates to food, is necessarily going to involve climate change and climate justice. Indigenous climate justice and food sovereignty movements could conceivably focus on one particular food. But such focus opens up issues of collective continuance that relate to culture, politics, economics, and many other issues tied to our collective capacities to survive and live well given the challenges we face.

Relating food with climate change from an Indigenous perspective also opens up the history of colonialism that perpetrating groups—such as the United States or major industries and corporations—still have not reckoned with. The St. Regis Mohawk Climate Change Plan, the Treaty Rights at Risk Initiative, and the Karuk TEK & Knowledge Sovereignty plans all speak to the continuing failure of neighboring communities, industries, and nations to respect Indigenous self-determination and the environmental conditions required for Indigenous peoples to practice their ways of life and enjoy economic sustenance. It is precisely these issues around conflict and disrespect that climate justice discourses outside of Indigenous perspectives do not address. For Indigenous peoples, food is a significant vector for bringing out the realities and histories that are part of Indigenous climate justice.

NOTES

1. Peter Robb, "Q and A: Sheila Watt-Cloutier Seeks Some Cold Comfort," *Ottawa Citizen*, March 27, 2015.

2. Candis Callison, *How Climate Change Comes to Matter: The Communal Life of Facts* (Durham, NC: Duke University Press, 2014), 74, 72.

3. Michael Witgen, *An Infinity of Nations: How the Native New World Shaped Modern North America* (Philadelphia: University of Pennsylvania Press, 2011); Susan Sleeper-Smith, *Indian Women and French Men: Rethinking Cultural Encounter in the Western Great Lakes* (Amherst: University of Massachusetts Press, 2001).

4. USGCRP, *The Impacts of Climate Change on Human Health in the United States: A Scientific Assessment* (Washington, DC: Government Printing Office, 2016), 253–54, https://health2016.globalchange.gov/.

5. USGCRP, *Impacts of Climate Change on Human Health*, 247–86.

6. Northwest Indian Fisheries Commission, *Treaty Rights at Risk*, http://www.treatyrightsatrisk.org.

7. Northwest Treaty Tribes, *Paddling to the Potlatch*, http://nwtreatytribes.org/paddling-to-the-potlatch/.

8. Billy Frank Jr., "Northwest Salmon, Tribal Cultures and Treaty Rights at Risk from Disappearing Habitat," *Seattle Times*, August 4, 2011, https://www.seattletimes.com/opinion/northwest-salmon-tribal-cultures-and-treaty-rights-at-risk-from-disappearing-habitat/.

9. Elizabeth Hoover, *The River Is in Us* (Minneapolis: University of Minnesota Press, 2017); Alice Tarbell and Mary Arquette, "Akwesasne: A Native American Community's Resistance to Cultural and Environmental Damage," in *Reclaiming the Environmental Debate: The Politics of Health in a Toxic Culture*, edited by Richard Hofrichter (Cambridge, MA: MIT Press, 2000).

10. St. Regis Mohawk Environmental Division, *Climate Change Adaptation Plan for Akwesasne*, 2013, St. Regis Mohawk Tribe, 24–26, http://srmtenv.org/web_docs/2013/09/SRMT_CCAP_08-30-13.pdf.

11. Jane Mt. Pleasant, *Traditional Iroquois Corn: Its History, Cultivation, and Use* (Ithaca, NY: Natural Resource, Agriculture, and Engineering Service, 2011).

12. St. Regis Mohawk Environmental Division, *Climate Change Adaption Plan*.

13. Karuk Climate Change Projects, *Climate Vulnerability Assessment*, 2016, https://karuktribeclimatechangeprojects.wordpress.com/climate-vulnerabilty-assessment/.

14. Karuk Climate Change Projects, *Karuk TEK & Knowledge Sovereignty*, 2016, https://karuktribeclimatechangeprojects.wordpress.com/chapter-2-its-illegal-to-be-a-karuk-indian-in-the-21st-century/.

15. Kari Marie Norgaard, *Karuk Traditional Ecological Knowledge and the Need for Knowledge Sovereignty: Social, Cultural, and Economic Impacts of Denied Access to*

Traditional Management (Happy Camp, CA: Karuk Tribe Department of Natural Resources, 2014), https://cpb-us-e1.wpmucdn.com/blogs.uoregon.edu/dist/c/389/files/2010/11/Final-pt-1-KARUK-TEK-AND-THE-NEED-FOR-KNOWLEDGE-SOVEREIGNTY-1phd94j.pdf.

16. Tebtebba: Indigenous Peoples' International Centre for Policy Research and Education, *Guide on Climate Change and Indigenous Peoples*, 2nd ed. (Quezon City, Philippines: Tebtebba Foundation, 2009), 12.

17. Victoria Tauli-Corpuz, "UN Special Rapporteur: Indigenous Peoples' Rights Must Be Respected in Global Climate Change Agreement," *Victoria Tauli-Corpuz: United Nations Special Rapporteur on the Rights of Indigenous Peoples*, March 3, 2015, http://unsr.vtaulicorpuz.org/site/index.php/en/press-releases/61-clima-change-hrc.

18. See, for example, Northwest Treaty Tribes, *Paddling to the Potlatch*.

19. Robin Wall Kimmerer, *Braiding Sweetgrass: Indigenous Wisdom, Scientific Knowledge and the Teachings of Plants* (Minneapolis, MN: Milkweed Editions, 2013); Ronald L. Trosper, "Northwest Coast Indigenous Institutions That Supported Resilience and Sustainability," *Ecological Economics* 41 (2002): 329–44.

20. Marlene R. Atleo, "The Ancient Nuu-chah-nulth Strategy of Hahuulthi: Education for Indigenous Cultural Survivance," *International Journal of Environmental, Cultural, Economic, and Social Sustainability* 2, no. 1 (2006): 153–62; Benedict J. Colombi, "Salmon and the Adaptive Capacity of Nimipuu (Nez Perce) Culture to Cope with Change," *American Indian Quarterly* 36, no. 1 (2012): 75–97.

21. Jesse Ford and Dennis Martinez, "Traditional Ecological Knowledge, Ecosystem Science, and Environmental Management," *Ecological Applications* 10, no. 5 (2000): 1249–50.

22. See, with regard to other traditions: Robert Melchior and Gordon Waitt, "Climb: Restorative Justice, Environmental Heritage, and the Moral Terrains of Uluru-Kata Tjuta National Park," *Environmental Philosophy* 7, no. 2 (2011): 135–63; Ian Werkheiser, "Community Epistemic Capacity," *Social Epistemology* 30, no. 1 (2015): 25–44.

23. Trudy Govier, *Social Trust and Human Communities* (Montreal, Que.: McGill-Queen's University Press, 1997); Margaret Urban Walker, *Moral Repair: Reconstructing Moral Relations after Wrongdoing* (New York: Cambridge University Press, 2006).

24. St. Regis Mohawk Environmental Division, *Climate Change Adaption Plan*.

25. E. Richard Atleo, "Discourse in and about the Clayoquot Sound: A First Nations Perspective," in *A Political Space: Reading the Global through Clayoquot Sound*, edited by Warren Magnusson and Karena Shaw (Montreal, Que.: McGill-Queen's University Press, 2002).

26. Kyle P. Whyte, "Is It Colonial Déjà Vu? Indigenous Peoples and Climate Injustice," in *Humanities for the Environment: Integrating Knowledges, Forging New Constellations of Practice*, edited by Joni Adamson and Michael Davis (New York: Routledge, 2016).

27. Heather Davis and Zoe Todd, "On the Importance of a Date, or, Decolonizing the Anthropocene," *ACME: An International Journal for Critical Geographies* 16, no. 4 (2017): 761–80.

ELIZABETH HOOVER
DEVON A. MIHESUAH

CONCLUSION
Food for Thought

Food is an inherent part of any society. Language, culture, identity, family relations, and human interactions with the environment, broadly defined, are all deeply connected to Native food systems. As Kyle Whyte explains, what constitutes "food" should not be thought of as independent from "food systems" that integrate ecological, cultural, social, and political dimensions of our lives as "relatives" to our societies, nations, and animate and inanimate inhabitants of the environment. When asked about his mentors across his culinary career, Indigenous food activist Karlos Baca (Diné/Tewa/Nuu-ciu), who founded the Taste of Native Cuisine collective, states, "the entire desert Southwest would be my mentor, because engaging with the landscape as a whole is the ultimate teacher. It's how our recent and ancient ancestors evolved their understanding of the edible landscape." Baca went on to describe human relatives—as well as rocks, plants, and aspects of the natural landscape—as having influenced his knowledge base and cuisine.[1] As the authors of the chapters in this book have described, settler colonial encounters, loss of land, assimilationist policies, and environmental change have forcibly separated many Native peoples from the land that mentored their ancestors. This has caused dramatic changes to traditional food systems, which in turn have contributed to immense suffering, including disproportionate rates of adverse metabolic health conditions, culture and language loss, and ongoing economic deprivation.

Yet community-based projects seeking to provide healthy food, restore the environment and people's relationship to it, and in many cases revive traditional seeds and horticultural practices, have been cropping up in Native American communities. The goals of these projects are to impart food production knowledge, help community members "decolonize" their diets, and work toward broader goals of food sovereignty.

Numerous themes run through this book, from health decline to health recovery initiatives; recovery of TEK; the importance of protecting and respecting the Natural World; the process of becoming an activist for social, economic, and political justice; and the discovery and transmittal of knowledge about precontact foods, seed saving, cultivation of crops, fishing, and hunting. Plus, how crucial it is for all of us to care of each other. The contributors to this book are seeking to establish a particularly Indigenous food sovereignty, one that builds on but goes beyond conventional understandings of the need for a just economy, equitable landholdings, and control over means of production, to recognize the importance of education around tribal history and culture, and a food system that connects and benefits human and nonhuman relatives.

At the same time, none of these authors shies away from the difficulties of achieving this goal, which require overcoming tribal politics, economic challenges, erratic weather, apathy, pests, and exhaustion. These chapters describe the hard work that has been done, and still needs to be done. As Devon Peña points out, in order to celebrate and revitalize the methods, practices, and ethics of Indigenous TEK, we need to understand the conditions of disruption and change specific to each place-based culture. Similarly, in "Searching for *Haknip Achukma* (Good Health)," Mihesuah points out that tribes need to understand how to negotiate and navigate the various challenges that outside forces will present in efforts to compromise treaty-ensured guarantees to clean water and hunting, fishing, and gathering rights. As all of these contributors have pointed out, a return to traditional ways of eating can be accomplished only with sufficient access to healthy ecosystems. But as Livingston, Gwin, and Mihesuah demonstrated, challenges to community food projects can come not only from the outside but from within tribal communities: elected tribal council members may choose to side with lobbyists rather than their community members or may defund a project associated with a previous administration; tribal citizens may object to the tribal government having authority over a cultural food program; or community members may vandalize a project out of jealousy or desperation.

On the other hand, as Kevin Chang and his co-authors from Hawaii describe, "we work from the inside out, and from the bottom up." Determined people commit to gathering and working hard in pursuit of a shared belief in the value of community stewardship, seed saving, and cultural reclamation efforts. Chang and his colleagues describe the restoration of community and ecosystem relations as "mending our net." There is a recognition, as Lindholm points out in the context of Alaska, that traditional foods represent cultural, spiritual, emotional, social, physical, and mental nourishment. As many of these contributors have established, these efforts can only be accomplished through the cooperation of community members across generations, the reclaiming of language, the recovery of seeds, and as Clark has described in the context of California, the reinstitution of traditional land management regimes. Mohawk seed keeper Rowen White describes the slow, intergenerational process that is seed work—the recovery of "Seedsongs" that represent the intergenerational memory of reciprocal relationships with food and seed plant relatives. As Pat Gwin details in reflecting on his ancestors' experience, "it was the plants that kept them alive and, in fact, what made them Cherokee. One could not exist without the other." In this way, the contributors to this book, and the community members they work with, have demonstrated that Indigenous food sovereignty is a method, a tool, and a process that is intergenerational, is interconnected, and symbolizes what Chef Nephi Craig describes as "a return to cultural pride."

Throughout this book, the authors have sought to examine the local histories that have led to the conditions that produce food injustices, but also to celebrate the actions that community members are taking in the present, and to highlight the potential for future growth. We all hope that this book will begin, or continue, discussions in your kitchens, gardens, classrooms, and community centers about how to define and enact food sovereignty in your own community. The study questions are meant to facilitate this.

NOTES

1. Karlos Baca, interview with Elizabeth Hoover, Providence RI, November 7, 2017.

STUDY QUESTIONS

Before reading the book, define the following terms as you understand them. After reading the book, define them again. Did the readings give you new insights?
- Sovereignty
- Indigenous food sovereignty
- Food sustainability
- Traditional indigenous foods
- Food activist
- Traditional ecological knowledge (TEK)
- Treaty rights

As you read the introduction and each chapter, keep in mind the following foundational questions:

1. What are the various definitions of food sovereignty?
2. Can you think of a better descriptor than "food sovereignty" to define the initiatives in this book?
3. What are the most significant differences between contemporary American foods and traditional Indigenous foods?
4. What are reasons for the decline in Indigenous health? Are you at risk for declines in your own health?
5. How does the destruction of the environment affect food production and traditional food culture?
6. How has the loss of traditional ecological knowledge (TEK) influenced the way Natives think about food and the environment?

7. Do any of these Indigenous food sovereignty initiatives inspire you to become a food activist?
8. What are the challenges to achieving food sovereignty?
9. If culture and traditional foods are closely intertwined, what could happen to a tribe that cannot recover its traditional resources?
10. What is the meaning of "decolonizing" foods and why is that concept important?
11. If you were to walk or hike a mile within your neighborhood, what types of ingredients would you forage?

CHAPTER-SPECIFIC QUESTIONS

Chapter 1: Voices from the Indigenous Food Movement

1. Are there particular voices in this chapter that resonate with you? Why do you think this is so: your tribal affiliation (if any), your personal background or ethical beliefs, teachings of family members or elders, your education, or other factors?
2. Are there voices you disagree with or cannot relate to? Why, and what differences would make you support them?

Chapter 2. "You can't say you're sovereign if you can't feed yourself"

1. Why would "culturally appropriate" foods, as opposed to perhaps more easily grown conventional vegetable varieties, be important for some food sovereignty projects?
2. What does it mean for food to be both "biological and spiritual nourishment?"
3. How might food and language be connected?
4. What are different relationships that might be important to define and preserve in determining what community food sovereignty would look like?
5. Would a community need to produce all its own food to be truly food sovereign?
6. How do issues of "choice" and "control"—at individual, community, and tribal levels—factor into food sovereignty?

Chapter 3. Searching for *Haknip Achukma* (Good Health)

1. What factors accounted for the decline of health among the Five Tribes and other Native peoples in Indian Territory?

2. What challenges do tribes face currently in achieving food sovereignty in Oklahoma?
   3. How do the various definitions of "traditional foods" affect Natives' health?
   4. What are some food sovereignty and health initiatives in Oklahoma?

Chapter 4. Kua'āina Ulu 'Auamo

   1. What is the Hawaiian name for fishponds? For seaweeds? How were, and are, these resources important to Native Hawaiians and to ecological preservation on the islands?
   2. Why does Hawai'i import the vast majority of its food? How do the boundaries and limitations of island living contribute lessons and opportunities that the rest of the world can learn from?
   3. In Hawai'i, how does the health of the land rely on the health of the ocean, and vice versa? How do you think the land and ocean are interconnected where you live, if at all?
   4. How important is water for Indigenous seafood and food cultivation in Hawai'i?
   5. What might 'āina momona look like in the place where you live? What are some things you and your community can do to achieve 'āina momona in your place?
   6. How could the practice of "giving before taking" on all levels result in abundance in the system?

Chapter 5. Alaska Native Perceptions of Food, Health, and Community Well-Being

   1. How might Alaska Native perceptions and worldviews contribute to more informed policy and decision-making on state, national, and international levels?
   2. In what ways does the dominant capitalist culture run contrary to Alaska Native culture?
   3. How does nutritional colonialism impact Native lifeways and subsistence culture?
   4. What is included in the Indigenous definition of nourishment?
   5. How do Alaska Natives find balance between two different cultures with often incompatible values?

Chapter 6. Healthy Diné Nation Initiatives

   1. Why is there a debate among Navajos over the meaning of "junk food"?

2. Who is responsible for the health of tribal members? Explain why.
3. Which of your favorite foods would be taxed under the unhealthy foods 2 percent sales tax? How would you respond to the tax?
4. What do you think are the most effective initiatives funded by the tax?
5. One nation does not mean one point of view. If you were a food activist on the Navajo Nation, how would you rally citizens and leaders to your cause?

Chapter 7. Planting Sacred Seeds in a Modern World

1. Where do Native peoples find "lost" seeds?
2. What are "Seedsongs"?
3. Why are seeds central to Haudenosaunee collective identity?
4. How is protecting seed stock from patenting and bio-piracy important to ethical seed stewardship and Indigenous food sovereignty?
5. Do you think there are ethical ways of commercializing seeds?
6. For Rowen White "these seeds are my wise teachers." Who or what are the wise teachers who lead you on your life journey?

Chapter 8. What If the Seeds Do Not Sprout?

1. In this age of "improved cultivars" and genetically modified organisms, why are pre-1492 varieties important culturally? What about agriculturally?
2. What would happen to the Cherokee people if "the seeds did not sprout"? Can a lost heirloom ever be replaced?
3. What is the meaning of the saying, "No self-respecting Cherokee would ever be without a corn patch?" (Remember that "agriculture" cannot be spelled without "culture.")
4. If you were going to go on a quest to recover your community's lost agriculture, and culture, where would you start?

Chapter 9. People of the Corn

1. How is dry-farming effective for Hopis?
2. Describe three aspects of the relationship between corn and the Hopi people.
3. What are some aspects of life that corn represents to the Hopi people?
4. How were historic farming, hunting, and gathering tasks assigned by gender, and have those roles changed through time? If your tribe traditionally assigned farming to females, has anything changed? Why?

5. What modern-day demands threaten the relationship between Hopi people and corn?

Chapter 10. Comanche Traditional Foodways and the Decline of Health

1. How does the traditional Comanche diet compare to what the people consume today? What about traditional and modern lifestyles?
2. Why is it that Comanches do not have access to their traditional foods?
3. How do conflicting historical accounts—and modern-day interpretations of accounts—of Comanche diets and physicality influence those interested in recovering their traditional ways of eating and improving their health?
4. If Comanches have no seeds to recover, and if bison and game animals are not available, how can Comanches recover their health?
5. Historical writings by settler colonists contain information about how Comanches lived prior to confinement to a reservation, but these descriptions clearly convey whites' assumptions of racial and cultural superiority. Do these quotations contain legitimate information, and why or why not? Can the Comanches and other tribes recover food sovereignty through the lens of an outside culture?

Chapter 11. Bringing the Past to the Present

1. What does Gerald Clarke mean by saying "California was one big farm?"
2. In what ways was California's geography not conducive to large, densely populated settlements? To the development of farms as traditionally defined?
3. What edible plants are found throughout California, and how are they prepared?
4. How are Native Californians revitalizing their foodways traditions?
5. What is the importance of revitalizing Indigenous plant knowledge, and what benefits can come of that?

Chapter 12. On Intimacy with Soils

1. What roles do companion plants play in Indigenous agroecosystems? Name a native companion plant that is used in your bioregion as a medicine or as healer of soil quality.
2. Does Devon Peña provide sufficient evidence for the case that Indigenous farmers have practiced permaculture and biodynamics for centuries before Europeans and Americans described and promoted these practices?

3. Why does TEK matter? Why should Indigenous agroecological knowledge receive recognition, and what could be the consequences if sustainable agriculture discourses continue to obscure and erase these Indigenous antecedents?
4. If you were to work on a project to restore soil health on damaged lands, which crops (favored locally) would you choose as companion plants for this regenerative agricultural practice?

Chapter 13. Nephi Craig

1. Why might some Indigenous chefs feel conflicted when working in mainstream restaurants, even those that serve haute cuisine?
2. How is Nephi Craig using his knowledge and skills to assist his tribe?
3. Why are many stories about Indigenous chefs incomplete?
4. How does Craig live his "life in second sight?"
5. What are benefits of community-based Indigenous-only kitchens in tribal communities?

Chapter 14. Indigenous Climate Justice and Food Sovereignty

1. In what ways is climate change among the oldest issues that human societies have had to deal with, even in ancient times?
2. How is the concept of mutual responsibility tied to how some Indigenous peoples understand and work to address climate change?
3. Why is understanding the relationships between colonialism and capitalism important for also understanding how certain groups, including Indigenous peoples, are harmed more severely than others by the consequences of climate change?

After reading this book and considering the chapter questions, would you answer the opening questions any differently than you did before?

Possible Topics for Research Projects
- History of a tribe's health and food, how both of these changed over time, and what people are doing today to maintain or improve their health and access to traditional foods
- History of your culture's food (traditional foods and how they are used today)

- Environmental health issues related to food
- Local and tribal food organizations and activities in your region
- Compilation of foodways words and concepts in your tribal language
- Controversies over genetically modified seeds and resource-intensive monoculture agriculture
- Successful tribal health initiatives (profile at least five)
- Activism: How to challenge the industrial food industry
- School food and health curricula
- Ways in which traditional foods could be better incorporated as part of your community's health-care systems
- How diets of specific tribespeople were altered because of colonization
- Traditional dishes or meals in your tribal community: how those dishes are prepared now and were prepared in the past, the origins of some of the ingredients, and ways to access the original ingredients if the ingredients have changed
- Strategies for health recovery of a certain tribe (recovery of agricultural techniques, tribal health efforts, and so on)
- If price and convenience of traditional Indigenous foods are the biggest factors standing in the way of people eating them, discuss possible strategies that could address these issues
- The best ways to engage the public, corporations, and government in the conservation of species of aesthetic and biological importance
- How increasing temperatures, extensive droughts, acidification of the oceans, and rising sea levels can become the major drivers of the future of our food systems

Ideas for Activities Related to Indigenous Food Sovereignty
- Plan a school or community garden.
- Prepare one precontact meal a day for one week.
- Plant a small backyard or kitchen garden.
- Create a book of traditional recipes for your community.
- Learn about the wild foods your ancestors foraged.

CONTRIBUTORS

BRENDA F. ASUNCION focused on the marine environment in her formal studies (MS in marine science), and after gaining a few years of working experience in marine policy in Hawaiʻi, she currently serves as the Loko Iʻa Coordinator of Kuaʻāina Ulu ʻAuamo (KUA), facilitating a statewide network of kiaʻi loko called Hui Mālama Loko Iʻa. The source of her fishpond learning is Heʻeia fishpond, where she started as a volunteer. In addition, she is a student of hula in Hālau I Ka Wēkiu.

STEVEN BOND-HIKATUBBI is an enrolled member of the Chickasaw Nation and has served as the tribe's ethnobotanist since 2007. He has worked with the Intertribal Agriculture Council since 2011, facilitating the development of several agricultural ventures for tribes and tribal producers throughout the United States. He is a writer, photographer, consultant, and lecturer in the disciplines of ethnobotany, agriculture, and food safety. Currently, he is developing his own farm and continues to stay engaged in the pursuit, application, and sharing of knowledge among tribal communities and their neighbors.

CARRIE CALISAY CANNON is a member of the Kiowa tribe of Oklahoma and is also of Oglala Lakota descent. She has a BS in wildlife biology, and an MS in resource management. In 2005 she began working as a tribal biologist for the Hualapai Tribe of Peach Springs, Arizona, where she initiated an intergenerational ethnobotany program for the Hualapai community. Carrie is currently employed as an ethnobotanist for the Hualapai Department of Cultural Resources. She administers a number of department projects and programs that promote the intergenerational teaching of Hualapai ethnobotanical knowledge. She works to promote both preservation and revitalization, focusing her energy on ensuring that tribal ethnobotanical knowledge persists as a living practice and tradition.

KEVIN K. J. CHANG is a member of the band Kupaʻaina. An attorney and musician, he currently serves as the executive director of Kuaʻāina Ulu ʻAuamo (KUA) a movement-oriented, capacity-building, and network-facilitating organization for community-driven biocultural-resource-management initiatives in Hawaiʻi. Prior to his involvement with KUA, he worked for the Office of Hawaiian Affairs and the Trust for Public Lands Hawaiian Islands Program.

GERALD CLARKE is an enrolled member of the Cahuilla Band of Indians and currently lives on the Cahuilla Indian Reservation. An assistant professor of ethnic studies at the University of California, Riverside, he is a frequent lecturer on Native art, culture, and issues. He has served on the Cahuilla Tribal Council and participates in Bird Singing, a traditional form of singing that tells the cosmology of the Cahuilla people.

PAT GWIN is a citizen of the Cherokee Nation and a fourth-generation Oklahoman. His birth, upbringing, education, and career have all unfolded within northeastern Oklahoma. He graduated from Northeastern State University with a BS degree in biology, began his work with the Cherokee Nation in 1992, and has served in the Environmental, Natural Resources, Executive Administration, and Secretary of Natural Resources Departments. Currently, he is the Cherokee Nation's senior director of environmental resources and works on numerous Department of the Interior, Environmental Protection Agency, and US Department of Agriculture–related projects dealing with environmental issues and policies. Additionally, he directs the Cherokee Nation's Ethnobiology Program, which includes the seedbank and native plant initiatives. His nonprofessional life is consumed by his passion for the outdoors and sustainability issues.

ELIZABETH HOOVER is Manning Associate Professor of American Studies at Brown University, where she teaches courses on environmental health and justice in Native communities, Indigenous food movements, and community-engaged research. Her book *The River Is in Us: Fighting Toxics in a Mohawk Community*, an ethnographic exploration of Akwesasne Mohawks' response to Superfund contamination and environmental health research, was published by the University of Minnesota Press in fall 2017. Elizabeth also serves on the executive committee of the Native American Food Sovereignty Alliance (NAFSA), and the newly formed Slow Food Turtle Island regional association.

WALLACE K. ITO is the Limu Hui Coordinator of Kuaʻāina Ulu ʻAuamo (KUA). His job is to weave a network of kupuna (community elders) who retain knowledge of Hawaiian customary limu (seaweed) practices in order to help perpetuate them. A spearfisher since the age of eight, Wally has witnessed the decline of the nearshore fishery over a period of more than fifty years. He returned to school at age fifty-one to complete a college degree, finally accomplishing that goal in 2014, graduating from Hawaiʻi Pacific University with a BS in marine biology.

WINONA LaDUKE (Anishinaabe) is an internationally acclaimed author, orator, and activist who works on issues of environmental justice, sustainable development, renewable energy, and food systems. She helped found both the White Earth Land Recovery Project—a reservation-based nonprofit organization devoted to restoring the land base and culture of the White Earth Anishinaabeg—as well as Honor the Earth—a nonprofit organization that fights to create awareness and support for Native environmental issues and to develop needed financial and political resources for the survival of sustainable Native communities. Winona also served as Ralph Nader's vice-presidential running mate on the Green Party ticket in the 1996 and 2000 presidential elections. In addition to leading Love Water Not Oil tours across proposed pipeline routes in the Midwest, Winona is currently working to build and promote Winona's Hemp & Heritage Farm.

MELANIE M. LINDHOLM completed her AS at Dixie State College, her BS at the University of Nebraska–Lincoln, and her MA at the University of Alaska–Fairbanks.

Her research interests are related to causes and consequences of nutritional transitions among northern Indigenous populations. She enjoys teaching sociology, scrapbooking, and dancing.

DENISA LIVINGSTON (Diné, New Mexico) is an unapologetic food justice organizer, social entrepreneur, and volunteer community health advocate for the Diné Community Advocacy Alliance. Her mission is to empower others and improve their lives. She is committed to addressing the diabetes epidemic, the dominant culture of unhealthy foods, and the lack of access to healthy foods, not only on the Navajo Nation but nationally and internationally. Livingston is passionate about servant leadership, creating new roles for society, and motivating community members to purpose and innovation.

VIRGIL MASAYESVA was formerly director of ITEP and was a member of the Hopi Tribe, raised in the village of Hotevilla on Third Mesa on his family's farm located in a valley that his family calls Hopaq.

LATICIA MCNAUGHTON is an enrolled Six Nations Mohawk, a member of the Wolf Clan. She is a PhD candidate in (Native) American studies at the State University of New York at Buffalo. She holds an MA in Native American studies from the University of Oklahoma and a BA in English/anthropology from Buffalo State College. Her dissertation work examines Haudenosaunee (Iroquois) Indigenous foods, food sovereignty practices, health and wellness, and the revitalization of traditional diet. McNaughton shares research, recipes, stories, and Mohawk language on her blog site, "Indigenous Food Revolutionary" (www.indigenousfoodrevolution.blogspot.com).

DEVON A. MIHESUAH, an enrolled citizen of the Choctaw Nation of Oklahoma, is the Cora Lee Beers Price Professor in the Humanities Program at the University of Kansas. She is the author of numerous award-winning books on Indigenous history and current issues, including *Choctaw Crime and Punishment, 1884–1907*; *Recovering Our Ancestors' Gardens: Indigenous Recipes and Guide to Diet and Fitness*; *American Indigenous Women: Decolonization, Empowerment, Activism*; and *Ned Christie: The Creation of an Outlaw and Cherokee Hero*. She is former editor of the *American Indian Quarterly* and oversees the American Indian Health and Diet Project at the University of Kansas.

DEVON G. PEÑA (Chicano) is the founder and president of the Acequia Institute, a charitable foundation created to provide support for the environmental and food justice movements. He operates the institute's 181-acre farm in south central Colorado's San Luis Valley. Dr. Peña is also a professor of American ethnic studies and anthropology at the University of Washington. A well known and prolific author, he most recently published *Mexican-Origin Foods, Foodways, and Social Movements: Decolonial Perspectives* (University of Arkansas Press, 2017).

BRIT REED holds an MPA with a concentration in tribal governance from Evergreen State College. She is the founder of Food Sovereignty is Tribal Sovereignty and a member of the I-Collective—a group of Indigenous cooks, chefs, knowledge and seed keepers, and artists.

MARTIN REINHARDT is an Anishinaabe Ojibway citizen of the Sault Ste. Marie Tribe of Chippewa Indians from Michigan. A tenured professor of Native American studies at Northern Michigan University, he also serves as the president of the Michigan Indian Education Council. His current research focuses on revitalizing relationships between humans and Indigenous plants and animals of the Great Lakes Region. He is the primary investigator for the Decolonizing Diet Project and coauthor of the *Decolonizing Diet Project Cookbook*. He has a PhD in educational leadership from Penn State, where his doctoral research focused on Indian education and the law, with a special focus on treaty educational provisions. Reinhardt also serves as a panelist for the National Indian Education Study Technical Review Panel and as a co-primary investigator for the Indigenous Women Working Within the Sciences (IWWS) Project, sponsored by the National Science Foundation.

VALERIE SEGREST, BSN, MS, is a Native nutrition educator and activist who specializes in local and traditional foods. As an enrolled member of the Muckleshoot Indian Tribe, she serves her community by working to inspire and educate others about the importance of a nutrient-dense diet through a simple, commonsense approach to eating.

SEAN SHERMAN (Oglala Lakota) is the creator of the Indigenous culinary business the Sioux Chef and cofounder of the Tatanka Truck and the nonprofit organization North American Traditional Indigenous Foods (NATIFS.org). He is also coauthor (with Beth Dooley) of *The Sioux Chef's Indigenous Kitchen* (University of Minnesota Press, 2017), which was named among the best cookbooks of 2017 by the *Washington Post*, National Public Radio, the *Los Angeles Times*, the *San Francisco Chronicle*, and *Village Voice*, as well as being awarded the prestigious James Beard Award for best American cookbook of 2018.

WAYNE C. TANAKA is a former board member of Kuaʻāina Ulu ʻAuamo (KUA) and is a senior public policy advocate at the Office of Hawaiian Affairs, where he focuses on water and ocean resource management issues, including community-based fisheries management initiatives. An attorney and engineer by training, he currently serves as president of the board of directors for the Conservation Council for Hawaiʻi. He has been an avid fisher throughout his life.

CHIP TAYLOR is trained as an insect ecologist and has published papers on insect reproductive biology and population dynamics, hybridization, plant demography, and pollination. Starting in 1974, he established research sites and directed students studying Neotropical African honey bees (killer bees) in French Guiana, Venezuela, and Mexico. In 1992 Chip founded Monarch Watch, an outreach program focused on monarch butterflies with an emphasis on education, research, and conservation. To learn more about the monarch butterfly migration, Monarch Watch has enlisted the help of volunteers to tag monarchs during the fall migration. The tagging program has produced many new insights into the dynamics of this migration. Taylor created the Monarch Waystation program in 2005 and the Bring Back the Monarchs program in 2010, both in recognition that habitats for monarchs were declining. These program are intended to inspire individuals, schools, and others to create habitats for monarch butterflies and to sustain all the wildlife that shares these habitats with monarchs.

DENNIS WALL has been editor for the Institute for Tribal Environmental Professionals (ITEP) since May 1998. His work includes producing *Native Voices*, the institute's quarterly newsletter, and editing/assembling training material for the institute's air and climate-change courses. Before joining the ITEP staff, Wall made a living as a freelance writer, worked as a technical editor for Los Alamos National Laboratory in New Mexico and as a writer-editor with a publishing firm in Indiana, and produced a guide to the ecology of federal refuges, *Western National Wildlife Refuges: 36 Ecological Havens from California to Texas* (Museum of New Mexico Press, 1995). He holds a BA in psychology from the University of Arizona and an MA in English from Northern Arizona University. His leisure activities include fiction writing, birding, exploring nature, and traveling. Dennis shares a home with his wife and two children, along with goats, chickens, and assorted furred, finned, and feathered critters.

ELIZABETH KRONK WARNER is a citizen of the Sault Ste. Marie Tribe of Chippewa Indians. She joined the University of Kansas law faculty in June 2012. In 2010, Warner was selected to serve as an Environmental Justice Young Fellow through the Woodrow Wilson International Center for Scholars and the US-China Partnership for Environmental Law at Vermont Law School. She has also served as a visiting professor at Xiamen University in Xiamen, China, and Bahcesehir University in Istanbul, Turkey. Her scholarship focuses on the intersection of Indian law and environmental law. She is coauthor of the casebook *Native American Natural Resources* and she coedited *Climate Change and Indigenous People: The Search for Legal Remedies*. In addition to teaching, Warner serves as an appellate judge for the Sault Ste. Marie Tribe of Chippewa Indians Court of Appeals in Michigan and as a district judge for the Prairie Band Potawatomi Nation in Kansas. Warner previously served as chairwoman of the Kansas Advisory Committee to the US Civil Rights Commission. She received her JD from the University of Michigan Law School and her BS from Cornell University.

ROWEN WHITE, a seed keeper and farmer from the Mohawk community of Akwesasne, is a passionate activist for Indigenous seed and food sovereignty. She is the director and founder of Sierra Seeds, based in Nevada City, California, an innovative organic seed stewardship organization focusing on local seed and hands-on education. She is the current national project coordinator and advisor for the Indigenous SeedKeepers Network (ISKN), an initiative of the Native American Food Sovereignty Alliance, a nonprofit organization aimed at leveraging resources to support tribal food sovereignty projects. Having developed many curricula focused on a holistic, Indigenous permaculture-based approach to seed stewardship, White teaches and facilitates creative seed stewardship immersions around the country within tribal and small farming communities. She weaves stories of seeds, food, culture, and sacred Earth stewardship on her blog, *Seed Songs*. Follow her seed journeys at www.sierraseeds.org.

KYLE POWYS WHYTE is an enrolled member of the Citizen Potawatomi Nation. He is the Timnick Chair in the Humanities, associate professor of philosophy, and associate professor of community sustainability at Michigan State University. His research addresses moral and political issues concerning climate policy and Indigenous peoples, the ethics of cooperative relationships between Indigenous peoples and science organizations, and problems of Indigenous justice in public and academic discussions of food sovereignty, environmental justice, and the anthropocene.

KAWIKA B. WINTER, PhD, is a multidisciplinary ecologist who operates in the spheres of academia, conservation, and policy. His research and professional career have focused on large-scale biocultural restoration of social-ecological systems in Hawaiʻi, and he has particular interest in the theoretical, philosophical, and applied aspects of reviving traditional resource management. He is currently in charge of managing the Heʻeia National Estuarine Research Reserve (HeNERR) on Oʻahu, and holds positions at the University of Hawaiʻi at Mānoa in the Hawaiʻi Institute of Marine Biology, and in the Department of Natural Resources and Environmental Management.

BRIAN YAZZIE, "Yazzie the Chef," is from a small community called Dennehotso, Arizona, located on the northeastern part of the Navajo Nation and is Áshįįhí (Salt People Clan), born for Nóóda'i dine'é Táchii'nii (Ute People Division of the Red Running into the Water Clan). He has worked in venues from New York, Minnesota, North Dakota, and Oregon to France. As an activist chef, he is concerned with cultural food appropriation and focuses his culinary talents on placed-based foods and education about healthy eating.

CHARLES K. H. YOUNG is a retired businessman with a degree in mechanical engineering who worked in food supply chain management and manufacturing in Hawaiʻi. He is a cofounder of and remains active in two community-based nonprofit organizations. One organization, Kamaʻāina United to Protect the ʻĀina (KUPA) Friends of Hoʻokena Beach Park, works with the Hawaiʻi County Department of Parks and Recreation under a memorandum of understanding to manage daily activities at Hoʻokena Beach Park. He also serves on the West Hawaiʻi Fishery Council and the Kona Community Development Plan Action Committee.

INDEX

Abbott, Isabella, 142–43
Abbott, Thomas, Sr., 242
A Big Fat Fall by Tripping (Comanche), 232, 233
Abya Yala, 276
access (component of food sovereignty), 73–74
Acequia Association, 296n18
acequias, 279, 281, 288, 295n9, 298n23
Achieving Food Sovereignty Chart, 14
acorns, 28, 108, 231, 258–59, 260, 261, 302, 307, 316, 317
activism, 14, 33, 105, 128, 345; achieving food sovereignty, 4, 14; challenges, 105–6, 112; home activism, 310. *See also* protesters
Affiliated Tribes of Northwest Indians, 327
Affordable Care Act, 102
agricultural tradition, 17, 96, 109, 223, 244, 245
agro-biomimicry, 290
agroecology, 283, 284, 285
agroecosystems, 277
Aguilar, Miguel Escalona, 283
ailments related to diet, 241. *See also* diabetes; high blood pressure; obesity

Akwesasne, 6, 30, 31, 58,187, 188, 189, 191
Akwesasne Cultural Restoration Program, 31, 65, 191
Akwesasne Task Force on the Environment, 30
Alaska Department of Fish and Game, 164
Alaska Federation of Natives, 167
Alaska Native Claims Settlement Act, 163
Alaska Natives: effects of processed foods, 159; cancer rates, 156, 165; connection to land and sea, 156–57; diabetes, 156, 165; diet changes, 156; health problems, 156; health strategies, 167–68; subsistence resources, 156, 159–60; substance abuse, 166
Alcoa, 325
Alfred, Taiaiake, 111
allotments, 240
American Association of Retired Persons (AARP), 112
American Indian Health and Diet Project, 36
Anadarko, Okla., 227
ancestral health, 312

Ancestral Lands Program, 107
Anderson, M. Kat, 265
Andrade, Carlos, 131
animals: chemicals effect on, 164–65; climate change effects on, 164; exploitation of, 106, 161, 162; respect for, 160–61
Anishinaabes, lifestyle changes of, 323–24
Annual Apache Harvest Festival, 311
antelope, 17, 99, 224, 225, 227, 240, 256, 301
anthropomorphism, 208
anthrosols, 283, 291
Anzaldúa, Gloria, 277
Apaches, 19, 109, 226, 228, 235, 236, 310, 312, 316; Chiricahua, 301; Cibecue, 301; Coyotero, 301; Jicarilla, 301; Kiowa-Apaches, 235, 241, 301; Lipan, 301; Mescalero, 301; San Carlos, 301; Tonto, 301; White Mountain, 301, 302, 308, 315, 317
Apaches in the Kitchen (blog), 300
Apache-Sitgreaves National Forest, 301, 302
Appropriation Act (1889), 103
Arizona-Sonora Desert Museum, 304, 310
Army Topographical Corporation, 261
Arquette, Dave, 191
Arquette, Mary, 191
Arrowhead Café and Marketplace, 311
Á:se Tsi Tewá:ton Akwesasne Cultural Restoration Program, 31, 65, 82, 191
atocpan (land type), 284
atoctlalli (land type), 284
Atteo, Marlene, 328
Attocknie, Joe, 233
Avian flu, 97
Aztecs, 255

Baca, Angelo, 88
Baca, Karlos, 335
backyard/family gardens, 107–8, 109, 236; Choctaw, 108; connection to soil, 278; Mihesuahs', 109
Bacock, Alan, 67, 85, 87, 89
Bad River Gitiganing Community Garden, 63, 90
Bad River Gitiganing Project, 61, 63, 67, 87, 90
Baker, Lannesse, 67, 74, 89
Ballet White Mountains, 31
Barker, Joanne, 10
Barnes, Carl Leon "White Eagle" (Corn Man), 17, 200, 201
Barreiro, José, 189
Barton, Taelor, 37, 38
Bartram, William, 108
baskets, 42, 95, 258, 259, 260, 263, 272, 273
Battey, Thomas C., 226, 227, 228
Battle of Adobe Walls, 236
Baweting, 39
Bay de Wasie, 39
beans, 17, 30, 34, 35, 79, 108, 178, 188, 189, 202, 204, 206, 207, 226, 227, 234, 244, 276, 277, 279, 287, 291, 293, 296, 298, 299n32, 302, 312, 313, 325; common, 292; cranberry, 31; green, 109; kidney, 302; locust, 302; mescal, 237; pinto, 292; Scarlett runner, 31; screwbean, 263–64; soy, 30; speckled, 196; tepary, 4; zebra-striped, 191
bears, 34, 108, 213, 225, 225, 227, 301
beef, 6, 46, 97, 99, 100, 108, 227, 236, 239, 241, 244. *See also* cattle
beefalo, 100
Begaye, Russell, 181
Bella Coolas, 230
Benedict, Angela, 325
Bergen, Kristopher, 314
Berry, Wendell, 97
Berryhill, Stephanie, 66, 71, 74, 75, 78, 88, 89

Bighorse, Vann, 111
biocultural resource management, 130–32
biodynamic principles, 281
biodynamic soil treatments, 291
bison, 4, 35, 55, 94, 96, 99, 100, 116, 223, 226, 236, 237, 242, 244, 306; ranching, 100
Black Mesa Water Coalition, 61, 65, 84
Blanchard, Paulette, 327
"blundering intruders," 104
boarding schools, effects on student health, 5, 16, 35, 45, 96, 267
Boas, Frank, 234
Bogner, Angela "Tangie," 273
bois d'arc (Osage orange), 226, 240
Bold Oklahoma, 105
Bon Appétit, 309
bone marrow, 228
Boomer Sooners, 103
Boone and Crockett Club, 301
Bosque Redondo, 175, 239
Bourdain, Anthony, 302
Braiding the Sacred and Voices of Maize network, 296n18
Brant, Janice, 191
Brant, Terrylynn, 191
Brascoupe, Clayton, 68, 70, 79, 89
Breen, Sheryl D., 80
Bryce, Cheryl, 111
Buffalo Hump, 232
Bugbee, Richard, 266–67
buglers (elk teeth), 226
Bureau of Indian Affairs, 105, 198, 199, 267, 273
Byrd, Grace Ann, 70, 77, 82, 89

Café Gozhóó, 313–15
Caldwell, Chris, 327
CalFire, 273
California Indian Basketweavers Association, 272

California tribes: concern for land, 257–58; wild plant management, 265–66
Callison, Candis, 323
CalTrans, 273
cancer, 15, 100, 105, 156, 165, 169, 263, 311
Cannon, Carrie Calisay, 27–30
Carlson, Gustav, 229
Carlson, Leonard, 234
Carter, R. G., 227, 228
Catlin, George, 225, 226, 232, 233, 234
cattle, 4, 94, 99, 100, 104, 106, 108, 112, 226, 234, 236, 237, 238, 244, 316. *See also* beef
Cedar Box Teaching Boxes, 44
celebrity chefs, 307
Cestou Necestou (documentary series), 264
Changing Woman, 312
Charnon, Don, 75, 82, 89
"Chefy Phenomenon," 306
chemicals, 162, 164
"Cherokee Ceremonial Tobacco" seeds, 202
Cherokee Nation: bison ranching, 100; Communications Department, 205; heirloom seeds, 199, 208; hunting and fishing rights, 103; Inter-tribal Environmental Council, 105; Natural Resources Department, 199
Cherokee Nation Seed Bank and Native Plant Site, 199, 202–3, 207
Cherokee Nation v. Georgia (1831), 10
Cherokee wild plants, challenges in growing, 205, 207
Cheyenne River Youth Project, 61, 63, 71, 81
Cheyennes, 96, 100, 109, 236; foods used by 234; physicality, 234, 235
Chia Café Collective, 268, 269
Choctaws: definition of traditional foods, 98–99; financial assets, 100; hunting licenses, 103; loss of resources, 104–5; mining towns, 242; post-removal gardens, 108–9;

Choctaws (*continued*)
  poverty among, 100; Promise Zone awarded, 100–101; Regional Medical Clinic, 102; residents in Choctaw Nation, 98; roasting-ear patches, 109; Small Business Development Services, 101; traditional foods, 108
choices (important to food sovereignty), 75
Christianization, 170
Ciocco, Anthony "Chako," 107
City of Albuquerque v. Browner (1996), 52
Civil War, 96, 99, 241, 242
Clean Air Act, 51
Clean Water Act, 51
climate change: effects of, 5, 6, 9, 19, 51, 73, 97, 103, 162, 163, 164, 166, 206, 207, 288, 320–33, 344
Climate Change Strategic Plan of the Confederated Salish and Kootenai Tribes, 51
coal companies, in Indian Territory, 242
Cobb, Amanda, 87
Cochiti Youth Experience, 61, 69, 90
Códice Badiano ("Little Book of the Medicinal Herbs of the Indians," 1552), 283
collective continuance (way of understanding), 321
collective responsibility (Hawaiʻi), 130
College of Menominee Nation Sustainable Development Institute, 327
Colombi, Ben, 328
colonization, 5, 9, 12, 15, 27, 40, 45, 79, 94, 128, 162, 163, 225, 309
colonizers, desire for land, 257–58
comadrazgo (ritual kinship relations), 277
Comanche Nation Diabetes Awareness Program, 245
Comanche Nation News, 110, 245
Comanchería, 109
comancheros, 234

Comanches (Nʉmʉnʉʉ): bands, 223, 224; as "Cossacks of the Plains," 224; as farmers, 241; Fish Eater Band, 228; Kotsotaka Band, 224; lack food initiatives, 109–10; as "Lards of the Plains," 245; as "Lords of the Plains," 224; as "Lords of the Southern Plains," 224; modern diet of, 110; Nokoni Band, 224; Peneteka Band, 224; physical description of, 232–35; Quahada Band, 224, 228; raiders, 109, 224, 226, 232, 233, 234, 236; as "Spartans of the Plains," 224; Taninuu Band, 224; Tenewa Band, 224; as "Terror of the Southern Plains," 224; traditional diet, 223, 225; Yamparika Band, 224
Comanches: Lords of the South Plains, The (Wallace and Hoebel), problems with research, 227
commodities, 97, 98, 306
Community-Based Subsistence Fishing Area, 140–41
Community Wellness Development Projects (Navajo), 179, 181, 184, 185
community gardens, 30, 57, 63, 69, 77, 90, 107–8, 245. *See also* milpas; Three Sisters
companion planting, 282, 283, 286–87, 291
compost, 278–79
Conoco Phillips Refinery, 105
Continental Network of Indigenous Women of the Americas, 311
control (importance to food sovereignty), 75–76
Cook, Katsi, 189
Cook, Roger, 191
Cook, Tom, 65, 89
Coppola, Francis Ford, 307
corn, 4, 6, 16, 17, 30, 31, 34, 35, 49, 58, 59, 65, 66, 69, 70, 73, 74, 78, 79, 80, 81, 83, 97, 98, 99, 108, 109, 110, 115n25,

120n81, 179, 188, 189, 191, 192, 200, 201, 202, 204, 206, 208, 209–22, 226, 227, 233 236, 238, 240, 244, 258, 276, 277, 278, 279, 282, 283, 287, 291, 292, 296, 298, 299, 302, 307, 309, 312, 313, 316, 317. *See also* Haudenosaunee; Hopi corn
Cornelius, Dan, 73, 86
Corn Mothers (Hopi), 219
corn-squash-bean triad, 279, 283
Corntassel, Jeff, 111
corporate food systems, 158
Country Club at D.C. Ranch, 302
Coward, Rosalind, 307
Craig, Ari (daughter), 311
Craig, Bob E. (grandfather), 311
Craig, Jandi (spouse), 310, 311
Craig, Kaia (daughter), 311
Craig, Mariddie Ann (mother), 311
Craig, Nephi: children, 311; culinary education, 302–3; philosophies of, 304, 306, 307, 312, 313, 314–17; seeds, 313; signature desert, 304; sobriety, 310; work experience, 302–3
Craig, Rowdy Nephi (son), 311
Craig, Tawny Cassia (daughter), 311
Craig, Vincent (father), 311
Crespí, Fray Juan, 257
crop nutrient density, 281
Cross Timbers, 224, 226, 234
Cruikshank, Julie, 293n3
cuenchihu, 284
Culhua Mexica, 282
"culinary truths," 308
cultural genocide, 156
"culturally appropriate foods," 64, 65, 167, 170, 188, 340

Dahl, Michael, 77, 89
Davis, Anne Walendy, 243
Davis, Heather, 332
decolonization, 11, 33, 289, 308, 310
Declaration of Nyéléni, 7, 95

Decolonizing Diet Project, 40–42
deer, 28, 34, 35, 51, 99, 103, 104, 106, 108, 109, 203, 206, 225, 226, 227, 240, 241, 256, 257, 301, 326. *See also* venison
deforesting landscapes, 322
degradation of ecosystems, 159
de la Cruz-Badiano, Martín, Codex, 283
Deloria, Vine, Jr., 87
dentists, Indian Territory, 241
de Portolá, Gaspar, 257
de Sahagún, Bernardino, 282
Deyo Mission, 240
diabetes, 4, 6, 7, 15, 16, 17, 28, 32, 33, 37, 40, 44, 44, 63, 64, 70, 71, 97, 102, 109, 110, 156, 165, 169, 170, 173, 174, 223, 241, 242, 243, 244, 245, 252n110, 269, 309, 324, 350; false ads for cure, 241
Diabetes Multi-resource Task Force, 102
Diamond Pipeline, 105–6
"diet transition," among Alaskan Natives, 156. *See also* cultural genocide
Diné Community Advocacy Alliance, 15, 174, 181
Dow AgroSciences, 97
Downstream Casino, 100
Drake, Barbara, 267, 269
Drake, Sir Francis, 257
Dream of Wild Health (farm), 48, 64, 72, 76, 90
Dry, Bradley James, 38
dry farming, 17, 65, 209, 211, 342
Dumas, Jane, 267
DuPont/Pioneer Co., 97
Dutch elm disease, 106
Dził Łigai (White Mountain), 317

E Alu Pū (network), 15, 122, 123, 124, 125, 129, 130–31, 132, 133, 140, 141, 144, 145, 148, 149n16
earthquakes, 105
Eastern Oklahoma State College (Ada), 101

education (importance to food sovereignty), 14, 81–83
elk, 28, 99, 106, 225, 226, 227, 256, 301, 304, 313
El Three Points, 277
entrails, 228, 241; *quetucks* (Comanche), 241
environmental changes: Alaska, 164; Hawai'i, 135, 143, 146; Oklahoma, 104
Environmental Department of the Pala Band of Mission Indians, 272
environmental law, importance of, 51–53
"environmental racism," 105
ethanol, 49
ethnobotany, 281
Ethnobotany Project: Contemporary Uses of Native Plants (Ramirez and Small), 270
ethnoecology, 266, 272, 295
ethnoedaphology, 281
Euchee Language Project, 66
Evergreen State College's Reservation-Based, Community-Determined Program, 36
ethnopedology, 281, 284, 191
Euchee Butterfly Farm, 107

factionalism, 36, 95, 112
Fairbanks, Ala., 73, 89
Fallin, Mary, 105–6
family gardens, 107–8. *See also* backyard/family gardens
farm: as defined by Europeans, 254; defined by Native Californians, 254
farming, among Five Tribes, 101, 107, 108, 180
farming tools, 99
Feast of First Fruits, 108
Ferguson, Angela, 191
First Nations Development Institute, 57, 58, 296n18
First Nations Food Taster, Northern Michigan University, 40

fisheries management, in Hawai'i, 140, 144–45
fishing, 5, 11, 13, 14, 46, 81, 85, 99; in Akwesasne, 5–6; in Alaska, 161, 162; in Anishinaabemowin, 39, 40; in Arizona, 303; Hawaiian laws, 133, 139, 140, 145, 151n25; Hawai'i traditions, 152n39, 123; in Oklahoma, 103
fishponds, in Hawai'i, 144, 146
food: as foundational component of culture, 64–67; meanings of, 320–21
food desert, 20, 97, 173, 306
food distribution program, 6, 306
food safety laws, 116n31
food insecurity, 5, 6, 31, 326
food porn, 307
food security, 100
food sovereignty, 166, 264, 267, 274, 289, 308; achieving, 4, 14, 291, 292, 326, 329, 336; complexity of, 85–86, 307, 327; components of 62–85, 88; definitions of, 7–10, 11, 94–95, 320, 321, 332, 337; literal meaning of, 95; principles of, 12
Food Sovereignty is Tribal Sovereignty (FSiTS; Facebook group), 36
Food Symposium and Celebration of Basketry and Native Foods Festival (Tucson), 304
food systems, responsibilities of, 329
food writers: attraction to chefs, 308; ignore tribal issues, 308; romanticize Native foods, 18, 307
Fort Massachusetts (Fort Garland), 280
Fort Sill, 17, 223, 225, 227, 228, 234, 235, 236, 237, 238, 239, 240, 243
Fort Sill Indian School, 240
Forum for Food Sovereignty (2007), 7
fossil fuels, 322
fracking, 104–5
Frank, Lois Ellen, 304
Frank, Billy, Jr., 325

Freedom School (Mohawk), 82
Frémont, John Charles, 261
fry bread, 40, 53, 55, 98, 99, 173; myth of, 239
Fundamental Law of Diné, 177

Garcia, Michelle, 271
garden flowers, 291, 279
gardens: challenges to creating, 110; "Learn to Grow," 110; modern Cherokee, 198–99; vegetables, 109
Garreau, Julie, 63, 71, 75, 77, 81, 84, 89
General History of the Things of New Spain (Sahagún), 282
General Motors, 325
genetically modified organisms (GMOs), 8, 16, 32, 80, 84, 97, 111, 161, 208
genetic purity, 202
Gitche Gumee, 50
Glidewell, Keith, 74, 89
glyphosate, 49
Goldstein, Marcus, 233, 235
Goyanthlay (Geronimo), 301
Grand Canyon, 27, 28, 221
Great Mystery (Hopi), 209
greenhouse gases, 322
greenhouses, 35, 58, 101, 109, 180, 187, 312
Greenland, 322
Grounds, Richard, 66, 89
Great Depression, 241
Great Salt Lake, 224

Hahamongna Nursery, 270
Hale, Jonathan, 181
Hämäläinen, Pekka, 232
Hamilton, Annie, 268
Hannagan Meadow, 302
Harvey, Clayton, 312
Haskell, Charles N., 103
Haudenosaunee: community, 33, 79, 187, 188, 189,190, 194; Confederacy, 32, 189, 190; conflicts with Euro-Americans, 190–91; cornfields, 190, 196; Seedkeepers Society, 16, 191; Task Force on Environment, 191; and Three Sisters, 30, 281, 292–93n1, 325
Hawai'i: Department of Land and Natural Resources, 132, 139; language revitalization, 129; movement building, 127; Natives' origin stories, 136; Office of Hawaiian Affairs, 129; overthrow of kingdom, 128; post-overthrow achievements, 150n19; resource management, 138, 139; social-ecological zones, 134–35; and traditional ecological knowledge (ETK), 129, 134
health problems: among American Indians, 6–7; concerns about, 62–63. *See also* diabetes; heart disease; high blood pressure; obesity
Healthy Dine Nation Act (2014), 174, 178, 182; opposition to, 177–82
heart disease, 37, 40, 64, 97, 100, 151, 156, 165, 169
heirloom seeds, challenges to integrity of, 111
heritage seeds, 13, 30, 62, 79, 196
high blood pressure, 15, 17, 223, 243, 245, 309
Hill, Lilian, 71, 76, 80, 89
Hinton, Amos, 70, 74, 76, 78, 83, 89
Hoebel, E. Adamson, 227, 228, 233, 237, 247n22
Hōkūle'a, voyage of, 129
Holt-Giménez, Eric, 100
Hoover, Elizabeth, 10, 30–31
Hopaq (valley), 210
Hopi corn: in ceremonies, 220–22; challenges to growing, 218; harvest of, 218–19; preparation of 219–20; sacredness, 211–20

Hopis: clans of, 210, 214, 213, 220, 221, 222; corn farming, 214–18; Farmers Market and Exchange, 71; kiva chiefs, 215–16; mesas, 210, 211; migration period, 213; Natwani Coalition, 71; origin story, 210–11; women's roles, 215–20
Hopi Tutskwa Permaculture, 61, 71, 76
Hopi villages, 212, 214, 218, 220; Hotevilla, 210, 212, 216; Kykotsmovi, 211; Oraibi, 213; Paaqavi, 211, 217
Horinek, Mekasi, 105
horses, 99, 108, 217, 224, 225, 226, 232, 236, 240, 278
Hoskins, Chuck, 70, 76, 84, 89
House Bill 1123 (Okla.), 105
Hualapais: Cultural Center, 29; Ethnobotany Youth Project, 27, 29
Huambachano, Mariaelena, 327
Hocabá, Yucatan, Maya farmers in, 288
Hugelkultur ("hill/mound culture"), 283
Hummingbird, Nicholas, 270
hunter-gatherers, 156, 163
hunters, 15, 58, 96, 99, 161, 225, 236, 243–44, 245, 301, 312, 326
hunting: regulations, 163, 164; rights, 103–4
hybridization, 352

ICE (US Immigration and Customs Enforcement), 38
I-collective (organization), 38
independence (component of food sovereignty), 13, 62, 69–73
"Indian food," 98
Indian Health Service, 102
Indian removal policy, 95
Indigenous agroecologies, 289–91
Indigenous climate justice, 331–32
Indigenous Eating (Facebook), 36, 304
Indigenous Farming Conference at White Earth, 31, 39

Indigenous Food and Agriculture Initiative (University of Arkansas), 112
Indigenous (traditional) foods/animals, fauna: armadillo, 225; bear, 34, 108, 213, 225, 226, 227, 301; bighorn sheep, 301; bream, 108; catfish, 34, 106, 108, 109, 113, 301; crow, 235; earthworm, 279; frog, 225; goose, 108; ladybug, 279; lizard, 225; milkfish, 146; mountain lion, 301; mule deer, 301; mullet, 143, 146; quail, 35, 106, 109, 241, 256; rabbit, 33, 46, 104, 108, 240, 256, 262, 266; raccoon, 108; rat, 225; salmon, 325, 328, 329; seabirds (Hawai'i), 135; skunk, 225; snakes, 225, 241, 287; squirrel, 108; sulfur butterfly, 48; sunfish, 301; trout, 108, 301; turtles, 65, 106, 108, 228; white tailed deer, 301; wolf, 227, 241, 301. *See also* antelope; deer; elk; turkey
Indigenous (traditional) foods/plant tools, flora: amaranth, 304; American lotus, 231; blackberries, 108; bois d'arc, 226, 240; breadroot, 231; calabacita, 292; camas lily, 229; Cherokee purple tomato, 199; Chickasaw plum, 230; chinquapin, 108; chokecherries, 230; cholla cactus bud, 269, 302; clover, 231; cottonwood bud, 302; crappie, 106; dandelion, 302; Devil's claw, 277; Devil's shoestring, 113; datura, 286; elderberry, 262–63; fall plum, 230; grapes, 82, 99, 108, 225, 226, 229, 257; ground cherry, 302; hackberry, 229; hawthorn, 230; hickory nut, 34, 99, 108; hollyleaf cherry, 259–60; huckleberry, 42; Indian bread root, 231; Indian potato, 264; Indian rice grass, 302; junka, 273; kipuka (Hawaiian tree), 132; koa (Hawaiian tree), 135; locust beans, 302;

mangrove (Florida), 146; Manzanita berry, 28; marigold, 287; mesquite, 28, 226, 262–64, 277, 302; Mexican marigold, 279; milkweed, 49, 50, 107, 302; mulberries, 99, 108, 109, 225, 230; muscadines, 229; mushrooms, 45, 108; nopalitos, 277, 278; oak, 258; pandanus, 136; papalo, 287; pecans, 34, 99, 108, 110, 231; perch, 106; persimmons, 108, 225, 230; piki bread, 219; pineapple, 143; pine nut, 260–61; piñon, 179, 304; poi, 142; possum grape, 229; potato, 32, 34, 99, 108, 109, 178, 264, 286; prairie turnip, 231; 135, 229; prickly pear, 28, 225, 231, 277; pumpkin, 108; rattlesnake master, 231; red sumac berry, 179; rivercane, 207; sand plum, 230; screwbean, 263–64; serviceberry, 229; snakeroot, 113; stinging nettle, 282; sumac, 28, 178, 179, 231, 273, 302; sunchoke, 231; sunflower, 31, 108, 188, 231, 302, 317; sweet acacia, 277; sweet potato, 98, 108, 229; timpsala, 45; tree bark, 225; walnut, 34, 99, 108, 231, 302; water nettle, 283, 287; water parsnip, 302; white sage, 264; wild cherry, 259–60; wild plum, 108, 225, 226, 230; yarrow, 283, 287; yucca, 261–62; yucca pods, 302. *See also* acorns; beans; corn; squash

Indigenous Food Lab, 47
Indigenous Foods TEKnology in Great Lakes Region, 41–42
"Indigenous food systems," 6, 11, 47
Indigenous Healthcare Improvement Act, 102
Indigenous Permaculture Convergence at Woodbine Institute, 296n18
Indigenous Planning and Design Institute, 327
Indigenous Seed Keepers Network, 65, 91n7

Institute for Tribal Environmental Professionals, 209, 210
intercropping, 291
Intertribal Agriculture Council, 57, 58, 112
Inter-tribal Environmental Council, 105
invasive species: bark beetle, 271, 302; bighead carp, 106; Bradford pear, 106; eastern red cedar, 106; golden algae, 106; golden spotted oak borer beetle, 271; killer bees, 48; mangrove (Hawai'i), 146; musk thistle, 106; pigweed, 302; poison hemlock, 106; *Sericea lespedeza* (legume), 106; wild boars, 106; zebra mussels, 106
Ivins, Joseph C., 302

Johnson, Terrol, 7

Kaleohano, Kupuna Smith, 138
Kamal, Asfia Gulrukh, 12
Kanatsiohareke Mohawks, 30
Kanenhi:io Ionkwaienthon:hakie, 30, 58, 63, 73, 191
Karuk TEK and Knowledge Sovereignty, 332
katsinas, 214, 220
Kawagley, Oscar, 163
Ke'anae (community), 143
Kiamichi Technology Center, 101
Kiiwetinepinesiik Stark, Heidi, 86
Kimmerer, Robin, 328
Kingi, Wikuki, 327
Kiowas, 96, 109, 226, 227, 235, 236, 237, 241
kitchen gardens, 278. *See also* backyard/family gardens
Kloppenburg, Jack, 78, 80
Knight, Edith, 38
Komlos, John, 234
Krenn, Caitlin, 77, 89
Kroeber, Alfred, 255, 259
Kully Chaha, 35

Kumeyaay Community College, 272
Kuwanwisima, Leigh, 217–18, 222

LaDuke, Winona, 10, 68, 71, 87, 89
Laffrey, Shirley C., 243
LaFrance, Brenda, 188
Lake Casa Blanca State Park, 277
Lakotas, 45, 46, 69, 71, 73, 235, 323
landscape restoration, 4
lard, 34, 55, 98, 110, 241, 242, 245
Laredo, Texas, 277
Largo, Donna, 267
Latinized Nahuatl narratives, 282
La Via Campesina, 7
"Learn to Grow" (Cherokee project), 110
Lehmann, Herman, 228
Leupp, Francis E., 238
Little Beaver Creek, 106
Little Earth Housing Project, 73
liver, 224, 228, 241
Llano Estacado, 224
Loewen, Dawn, 72
Long Walk, 55, 175, 196

Mackenzie, Ranald Slidell, 236
Magical Antelope Surround, 225
Maheux, Emily, 312
maize, 292–93. *See also* corn
Malki Museum, 269
Marcy, Randolph B., 226, 232, 237
Margarita (grandmother), 276, 277, 278
Mariana, Nancy, 311
Marshall, John, 10
Martinez, Dennis, 328
Martinson, Pati, 73, 77, 89
Mary Elaine's (restaurant), 302, 306
Masaw, 210, 211, 213, 214, 221–22
Masayesva, Victor, Sr., 211, 216, 217, 218
Masayesva, Virgil, 210
Masayesva, Zetta, 216, 219
Mashkiikii Gitigan Medicine Garden, 61, 67, 74, 76

Mayas, 255
McCarthy, Daniel, 265, 270, 271
McClain, 67, 70, 87, 89
McClellan, Sara, 201–2
McComber, Steve, 191
McGregor, Davianna, 126
McHorse, Angela, 69, 78, 89
McMichael, Philip, 9
McNaughton, Leticia, 31–33
Medicine Bluffs Creek, 240
Meriam Report, 240
Mesoamerica, 282
Metoxen, Jeff, 74, 81, 84, 88, 89
Mexica (Aztec) glyphs, 282
Michelin Star, 306
Mihesuah, Carrie, 241
Mihesuah, Devon, 33–36, 40
Mihesuah, Henry, 228, 233, 240, 251n91
Mihesuah, Joshaway, 240–41
Mihesuah, Topachi, 240
Mihesuah family allotment, poachers on, 106
milpas, 279, 281, 282, 283, 286, 287, 290, 293, 296, 298
Minnesota Museum of Science, 65
Mino Bimadiziiwin (Creator's instructions), 68–69
Mitchell, Donald Craig, 163
mites, 279
Mobile Farmers Market Reconnecting the Tribal Trade Routes Roadtrip (2014), 73
Mohawk, John, 33
monarch butterflies, 48–49; decline of, 49
Monarch Watch, 48–49
Monsanto, 76, 97, 80
Moratto, Michael J., 265
Morgan, Gayley, 77, 89
Morongo Reservation, 269
Morrison, Dawn, 12
mountain/mountains: Baldy, 301; Blanca (Tsisnaasjini'), 304; Dził

Łigai, 301; Hesperus (Dibé Nitsaa), 304; Rainier, 42; San Francisco Peaks (Doko'oosliid), 302, 304; Sugar Loaf (Nvnih Chufvk), 35, 104; Taylor (Tsoodzil), 304; Tehachapi Range, 260; White, 301; Wichita, 226, 229, 240

Mt. Pleasant, Jane, 191

Muckleshoots, 43

Mutton Man, 311

Mvskoke-Creeks: community gardens, 107–8; traditional foods, 108

Mvskoke Food Sovereignty Initiative, 61, 66, 107

Naabik'íyáti' (cordial discussion), 174

Nā Kua'āina: Living Hawaiian Culture (McGregor), 126

Narragansett Food Sovereignty Initiative, 61, 66

National Environmental Policy Act, 199

National Fish and Wildlife Foundation, 107

Native American Culinary Association (NACA), 300, 303–4

Native American Graves Protection and Repatriation Act (NAGPRA), 195

Native Food Systems Resource Center, 112

Native Infusion: Rethink Your Drink (campaign) 43–44

Native Rehabilitation Programs, 312

Navajo Code Talkers, 311

Navajo Nation: Council Chamber, 176; Division of Community Development, 179, 181; Tax Commission, 177, 181–82

Ndée Bikíyaa (People's Farm), 312

Nelson, Melissa, 73, 76, 83, 89

New Mexico Acequia Association, 296n18

Nez, Jonathan, 179

Nisqually Community Garden, 61, 70, 77, 82

Nocona, Peta, 233

Nohwike' Bagowa (cultural center and museum), 319n22

nontraditional foods/animals: apple, 108, 302; apricot, 302; banana, 136; bitter yam, 136; breadfruit, 136; chamomile, 286, 298n23; cherry, 108; chestnut, 108; chicken, 30, 108, 277; Chinese bush clover, 106; chitterling, 241; cilantro, 276; curdled milk, 228; elephant ear (taro), 135–36, 146; goat, 30, 99, 108, 237, 278, 353; hog, 88, 99, 100; honey bee, 48; kidney bean, 302; lamb's quarters, 283, 286; laurel, 286; milk thistle, 106; mule, 99; peach, 108; pear, 108, 302; pea, 108; plum, 99, 108, 302; sheep, 17, 99, 109, 226, 237; sorghum, 99; sugarcane, 143; sweet pea, 276; tamarisk, 106; Texas longhorn, 236; water chestnut, 302; yam, 136. *See also* cattle; beef; horses; wheat

NordGen, 199

North American traditional Indigenous food systems, 47

Northern Arizona University, 209, 302

Northwest Passage, 257

Nutlouis, Roberto, 65, 78, 81, 84, 90

nutritional colonialism, 161–62, 163, 166, 170

Nuu-chah-Nulth, 323

Nvnih Chufvk, 35. *See also* Sugar Loaf Mountain

Oceti Sakowin camp, 55

Ochantubby, 33

Office of the High Commissioner for Human Rights (UN), 311

Oglala Lakota Cultural and Economic Revitalization, 296n18

Oherokon Rites of Passage, 31
Oklahoma: Coalition to Defeat the Diamond Pipeline's Oka Lawa Camp, 106; Department of Environmental Quality, 105; Department of Health, 97; first game laws, 104; fracking, 104–5; invasive species, 106; poaching, 105–6; pollinators, 106, 107; protesters, 105–6; racism in, 103; ranchers, 106; socioeconomic complexities, 95, 103; water pollution, 105
Oklahoma State University, 101
Oklahoma tribes, challenges to revitalizing food traditions, 95–113
Oleksa, Michael, 167
Omega-3s, 156, 165, 287
onions, 34, 99, 108, 225, 229, 230, 231, 262, 302, 309, 315; sweet, 230; twin crest, 302; wild, 108, 225, 229, 230; wild onion dinners, 230
O'odham Solidarity Project, 4
Oraibi (village), 213
Original Woman, 188
Osages: colonial foods, 98; environmental damage, 105; farms, 110–11; food desert, 97; health conditions, 97; oil discovered on land, 103; physicality, 234; poverty among, 97

Pahayuko (Comanche), 232
Paige, Zach, 75, 84, 90
Palestine, 38
Palmer, Parker, 127
Papakilo Database, 150n22
Parker, Cynthia Ann, 233
Parker, Quanah, 233
Parry Pinyon Pine Protection Project, 271
Patel, Raj, 94
Paya, Jorigine, 28
Peaches, Shalitha, 313

Pele (goddess), 126
Perkins, Kenny, 63, 65, 82, 90
permaculture, 46; principles, 289–90
Permaculture Guild, 296n18
Permaculture Project, 61, 67, 85
pesticides, 107, 162, 272, 297, 230
pests, 97, 108, 205, 218, 336. *See also* invasive species
Phoenix Indian Center Silver and Turquoise Ball, 303
Phoenix Indian School, 311
Pine Ridge Reservation, 45
Pink, Willie, 259
Pinyon Flat Campground, 271
Plains tribes, 233–34
plant polyculture, 282, 283; milpas, 281. *See also* Three Sisters
Poepoe, Uncle Kelson "Mac," 125, 133, 140
politics, as challenge to food sovereignty, 9, 95, 112, 202, 204, 205, 207–8, 280, 330, 332, 336
pollinators, 107; gardens, 14, 48–50, 104, 106, 107, 108, 109, 111, 206, 277, 291, 313
porcine epidemic diarrhea virus, 97
pork, 46, 97, 98, 99, 108, 230, 236, 241
potlatches, 323
Powamu (winter solstice), 220
Powless, Dan, 63, 90
Powskey, Malinda, 29
Prechtel, Martin, 196
prediabetes, 242
produce, rising prices of, 97
Promise Zone Award, 100–101
protesters, potential punishment of in Oklahoma, 105–6
Pruitt, Scott, 105
Pueblo Revolt (1680), 224
Puget Sound, 42
P'urhépecha campesinos, 285

Quakers, 226

Quapaw Cattle Co., 99–100
Quapaw mercantile, 99–100
Quapaws: and restaurants, 100; traditional foods, 99; food initiatives, 99–100
quauhtlalli (rotten-wood soil), 282
Quintano, Ester, 267

Radford, Hope, 112
rain barrels, 109
Rainbow Treatment Center, 312, 317
rain-fed agriculture, 328
Randlett, Agent James F., 238
Reader, Tristan, 7
Red Lake Reservation, 48
Red Willow Farm, 61, 69
Red Wind Foundation, 267
Reed, Brit, 36–39
Reinhardt, Martin, 39–42
relationships, 328–30; in Hawai'i natural world, 128; Three Sisters, 330; traditional in Hawai'i, 136–37
relocation, climate change induced, 331
rematriation, 186, 194, 195
repatriation, 194, 195
reptiles, 226, 227, 262
Reservation-Based, Community-Determined Program, Evergreen State College, 36
resilient co-inhabitation, 292
rewilding ecosystems, 290
Rickard, Norton, 191
Rico, David, 38
"Rights of Nature," 105
Rio Culebra acequia commonwealth, 280
rivers, 42, 77, 128, 188, 255; Arkansas, 99, 105, 224, 226; Canadian, 224, 266; Colorado, 213; Illinois, 105; Pecos, 226; Red, 224, 226, 235; Rio Grande, 213; Sabine, 224; in Sangre de Cristos, 279; San Luis Rey, 257; Spirit Mountain, 28; St. Lawrence, 325; St. Marys, 39; Trinity, 224; Washita, 226, 235
Riverside County Fire Department, 273
Riverside–San Bernardino County Indian Health Service, 273
Roads of My Relations (2000), family gardens in, 109
roasting-ear patches, 108–9
Romero, Jayson, 69, 90
Rosemont Mine, 4
Rush Springs, Okla., 106

Saint Paul College (St. Paul, Minn.), 53
Sampson, Don, 327
San Bernardino National Forest, 264
Sangre de Cristo Acequia Association, 296n18
Sargent, Josh, 73, 90
Schaefer, Marie, 327
Schumacher, Paul, 262
Science Museum of Minnesota Native American exhibition, 17, 202
Scottsdale Community College, 320
Seattle Culinary Academy, 37
"Seed Commons," 194
seeds: pepitas, 292; personal connections to, 277–78; plants' memories, 276; as "Vibrant Matter," 293–94n3, 276
seed saving, 14, 16, 17, 95, 112, 186–97, 294, 336, 337
Seeds of Native Health (institute), 112
seed songs, 197
Seed Sovereignty Assessment Toolkit, 193
Segrest, Valerie, 42–44, 64, 66, 74, 75, 81, 84, 85, 90
Senior Farmers' Market Nutrition Program, 101–2
Sequoyah Fields Fuels Corporation, 105
Sevier Complex, 224
Shánah daniidlįįgo as'ah neildeehdoo (Navajo slogan), 174

sharecroppers, 240
Shelly, Ben, 177, 178, 184
Sheridan, Philip, 235
Sherman, Sean, 44–48, 53, 71, 90
Sherman Indian School, 267
Shimek, Bob, 65, 68, 77, 90
Shirley, Michaela, 327
Shoemaker, Scott, 61, 64, 72, 90
Show Low, 308
Sierra Seeds Cooperative, 61, 73, 79
Simpson, Danny, 174, 181, 184
"Sioux Chef," 46
Sioux Chef's Indigenous Kitchen (Sherman and Dooley), 46–47
Sisquoc, Lorene "Lori," 267
Siva, Alvino, 269
sixth mass extinction, 288
Skibane, Alex Tallchief, 52
Sleeper-Smith, Susan, 324
Slim Buttes Agricultural Development Program, 61, 65, 69, 76
Sky Harbor International Airport, 208
Sneed, R. A., 227, 238
soil: mindfulness toward, 276–77; types, 285
Somiviki (Hopi), 219
sovereignty, 4, 185, 254, 272, 322, 324; meaning of, 10–11, 111; seed, 187, 188, 190, 193, 194, 195, 196
spearfish, 45
Spears, Cassius, Sr., 66, 82, 90
squash, 4, 17, 30, 34, 35, 108, 109, 110, 115n25, 120n81, 188, 189, 201, 202, 204, 206, 226, 234, 244, 276, 279, 283, 291, 292–93n1, 302, 304, 305, 315, 325
Standing Rock, 54, 55
St. Arnold, Jim, 40
Steckel, Richard, 234, 235
Steiner, Rudolf, 281
storied landscapes, 299
St. Regis Mohawk Climate Change Plan, 332
subsistence plots, 288

sugar, 6, 7, 15, 17, 32, 40, 44, 46, 47, 77, 96, 98, 99, 110, 143, 175, 176, 178, 227, 233, 236, 241, 242, 244, 263, 269, 277
Sugar Island, 39
Sullivan Expedition, 190, 196
Summit Restaurant, 303
Sunrise Park resort, 303
"survivance," 295n12
Svalbard Global Seed Vault, 199
Swinomish Indian Tribal Community, 51–52
Sycuan Band of the Kumeyaay Nation Community College, 272
Syngenta, 97

Talking Stick Resort, 303
Talley, Jared, 323
tamfula (Choctaw dish), 34
Taste of Native Cuisine (collective), 335
Tatanka Truck, 46
Tauli-Corpuz, Victoria, 327
Tawahquenah, 232, 233
Taylor, Chip, 48–50
Temporary Assistance to Needy Families (TANF), 269
Tending the Wild (Anderson), 256
Tesuque Farms Agricultural Initiative, 296n18
Thomas, Romajean, 65, 72, 77, 82, 90
Three Sisters, 30, 115n25, 120n81, 281, 292–93n1, 325, 328, 330
tiangüis (markets), 283
Tocabe (restaurant), 72, 89
Todd, Zoe, 332
Tohono O'odham Community Action, 3, 60, 61, 67, 89
Tomcat Bakery, 38
topography, 292–93; and frost, 292; infiltration, 292; saturation, 292
Torres, Craig, 269
Tova, George, 63, 90
Traditional Ecological Knowledge (TEK), 42, 43, 51, 129, 256, 272

traditional foods, 156–61; definitions of, 98–99; as modern restaurant "trends," 307
Traditional Foods of the Muckleshoot Kitchens, 43
Traditional Native American Farmers Association, 16, 61, 68, 296n18
Trail of Tears, 37, 196, 199
treaties: broken, 103, 104; as tools for environmental protection, 51–52
Treatment as State (TAS) provisions, 52
Treaty of Dancing Rabbit Creek, 33
Treaty of Medicine Creek, 44
Treaty of Point Elliot, 44
treaty rights, 103
Treaty Rights at Risk Initiative, 324–25, 328, 332
Tribal Alliance for Pollinators (TAP), 50
Tribal Environmental Action for Monarchs (TEAM), 49–50, 107
Tribal Forest Protection Act, 270
"tribelet," 255
tribes/nations: Acjachemen, 268; Arikara, 194; Big Pine Paiute, 35, 61, 85; Blackfeet, 235; Cahuilla, 255, 256; Cheyenne and Arapaho, 100; Chickasaw, 98, 107; Citizen Band Potawatomi, 107, 331; Delaware, 99; Eastern Band of Cherokee Indians, 17, 201; Eastern Shawnee, 107; Huichol, 45; Iowa, 100; Isleta, 52; Karuk, 325–26, 329; Kumeyaay, 255; Luiseño, 255; Lummi, 37; Maricopa, 116n31; Miami, 107; Modoc, 100; Nambé, 63, 90; Oneida, 31; O-Pipon-Na-Piwin Cree, 12; Paipai, 28; Paiute, 194; Pascua Yaqui, 4; Pawnee, 105, 194, 226; Pawnee Pict, 226, 232; Pima, 116n31; Pomo, 255; Ponca, 105; Pueblos, 194, 245; Sault Ste. Marie Tribe of Chippewa Indians, 50; Seminole, 96; Shoshone, 224, 235; Standing Rock, 5, 87; St. Regis Mohawk, 325; Swinomish Indian Tribal Community, 51–52; Sycuan Band of the Kumeyaay Nation, 272; Tesuque, 16, 61, 70, 77, 89; Tongva, 268; Ute, 226
"Triple Crisis," 9
Trosper, Ronald, 328
Trump, Donald, 102, 105
Tsyunhehkwa (Life Sustenance), 57
Tulalip Health Clinic, 37
turkeys, 30, 34, 35, 46, 51, 99, 103, 104, 106, 108, 109, 227, 240, 241, 259, 301
Turner, Nancy, 72, 104
Turtle Island (North America), 276
Tutsayatuhovit, 232, 233

"uncle," in Hawaiian, 148n3
unhealthy foods, promotion of, 98, 101
United Nations Food and Agriculture Organization, 8
United Nations Special Rapporteur on the Rights of Indigenous Peoples, 326–27
United States Global Change Research Program, 320, 324
United States v. Washington, 52
University of Kansas, 110

venison, 33, 40, 46, 108. *See also* deer
"vibrant matter," 276, 293–94n3. *See also* seeds

wage economy, as counter to Native values, 159, 168
Wah-Zha-Zhi (Osage), 110–11
Wallace, Ernest, 227, 228, 233, 237
Walmart Foundation, 112
Walters, Karina, 37
wao akua (sacred forest), 135
Warner, Elizabeth Kronk, 50–53
Washington, George, 190
water nettle (*Atzitzicaztli*), 283
water pollution, in Oklahoma, 105

Watso, Lori, 74, 90
Watt-Cloutier, Sheila, 322–23
Welch, Kevin, 201, 202
Werowocomoco (restaurant), 307
West, Kelly M., 241–45
wheat flour, 32, 55, 97, 98, 99, 242, 258, 282, 302
Whirling Thunder Garden Project, 61, 69, 82
White Earth Land Recovery Project, 58, 61, 65, 71, 75
White Earth Reservation, 48
White Mountain Apache Reservation, 301–2
White Mountain Apaches, traditional foods of, 302
White, Rowen, 16, 17, 73, 79, 80, 84, 88, 90
Whiteriver, Ariz., 308
Whitewater, Walter, 304
whooping cough, 237
Whyte, Kyle, 4, 11
Wichita Mountains National Wildlife Refuge, 226
Wiggins, Mike, 77, 90
Wilson, Diane, 64, 72, 76, 90
wimi (specialized knowledge), 210
Winters v. United States (1908), 52
Wise, Michael, 113
Wissler, Clark, 235
Witgen, Michael, 324
W. K. Kellogg Foundation, 112
Wo, Uncle Henry Chang, Jr., 132–33
Wolfgramm, Tania, 327
Working Group on Indigenous Food Sovereignty, 12

Yappalli Choctaw Road to Health, 37
Yazzie, Brian, 53–56
Yellowhair, Milo, 69, 70, 90
Yonkhinisténha Owéntsyia (Mother Earth), 32

Zumba, 176, 179, 245

CPSIA information can be obtained
at www.ICGtesting.com
Printed in the USA
LVHW090751301220
675344LV00003B/140